Power Sources and Supplies

World Class Designs

Newnes World Class Designs Series

Analog Circuits: World Class Designs

Robert A. Pease

ISBN: 978-0-7506-8627-3

Embedded Systems: World Class Designs

Jack Ganssle

ISBN: 978-0-7506-8625-9

Power Sources and Supplies: World Class Designs

Marty Brown

ISBN: 978-0-7506-8626-6

*For more information on these and other Newnes titles, visit: **www.newnespress.com***

Power Sources and Supplies

World Class Designs

Marty Brown

with

Nihal Kularatna
Raymond A. Mack, Jr.
Sanjaya Maniktala

AMSTERDAM • BOSTON • HEIDELBERG • LONDON
NEW YORK • OXFORD • PARIS • SAN DIEGO
SAN FRANCISCO • SINGAPORE • SYDNEY • TOKYO

Newnes is an imprint of Elsevier

Cover image by iStockphoto

Newnes is an imprint of Elsevier
30 Corporate Drive, Suite 400, Burlington, MA 01803, USA
Linacre House, Jordan Hill, Oxford OX2 8DP, UK

∞ Recognizing the importance of preserving what has been written,
Elsevier prints its books on acid-free paper whenever possible.

Library of Congress Cataloging-in-Publication Data
Brown, Marty, 1951-
   Power sources and supplies/Marty Brown.
        p. cm.—(Newnes world class designs series)
   Includes bibliographical references and index.
   ISBN-13: 978-0-7506-8626-6 (pbk. : alk. paper)
   1. Electric power supplies to apparatus—Design and construction.   2. Electric apparatus and
appliances—Power supply. I. Title.
   TK7868.P6B758 2007
   621.31′7—dc22   2007043748

British Library Cataloguing-in-Publication Data
A catalogue record for this book is available from the British Library.

ISBN: 978-0-7506-8626-6

For information on all Newnes publications
visit our web site at www.books.elsevier.com

07 08 09 10 11 12   10 9 8 7 6 5 4 3 2 1

Typeset by Charon Tec Ltd (A Macmillan Company), Chennai, India
www.charontec.com
Printed in the United States of America

Contents

Preface

This book is an assemblage of four books from different authors on topics from linear and switching power supply design. The book covers a wide range of different aspects of power supply design with differing perspectives from these authors; all this combines to provide a more rounded view of the power field.

There are many different stories as to how the designers in the field came to be "power supply engineers." For me it was quite by accident. I had been an engineer for 5 years, using nothing more than 780X linear regulators. I was working on an avionics product designing the digital sections. I went to my manager when I had finished and said, "We need to fly this product, and we cannot take these bench supplies along with us. We need some power supplies designed."

He responded, "I was just waiting to see who finished first." One year and four different versions later, the unit flew.

And that was the start of my involvement in switching power supplies. In 1974, there were no PWM controller ICs, so by providing positive feedback around a linear regulator or by using a 556, one could generate the needed function. Discrete bipolar transistors and fast-recovery diodes were the only power semiconductor devices of the day.

The technology has come a long way since then. Semiconductor suppliers are providing virtually "plug and play" solutions such as National Semiconductor's Simple Switcher™. The added help of the many online design and simulation tools makes the design task seemingly quick and painless. Other suppliers too, such as the magnetic component suppliers, have made it easier than ever to produce a switching power supply design. They produce many families of standard shapes and values of inductors and transformers.

However, once you go beyond the simple printed circuit board level switching power supply, the online tools quickly lose their usefulness. The majority of graduating engineers today are in the digital hardware and software fields, where only their basic electronics courses involved physical electronics. So when a power supply needs to be designed, it becomes a matter of who can run the fastest! Of course, the job usually falls to the most junior engineer and it then becomes his or her next assignment.

The power supply design then usually follows that engineer throughout their employment with that company because of the "millstone effect." For me, this turned out to be fortunate. I had always been intrigued by the unknown and loved the field's curious combination of RF, digital, analog and power. My reward is in realizing that I intuitively understand an area that few engineers do, and my power supply designs are in many of the products in use today.

The contributing authors of this book are very respected engineers coming from very different experiential backgrounds. Their differing perspectives should provide a good appreciation of the power supply field.

So, good learning and good designs.

Marty Brown
September 2007

About the Editor

Marty Brown (Chapters 1, 9, and 12) is the author of the *Power Supply Cookbook* and *Practical Switching Power Supply Design*. He earned his amateur radio license at the age of 11 and has had electronics as a hobby throughout his life. He graduated cum laude from Drexel University in 1974. His electronic design history includes underwater acoustics with the department of the Navy, airborne weather radar design (digital and SMPS), a satellite CODEC, and process control equipment. He was previously with Motorola Semiconductor as a principal application engineer, where he defined more than eight semiconductor products in the power conversion market and received two patents. He later started his own electronics consulting firm where he designed products from satellite power systems to power-related integrated circuits for many semiconductor companies. He is presently working in the field of digitally controlled power supplies with Microchip Technologies. He has eight children, five of whom are adopted. His wife is an internationally known writer and speaker in the area of inter racial adoption and related issues. He presently lives in Scottsdale, Arizona.

About the Contributors

Nihal Kularatna (Chapter 7) is the author of *Power Electronics Design Handbook*. He is an electronics engineer with over 30 years of experience in professional and research environments. He is a Fellow of the IEE (London), a Senior Member of IEEE (USA) and an honors graduate from University of Peradeniya, Sri Lanka. Presently, he is a Senior Lecturer in the Department of Engineering, the University of Waikato, New Zealand. He worked at the Arthur C. Clarke Institute for Modern Technologies (ACCIMT) in Sri Lanka as a Research and Development Engineer until 1990 when he reached Principal Researcher Engineer status. He was then appointed as CEO of ACCIMT in 2000. From 2002 to 2005 he was a Senior Lecturer at the Department of Electrical and Electronic Engineering, University of Auckland. He is currently active in research in transient propagation and power conditioning in power electronics, embedded processing applications for power electronics, and smart sensor systems. He has authored five books and is currently working on his sixth. His hobby is gardening cacti and succulents.

Raymond A. Mack, Jr. (Chapters 2, 4, 5, 6, and 11) is the author of *Demystifying Switching Power Supplies*.

Sanjaya Maniktala (Chapters 3, 8, and 10) is the author of *Switching Power Supplies A to Z*. He is a Senior Applications manager and Systems Architect at Fairchild Semiconductor. A post graduate in Physics from the Indian Institute of Technology, Bombay, and Northwestern University in Evanston, Illinois, he has worked on several continents, holding lead engineering positions in companies such as Artesyn Technologies (now part of Emerson Electric), Siemens AG, Freescale Semiconductor, and Power Integrations. More recently, he was the most prolific author/writer at National Semiconductor for five years in a row, during which time he wrote what are considered by many to be their most widely consulted and popular reference, Application Notes. In his spare time, he also wrote several articles for major electronics publications such as *EDN, Electronic Design, Power Electronics Technology*, and *Planet Analog*. He holds several patents in the area of power conversion and control, including the "Floating Buck Regulator Topology."

CHAPTER 1

An Introduction to the Linear Regulator

Marty Brown

Linear power supplies are the simplest of the DC/DC converters, but don't be fooled by the apparent simplicity of them. There are several factors in every application of linear supplies that are important for their reliable operation. These are thermal design, output regulation, stability considerations and its transient response, any of which could cause the system to behave badly.

Linear regulators are used much more often than switching regulators. One finds them distributed throughout products as POL (point of load) supplies, where local circuit regulation is needed, voltage bus quieting for noise sensitive circuits, and inexpensive voltage bus generation.

If you have done a design completely using linear regulators, you may technically call yourself a "power supply designer," but you will not fully appreciate the complexities of the field until you have experienced a switching power supply design. You have only reached the "tenderfoot" level of experience.

I've attempted to cover the material in a succinct and intuitive manner showing how flexible the humble linear regulator can be. The design examples can be scaled and adapted to many other applications. Related topics such as thermal design can be found in chapter 12.

—Marty Brown

The linear regulator is the original form of the regulating power supply. It relies upon the variable conductivity of an active electronic device to drop voltage from an input voltage to a regulated output voltage. In accomplishing this, the linear regulator wastes a lot of power in the form of heat, and therefore gets hot. It is, though, a very electrically "quiet" power supply.

The linear power supply finds a very strong niche within applications where its inefficiency is not important. These include wall-powered, ground-base equipment where forced air cooling is not a problem; and also those applications in which the instrument is so sensitive to electrical noise that it requires an electrically "quiet" power supply—these products might include audio and video amplifiers, RF receivers, and so forth. Linear regulators are also popular as local, board-level regulators. Here only a few watts are needed by the board, so the few watts of loss can be accommodated by a simple heatsink. If dielectric isolation is desired from an AC input power source, it is provided by an AC transformer or bulk power supply.

In general, the linear regulator is quite useful for those power supply applications requiring less than 10 W of output power. Above 10 W, the heatsink required becomes so large and expensive that a switching power supply becomes more attractive.

1.1 Basic Linear Regulator Operation

All power supplies work under the same basic principle, whether the supply is a linear or a more complicated switching supply. All power supplies have at their heart a closed negative feedback loop. This feedback loop does nothing more than hold the output voltage at a constant value. Figure 1.1 shows the major parts of a series-pass linear regulator.

Linear regulators are step-down regulators only; that is, the input voltage source must be higher than the desired output voltage. There are two types of linear regulators: the *shunt regulator* and the *series-pass regulator.* The shunt regulator is a voltage regulator that is placed in parallel with the load. An unregulated current source is connected to a higher voltage source; the shunt regulator draws output current to maintain a constant voltage across the load given a variable input voltage and load current. A common example of this is a Zener diode regulator. The series-pass linear regulator is more efficient than the shunt regulator and uses an active semiconductor as the series-pass unit, between the input source and the load.

The series-pass unit operates in the linear mode, which means that the unit is not designed to operate in the full on or off mode but instead operates in a degree of "partially

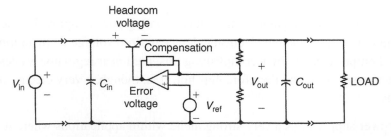

Figure 1.1: The basic linear regulator

on." The negative feedback loop determines the degree of conductivity the pass unit should assume to maintain the output voltage.

The heart of the negative feedback loop is a high-gain operational amplifier called a *voltage error amplifier*. Its purpose is to continuously compare the difference between a very stable voltage reference and the output voltage. If the output differs by mere millivolts, then a correction to the pass unit's conductivity is made. A stable voltage reference is placed on the noninverting input and is usually lower than the output voltage. The output voltage is divided down to the level of the voltage reference. This divided output voltage is placed into the inverting input of the operational amplifier. So at the rated output voltage, the center node of the output voltage divider is identical to the reference voltage.

The gain of the error amplifier produces a voltage that represents the greatly amplified difference between the reference and the output voltage (error voltage). The error voltage directly controls the conductivity of the pass unit thus maintaining the rated output voltage. If the load increases, the output voltage will fall. This will then increase the amplifier's output, thus providing more current to the load. Similarly, if the load decreases, the output voltage will rise, thus making the error amplifier respond by decreasing pass unit current to the load.

The speed by which the error amplifier responds to any changes on the output and how accurately the output voltage is maintained depends on the error amplifier's *feedback loop compensation*. The feedback compensation is controlled by the placement of elements within the voltage divider and between the negative input and the output of the error amplifier. Its design dictates how much gain at DC is exhibited, which dictates how accurate output voltage will be. It also dictates how much gain at a higher frequency and bandwidth the amplifier exhibits, which dictates the time it takes to respond to output load changes or *transient response time*.

The operation of a linear regulator is very simple. The very same circuitry exists in the heart of all regulators, including the more complicated switching regulators. The voltage feedback loop performs the ultimate function of the power supply—the maintaining of the output voltage.

1.2 General Linear Regulator Considerations

The majority of linear regulator applications today are board-level, low-power applications that are easily satisfied through the use of highly integrated three-terminal regulator integrated circuits. Occasionally, though, the application calls for either a higher output current or greater functionality than the three-terminal regulators can provide.

There are design considerations that are common to both approaches and those that are only applicable to the nonintegrated, custom designs. These considerations define the operating boundary conditions that the final design will meet, and the relevant ones must be calculated for each design. Unfortunately, many engineers neglect them and have trouble over the entire specified operating range of the product after production.

The first consideration is the *headroom voltage*. The headroom voltage is the actual voltage drop between the input voltage and the output voltage during operation. This enters predominantly into the later design process, but it should be considered first, just to see whether the linear supply is appropriate for the needs of the system. First, more than 95 percent of all the power lost within the linear regulator is lost across this voltage drop. This headroom loss is found by

$$P_{HR} = (V_{in(max)} - V_{out})I_{load(rated)} \tag{1-1}$$

If the system cannot handle the heat dissipated by this loss at its maximum specified ambient operating temperature, then another design approach should be taken. This loss determines how large a heatsink the linear regulator must have on the pass unit.

A quick estimated thermal analysis will reveal to the designer whether the linear regulator will have enough thermal margin to meet the needs of the product at its highest specified operating ambient temperature. One can find such a thermal analysis in Chapter 12.

The second major consideration is the minimum *dropout voltage* of a particular topology of linear regulator. This voltage is the minimum headroom voltage that can be experienced by the linear regulator, below which it falls out of regulation. This is predicated only by how the pass transistors derive their drive bias current and voltage. The common positive linear regulator utilizes an NPN bipolar power transistor (see Figure 1.2a). To generate the needed base-emitter voltage for the pass transistor's operation, this voltage must be derived from its own collector-emitter voltage. For the NPN pass units, this is the actual minimum headroom voltage. This dictates that the headroom voltage cannot get any lower than the base-emitter voltage ($\sim$0.65 VDC) of the NPN pass unit plus the drop across any base drive devices

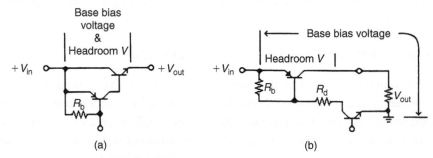

(a)   (b)

Figure 1.2: The pass unit's influence on the dropout voltage: (a) NPN pass unit; (b) PNP pass unit (low dropout)

(transistors and resistors). For the three terminal regulators such as the MC78XX series, this voltage is 1.8 to 1.5 VDC. For custom designs using NPN pass transistors for positive outputs, the dropout voltage may be higher. For applications where the input voltage may come even closer than 1.8 to 1.5 VDC to the output voltage, a *low dropout regulator* is recommended. This topology utilizes a PNP pass transistor, which now derives its base-emitter voltage from the output voltage instead of the headroom or input voltage (see Figure 1.2b). This allows the regulator to have a dropout voltage of 0.6 VDC minimum. P-Channel MOSFETs can also be used in this function and can exhibit dropout voltages close to zero volts.

The dropout voltage becomes a driving issue when the input to the linear regulator during normal operation is allowed to fall close to the output voltage. If operating from an AC wall transformer, this would occur at brown-out conditions (minimum AC voltages). The low dropout regulator (e.g., LM29XX) would allow the regulator to operate to a lower AC input voltage. Low dropout regulators are also widely used as *post regulators* on the output of switching power supplies. Within switching regulators, the efficiency is of great concern, so the headroom drop needs to be kept to a minimum. Here, the low dropout regulator will save several watts of loss over a conventional NPN-based linear regulator. If the application will never see headroom voltages less than 1.5 V, then use the conventional linear regulators (e.g., MC78XX).

Another consideration is the type of pass unit to be used. From a headroom loss standpoint, it makes absolutely no difference whether a bipolar power transistor or a power MOSFET is used. The difference comes in the drive circuitry. If the headroom voltage is high, the controller (usually a ground-oriented circuit) must pull current from the input or output voltage to ground. For a single bipolar pass transistor this current is

$$I_B = I_{Load}/h_{FE} \qquad (1\text{-}2)$$

The power lost just in driving the bipolar pass transistor is

$$P_{drive} = V_{in/max} \cdot I_B \text{ or } V_{out} \cdot I_B \qquad (1\text{-}3)$$

This drive loss can become significant. A driver transistor can be added to the pass transistor to increase the effective gain of the pass unit and thus decrease the drive current, or a power MOSFET can be used as a pass unit that uses magnitudes less DC drive current than the bipolar power transistor. Unfortunately, the MOSFET requires up to 10 VDC to drive the gate. This can drastically increase the dropout voltage. In the vast majority of linear regulator applications, there is little difference in operation between a buffered pass unit and a MOSFET insofar as efficiency is concerned. Bipolar transistors are much less expensive than power MOSFETs and have less propensity to oscillate.

The linear regulator is a mature technology and therefore can usually be accommodated by the integrated solutions provided by the semiconductor manufacturers. For applications beyond the limits of these integrated linear regulators alone, usually adding more components around the IC will satisfy the requirement. Otherwise, a completely custom approach would need to be utilized. These various approaches are overviewed in the design examples in the following section.

1.3 Linear Power Supply Design Examples

Linear regulators can be designed to meet a variety of cost and functional needs. The design examples that follow illustrate that linear regulator designs can range from the very elementary to the more complex. Designs for enhanced three-terminal regulator designs will be abbreviated, since the integrated circuit datasheets usually contain great detail. Due to the relatively large power loss of linear regulators, the thermal considerations typically represent a significant problem. Some thermal analysis and design is done in the examples. For further insight on this please refer to Chapter 12.

1.3.1 Elementary Discrete Linear Regulator Designs

These types of linear regulators were commonly built before the advent of operational amplifiers and they can save money in consumer designs. Some of their drawbacks include drift with temperature and limited load current range.

1.3.1.1 The Zener shunt regulator

This type of regulator is typically used for very local voltage regulation for less than 200 mW of a load. A series resistance is placed between a higher voltage and is used to limit the current to the load and Zener diode. The Zener diode compensates for the variation in load current. The Zener voltage will drift with temperature. The drift characteristics are given in many Zener diode datasheets. Its load regulation is adequate for most supply specifications for integrated circuits. It also has a higher loss than the series-pass type of linear regulator, since its loss is set for the maximum load current, which for any load remains less than that value. A Zener shunt regulator can be seen in Figure 1.3.

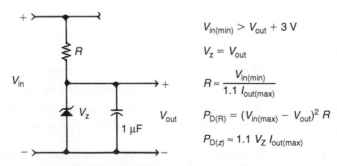

$$V_{in(min)} > V_{out} + 3 \text{ V}$$

$$V_z = V_{out}$$

$$R \approx \frac{V_{in(min)}}{1.1\, I_{out(max)}}$$

$$P_{D(R)} = (V_{in(max)} - V_{out})^2\, R$$

$$P_{D(z)} \approx 1.1\, V_z\, I_{out(max)}$$

Figure 1.3: A Zener shunt regulator

1.3.1.2 The one-transistor series-pass linear regulator

By adding a transistor to the basic Zener regulator, one can take advantage of the gain that the bipolar transistor offers. The transistor is hooked up as an emitter follower, which can now provide a much higher current to the load, and the Zener current can be lowered. Here the transistor acts as a rudimentary error amplifier (refer to Figure 1.4). When the load current increases, it places a higher voltage into the base, which increases its conductivity, thus restoring the voltage to its original level. The transistor can be sized to meet the demands of the load and the headroom loss. It can be a TO-92 transistor for those loads up to 0.25 W or a TO-220 for heavier loads (depending on heatsinking).

1.3.2 Basic Three-Terminal Regulator Designs

Three-terminal regulators are used in the majority of board-level regulator applications. They excel in cost and ease of use for these applications. They can also, with care, be used as the basis for higher functionality linear regulators.

The most-often ignored consideration is the overcurrent limiting method used in three-terminal regulators. They typically use an overtemperature cutoff on the die of the regulator, which is typically between +150°C and +165°C. If the load current is passed through the three-terminal regulator, and if the heatsink is too large, the regulator may fail due to overcurrent (bondwire, IC traces, etc.). If the heatsink is too small, then one may not be able to get enough power from the regulator. Another consideration is if the load current is being conducted by an external pass-unit, the overtemperature cutoff will be nonfunctional, and another method of overcurrent protection will be needed.

1.3.2.1 The basic three-terminal positive regulator design

This example will illustrate the design considerations that should be undertaken with each three-terminal regulator design. Many designers view only the electrical specifications of the regulators and forget the *thermal derating* of the part. At high headroom voltages,

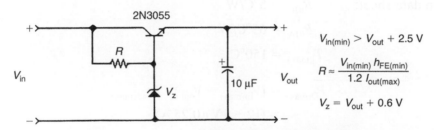

$$V_{in(min)} > V_{out} + 2.5\ V$$

$$R \approx \frac{V_{in(min)}\ h_{FE(min)}}{1.2\ I_{out(max)}}$$

$$V_z = V_{out} + 0.6\ V$$

Figure 1.4: A discrete bipolar series-pass regulator

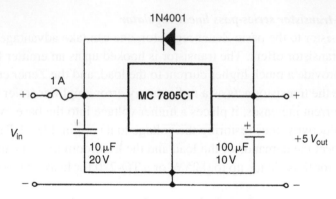

Figure 1.5: A three-terminal regulator

and at high ambient operating temperatures, the regulator can only deliver a fraction of its full-rated performance. Actually, in the majority of the three-terminal applications, the heatsink determines the regulator's maximum output current. The manufacturer's electrical ratings can be viewed as having the part bolted onto a large piece of metal and placed in an ocean. Any application not employing those unorthodox components must operate at a lower level. The following example illustrates a typical recommended design procedure (see Figure 1.5).

Design Example 1. Using three-terminal regulators

Specification	Input:	12 VDC (max)
		8.5 VDC (min)
	Output:	5.0 VDC
		0.1A to 0.25A
	Temperature:	−40 to +50°C

Note: The 1N4001 is required for discharging the 100 μF capacitor when the system is turned off.

Thermal Design (refer also to Appendix A)

Given in data sheet:

$$R_{\theta JC} = 5°C/W$$

$$R_{\theta JA} = 65°C/W$$

$$T_{j(max)} = 150°C$$

$$
\begin{aligned}
P_{D(max)} &= (V_{in(max)} - V_{out}) \cdot I_{load(max)} \\
&= (12 - 5\,V)(0.25\,A) \\
&= 1.75\,W \text{ (headroom loss)}
\end{aligned}
$$

Without a heatsink the junction temperature will be:

$$T_j = P_D \cdot R_{\theta JA} + T_{A(max)}$$
$$= (1.75\,\text{W})(65°\text{C/W}) + 50$$
$$= 163.75°\text{C}.$$

A small "clip-on" style heatsink is required to bring the junction temperature down to below its maximum ratings. Refer to Chapter 12 for aid in the selection of heatsinks.

Selecting the heatsink—Thermalloy P/N 6073B

Given in heatsink data: $R_{\theta SA} = 14°\text{C/W}$

Using a silicon insulator $R_{\theta CS} = 65°\text{C/W}$

The new worst-case junction temperature is now:

$$T_{j(max)} = P_D(R_{\theta JC} + R_{\theta CS} + R_{\theta SA}) + T_A$$
$$= (1.75\,\text{W})(5°\text{C/W} + 65°\text{C/W} + 14°\text{C/W}) + 50°\text{C}$$
$$= 84.4°\text{C}$$

1.3.2.2 Three-terminal regulator design variations

The following design examples illustrate how three-terminal regulator integrated circuits can form the basis of higher-current, more complicated designs. Care must be taken, though, because all of the examples render the overtemperature protection feature of the three-terminal regulators useless. Any overcurrent protection must now be added externally to the integrated circuit.

The current-boosted regulator

The design shown in Figure 1.6 adds just a resistor and a transistor to the three-terminal regulator to yield a linear regulator that can provide more current to the load. The current-boosted positive regulator is shown, but the same equations hold for the boosted negative regulator. For the negative regulators, the power transistor changes from a PNP to an

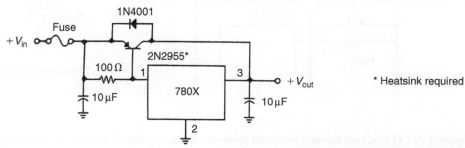

Figure 1.6: Current-boosted three-terminal regulator without overcurrent protection

NPN. Beware: there is no overcurrent or overtemperature protection in this particular design.

The current-boosted three-terminal regulator with overcurrent protection

This design adds the overcurrent protection externally to the IC. It employs the base-emitter (0.6 V) junction of a transistor to accomplish the overcurrent threshold and gain of the overcurrent stage. For the negative voltage version of this, all the external transistors change from NPN to PNP and vice versa. These can be seen in Figures 1.7a and b.

1.3.3 Floating Linear Regulators

A floating linear regulator is one way of achieving high-voltage linear regulation. Its philosophy is one in which the regulator controller section and the series-pass transistor "float" on the input voltage. The output voltage regulation is accomplished by sensing the ground, which appears as a negative voltage when referenced to the output voltage. The output voltage serves as the "floating ground" for the controller and the power for

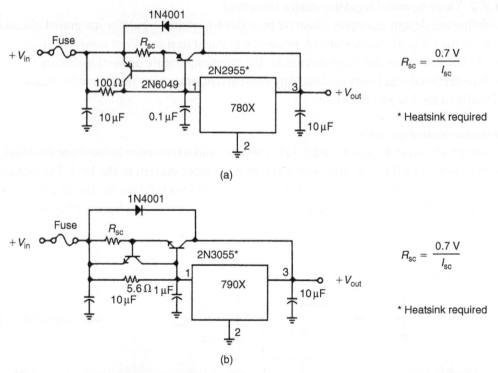

$$R_{sc} = \frac{0.7\ V}{I_{sc}}$$

* Heatsink required

(a)

$$R_{sc} = \frac{0.7\ V}{I_{sc}}$$

* Heatsink required

(b)

**Figure 1.7: (a) Positive current-boosted three-terminal regulator with current limiting
(b) Negative current-boosted three-terminal regulator with current limiting**

the controller and series-pass transistor is drawn from the headroom voltage (the input-to-output difference) or is provided by an auxiliary isolated power supply.

The power transistor still needs to have a breakdown voltage rating greater than the input voltage, since at start-up it must see the entire input voltage across it. Other methods such as a *bootstrap Zener diode* can also be used in order to shunt the voltage around the pass transistor, but only when the input voltage itself is switched on and off to activate the power supply. Also, caution must be taken to ensure that any controller input or output pin never goes negative with respect to the floating ground of the IC. Protection diodes are usually used for this purpose. One last caution is the little-known breakdown voltage of common resistors. If the output voltage exceeds 200 V, more than one sensing resistor must be placed in series in order to avoid the 250 V breakdown characteristic of 1/4 W resistors.

A common low-voltage positive floating regulator is the LM317 (the negative regulator complementary part is the LM337). The MC1723 can also be used to create a floating linear regulator, but care must be taken to protect the IC against the high voltage.

The first example shows how an LM317 can be modified to create a 70 V linear regulator from a 100 V input voltage. Several design restrictions must be strictly followed; for example, the operational headroom voltage must not exceed the voltage rating of the bootstrap Zener diode or regulation will be lost. Also the use of the protection diode on the error amplifier is mandatory. This regulator can be seen in Figure 1.8.

The second example illustrates a 350 V floating linear regulator that can provide up to 10 mA of load current from a 400 to 450 V unregulated source. The TIP50 provides the bias supply for the controller, which must withstand the full input voltage during start-up and power supply foldback. The controller is "grounded" on the output voltage and the

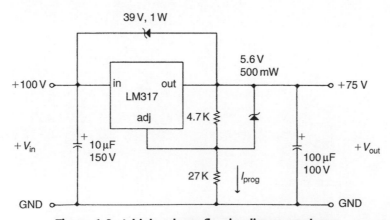

Figure 1.8: A high voltage floating linear regulator

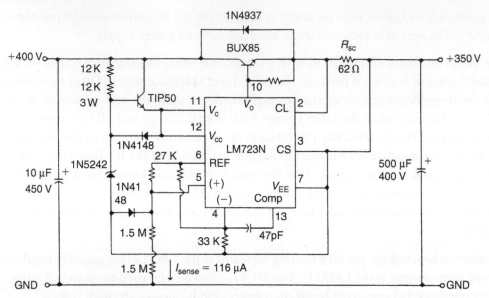

Figure 1.9: A 350 V, 10-mA floating linear regulator

minimum headroom voltage is 15 V. To readjust the output voltage, one changes the value of the two series resistors in the voltage sensing branch and this is set by

$$R_{\text{sense}} = (V_{\text{out}} + 4.0\text{V})/I_{\text{sense}} \qquad (1\text{-}4)$$

Floating linear regulators are particularly suited for high-output voltage regulation, but may be used anywhere. This regulator can be seen in Figure 1.9.

CHAPTER 2

Basic Switching Circuits

Raymond Mack

Understanding the fundamental operation of switching power supplies is an introduction into the dark arts of engineering. On paper, they are no more than the introductory college courses of electronics, but when you add in the parasitic behaviors of the components, then the challenge becomes much greater. For me, it is exciting.

I remember the first switching power supply I ever attempted. It was a 30-kHz flyback supply. I designed the control section using an NE556 and wound my own transformer on an MPP torroid core. When I powered it up, I made a shocking discovery. I had made a wonderful 3-MHz, AM modulated RF transmitter. At which I said, "Oh my—(fill in your preferred deity). I need to do a lot more learning."

To do a thorough design and analysis of a switching power supply you will need a new set of instruments, such as: oscilloscope voltage and current probes, a spectrum analyzer and a network analyzer. Welcome to a world where few engineers have ventured.

Switchmode power supplies first became practical in the 1970s, where one used improvised control circuits, bipolar transistors and slow diodes. They operated at less than 50 kHz. The first power MOSFETs emerged in the late 1970s. They were easier to drive and switched much faster than bipolar transistors. This allowed the switchmode power supplies to now exceed 100 kHz and even go as high as 300–500 kHz. Today, with the much improved diode performances, better magnetic materials, resonant techniques, and surface mount packaging, the switching power supply can easily operate over 1 MHz and have a much smaller size and high efficiency.

Switching power supply design will always involve a certain amount of "customization." Whether it is the number of outputs, size, height, efficiency or noise, some aspect of the design always needs to be tailored to the surrounding application. I have always said, "Once you put an AC flux into a magnetic core, you can do anything."

In this chapter Ray Mack provides an intuitive introduction to the basic operation of the basic building blocks of switching power supplies. I think you will find it very informative.

—Marty Brown

In this chapter, we will look at the time domain description of ideal inductors and capacitors and review ideal versions of each type of switching supply. In later chapters, we will look at the magnetic, electrical, and parasitic properties of inductors and capacitors and their effect on the design of individual components.

2.1 Energy Storage Basics

Equation (2-1) contains the definition of inductance. An inductor has an inductance of one henry if a change of current of one ampere/second produces one volt across the inductor.

$$V = L\frac{di}{dt} \tag{2-1}$$

This is *Lenz's law*. The first consequence of Eq. (2-1) is that the current through an inductor cannot change instantaneously. To do so would generate an infinite voltage across the inductor. In the real world, things such as an arc across switch contacts will limit the voltage to very high, but not infinite, values. The other consequence of Eq. (2-1) is that the voltage across an inductor changes instantaneously from positive to negative when we switch from storing energy in the inductor (*di/dt* is positive) to removing energy from it (*di/dt* is negative). Eq. (2-2) is the converse of Eq. (2-1) and is used to determine the current in the inductor when the voltage is known.

$$I = 1/L \int V dt + I_{\text{initial}} \tag{2-2}$$

Equation (2-3) contains the definition of a capacitor. It states that a capacitor is one farad if storing one coulomb of charge creates one volt.

$$Q = CV \tag{2-3}$$

Equations (2-4) and (2-5) describe a capacitor in terms of voltage and current (where charge is the integral of current and current is *dq/dt*).

$$V = 1/C \int i dt + V_{\text{initial}} \tag{2-4}$$

$$I = C\frac{dv}{dt} \tag{2-5}$$

The current waveform of the filter capacitor of a switching power supply is typically a sawtooth waveform. The goal of the capacitor is to limit the change in voltage (ripple voltage). There are two variables in Eq. (2-4) that can control the change in output voltage. We can either make the capacitance large or make *dt* small to control the voltage ripple. One of the major advantages of switching power supplies is that we can make *dt* very small (a high switching frequency), which allows the value of *C* to also be very small.

2.2 Buck Converter

Figure 2.1 shows an ideal buck converter regulator made of an ideal voltage source, an ideal voltage controlled switch, an ideal diode, an ideal inductor, an ideal capacitor, and a load resistor. It is called a buck converter because the voltage across the inductor "bucks" or opposes the supply voltage. The output voltage of a buck converter is always less than the input voltage. This ideal regulator is designed to use a 20 V source and provide 5 V to the 10-ohm load. The switch is opened and closed once every 10 µs. The switch produces a pulse width modulated waveform to the passive components. When the regulator is at steady state, the output voltage is:

$$V_{out} = V_{in} \cdot \text{Duty Cycle}$$

(2-6)

This equation is independent of the value of the inductor, the load current, and the output capacitor as long as the inductor current flows continuously. This equation assumes that the inductor voltage has a rectangular shape.

The diode acts as a voltage-controlled switch. It provides a path for the inductor current once the switch is opened. No current flows through the diode while the inductor is charging because it is reverse biased. When the control switch opens, the inductor current flows through the diode.

We design switching supplies with the simplifying assumption that the applied voltage to the inductor during charging is a perfect rectangular wave. Our example power supply has voltage output ripple of 20 mV. The perfect rectangle is a good approximation since the change in inductor voltage during charging is 0.02/15 or 0.13% and the variation on

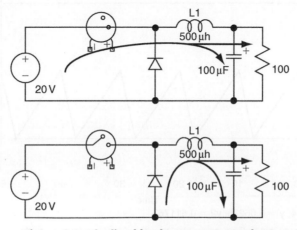

Figure 2.1: Idealized buck converter regulator

discharge is 0.02/5 or 0.4%. The constant voltage of the rectangular pulse causes *di/dt* in Eq. (2-1) to be a constant.

Figure 2.2 shows a plot of the output voltage (lower trace) and inductor current (upper trace) after the system is at steady-state providing 5 V and 500 mA to the load resistor.

Note that the change in output current is relatively small compared to the DC value of current in the inductor. In this case, the ripple current is 75 mA P-P. Another important point is that the ripple current is independent of load current when the system is steady-state. This is a consequence of the current through the inductor being controlled by the voltage across the inductor. The slope and duration of charging is controlled entirely by the difference ($V_{in} - V_{out}$). The average inductor current is equal to the output current.

It is also possible for the buck converter to work in discontinuous mode, which means the inductor current goes to zero during part of the switching period.

Equation (2-6) does not hold for discontinuous operation. The output ripple voltage is higher for a buck converter in discontinuous mode because the capacitor must supply the load current during the time that the inductor current is zero. Usually, a buck converter only runs in discontinuous mode when the load current becomes very small compared to the design current.

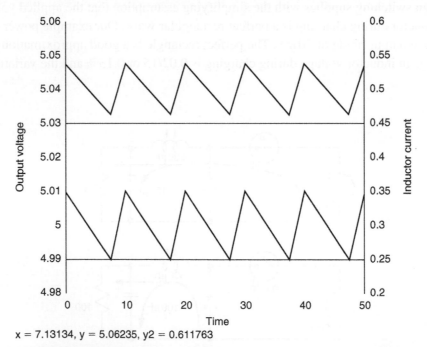

x = 7.13134, y = 5.06235, y2 = 0.611763

Figure 2.2: Output voltage and inductor current in a buck regulator

2.3 Boost Converter

Figure 2.3 shows an ideal boost converter regulator made of an ideal voltage source, an ideal switch, an ideal diode, an ideal inductor, a capacitor, and a load resistor. It is called a boost converter because the voltage across the inductor adds to the input supply voltage to boost the voltage above the input value. The output of a boost converter is always greater than the input voltage. This ideal regulator is designed to use a 5 V source and provide 20 V to the 1000-ohm load. The diode provides a path for the current once the switch is opened. The diode is off while the switch is closed. The switch is opened and closed once every 10 µs.

The switch and voltage source provide current to charge the inductor with energy while the switch is closed. While the inductor is charging, the current in the load is supplied by the capacitor because the diode is reverse biased. When the switch opens, the current in the inductor continues to flow, but now the inductor current forward biases the diode and flows through the load circuit. The voltage across the inductor reverses and adds to the voltage of the input supply. When the regulator is at steady-state, the output voltage is:

$$V_{\text{out}} = V_{\text{in}}/(1 - \text{Duty Cycle}) \tag{2-7}$$

This equation is independent of the value of the inductor, the load current, and the output capacitor for continuous mode operation.

Boost converters require much more capacitance than buck converters because the capacitor supplies all of the load current while the switch is closed.

Figure 2.4 shows a plot of the output voltage (lower trace) and inductor current (upper trace) after the system is at steady-state providing 20 V and 20 mA to the load resistor.

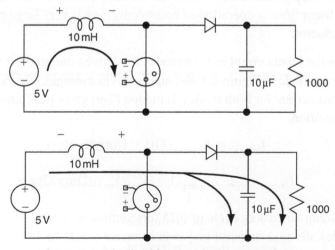

Figure 2.3: Idealized boost converter regulator

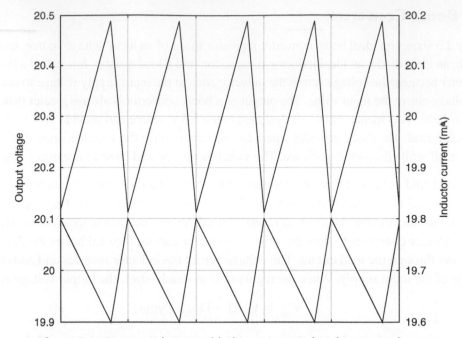

Figure 2.4: Output voltage and inductor current in a boost regulator

Just as in the buck converter, the ripple current in the inductor is independent of the output current for continuous mode operation. Typically, the peak inductor current is only slightly larger than the average inductor current.

It is also possible to run a boost converter in discontinuous mode. Discontinuous mode results in larger ripple current for boost converters, just as in the buck converter, because the capacitor must supply load current while the inductor current is zero. The other consequence of discontinuous operation of boost converters is very large peak current in the switch and inductor.

You can calculate the input current in both modes for a given output current. In our continuous mode example in Figure 2.3, the input current averages 80 mA. Equation (2-8) gives average input current for both modes. Equation (2-9) gives peak input current for discontinuous operation.

$$I_{\text{in-avg}} = I_{\text{out-avg}}(1/(1 - \text{Duty Cycle})) \tag{2-8}$$

$$I_{\text{in-peak}} = 2 \cdot I_{\text{out-avg}}(1 - (V_{\text{out}}/V_{\text{in}}))/\text{Duty Cycle} \tag{2-9}$$

If our example circuit had a duty cycle of 0.25 (discontinuous mode) instead of 0.75 (continuous mode), the peak inductor and switch current would be 480 mA instead of 81.75 mA.

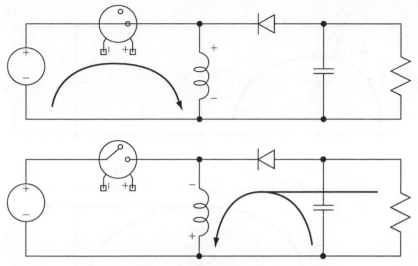

Figure 2.5: Idealized inverting boost converter

2.4 Inverting Boost Converter

Figure 2.5 shows the circuit of an ideal inverting boost converter. The switch and voltage source provide current to charge the inductor with energy while the switch is closed. While the inductor is charging, the current in the load is supplied by the capacitor because the diode is reverse biased. When the switch opens, the current in the inductor continues to flow, but now the inductor current forward biases the diode and flows through the load circuit. Since one side of the inductor is tied to the common point, the current flow when the switch opens causes a negative output voltage.

When the regulator is at steady-state, the output voltage is determined by Eq. (2-10) for continuous mode operation. Just as in the positive boost converter, the output voltage will be larger in magnitude than (or equal to) the input voltage.

$$V_{\text{out}} = -V_{\text{in}} \cdot (\text{Duty Cycle})/(1 - \text{Duty Cycle}) \qquad (2\text{-}10)$$

2.5 Buck-Boost Converter

If we add an additional switch and an additional diode to the boost converter as in Figure 2.6, we can create a buck-boost converter that will allow us to create a positive voltage that is either above or below the input voltage. Both switches close and open at the same time in this circuit. Again, the inductor is charged while the switches are closed and energy is delivered to the load when the switches open, just as it is in the boost converter.

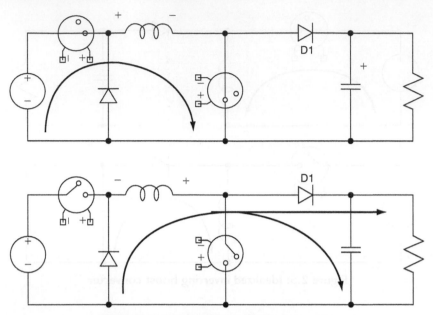

Figure 2.6: Idealized buck-boost converter

Diode D1 connects one end of the inductor to the common point so the voltage across the inductor can be either above or below the input voltage.

2.6 Transformer Isolated Converters

Power supplies that are intended to run directly from the AC power lines (offline supplies) require a transformer to isolate the load side from the AC lines. Transformers can also be used in power supplies where isolation is required for other reasons, such as medical equipment use. Table 2.1 lists power range and complexity versus appropriate converter type. This table gives a generally accepted range for each converter type. Each type can be used above or below these ranges, but the design problems to create an efficient supply become greater.

An off-line power supply is really a DC power supply that feeds a transformer isolated DC–DC converter. The rest of this section will focus on the DC–DC converter circuits.

Figure 2.7 shows a single switch flyback converter. It appears that this supply uses a transformer, but, in fact, the magnetic component is an inductor with two windings. This supply uses the primary winding of the inductor to store the magnetic energy in the same way that the boost converter works. Note that the phasing of the windings is opposite to normal transformer use. While the switch is closed, the energy is stored in the core and no current flows in the secondary. When the switch opens, current flows in the secondary

Table 2.1: Power Range and Complexity vs. Appropriate Converter Type

Circuit	Power Range	Relative complexity
Flyback	1 W–100 W	Low
Forward	1 W–200 W	Medium
Push-Pull	200 W–500 W	Medium
Half Bridge	200 W–500 W	High
Full Bridge	500 W–2000 W	Very high

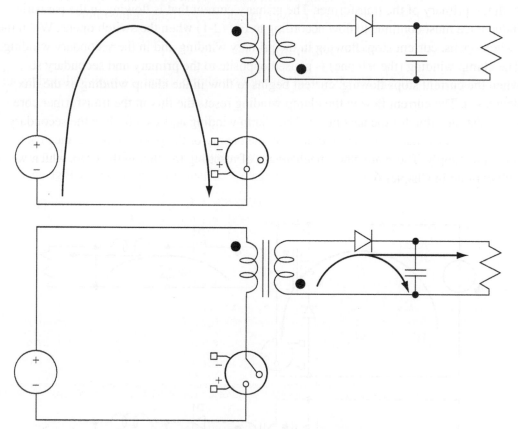

Figure 2.7: Idealized single switch flyback converter

[as required by Eq. (2-1)] and delivers energy to the load. The voltage on the output is determined by the turns ratio, just as in an actual transformer. The flyback converter is the only off-line converter that uses an inductor; all others use a transformer. One advantage of the flyback converter is that there is no need for an additional smoothing choke. The energy stored in the inductor is dumped directly into the capacitor and the load. This is also a disadvantage because the current for the load is supplied by the capacitor alone while the inductor is charging. The ripple voltage is larger for the flyback converter unless a larger output capacitor is used.

Figure 2.8 shows a single switch forward converter. During the time the switch is closed, current flows in the primary and in the secondary. The secondary current charges the filter choke just as in a buck converter. When the switch opens, current must continue to flow in the choke, as described in Eq. (2-1). The commutating diode (D2) in the secondary acts just as it does in the buck converter and allows inductor current to continue to flow.

Real transformers also have parasitic inductance that looks like an inductor in series with the primary of the transformer. The primary current that is flowing in the parasitic inductance must continue to flow according to Eq. (2-1) when the switch opens. When the switch opens, current stops flowing in the primary winding and in the secondary winding. The clamp winding (the left one) is phased opposite to the primary and secondary so when the current stops flowing, current begins to flow in the clamp winding as the flux decreases. The current flow in the clamp winding resets the flux in the transformer core to its resting value for the next pulse. The clamp winding acts exactly like the secondary winding of a flyback converter and delivers the energy of the parasitic inductance back to the input supply. There are other mechanisms of resetting the flux in the core, which we will explore in Chapter 6.

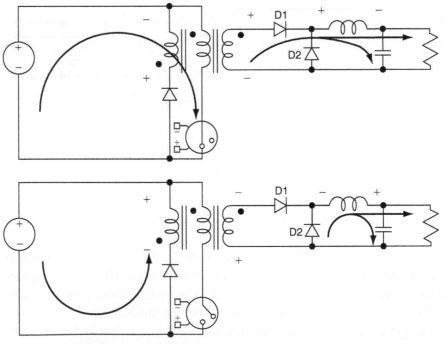

Figure 2.8: Idealized single switch forward converter

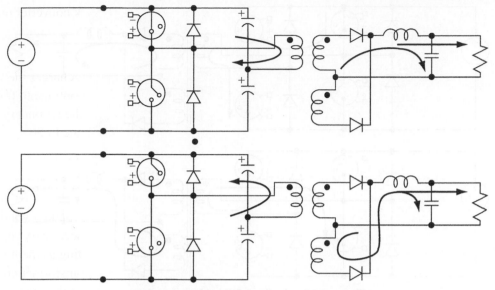

Figure 2.9: Idealized half bridge converter

Figure 2.9 shows a half bridge converter. This circuit is a high voltage equivalent of a TTL totem-pole output. The switches conduct alternately, which produces a bipolar voltage across the transformer primary. This requires that we have a full wave rectifier for the output. A clamp winding is not necessary since the opposite phase output diode will allow the current to flow in the secondary winding. We can add freewheeling diodes to the primary to control the voltage present on the secondary when the switches open. The capacitors provide a voltage divider that sets one end of the primary winding to one-half the input voltage. These capacitors are almost always part of the input DC power supply, so they perform the dual functions of voltage divider and input charge reservoir.

Figure 2.10 shows a full bridge converter. This design uses four switches to alternate the direction of current through the core.

Figure 2.11 shows a push-pull converter. The switches open and close 180 degrees out of phase, just as in a class B push-pull audio amplifier. Push-pull converters are rarely used in off-line supplies because they require high voltage transistors and it is very difficult to control the flux in the transformer. Modern current mode PWM controllers have made using push-pull circuits practical in low voltage circuits.

2.7 Synchronous Rectification

In all of the circuits we have reviewed in this chapter, we have used diodes as voltage-controlled switches. When they are reverse biased, they act as open switches. When they are forward biased, they act as closed switches. Power MOSFETs also work as switches.

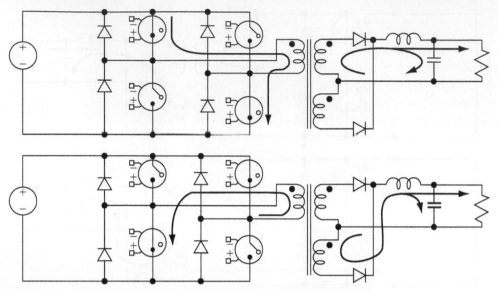

Figure 2.10: Idealized full bridge converter

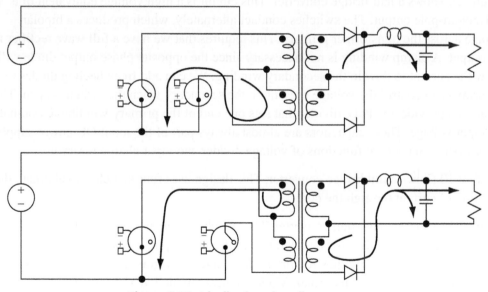

Figure 2.11: Idealized push-pull converter

When the gate to source voltage is sufficient to turn on a MOSFET, current can flow in either direction through the transistor. Power MOSFETs that are used as switches can have on resistance of 0.01 ohm or less. A Schottky diode that is conducting 5A will drop approximately 0.4V and dissipate 2W. A power MOSFET with 0.01 ohm on resistance

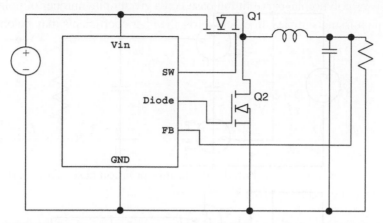

Figure 2.12: Buck converter using power MOSFETs as switches instead of diodes

will dissipate 0.25 W while conducting 5 A. This is a sizeable increase in efficiency. Figure 2.12 shows a buck regulator using synchronous rectification and ideal passive components. This circuit uses an ideal buck converter controller that sequences the MOSFETs and provides the voltage feedback control. When Q1 is on, the circuit turns off Q2. When Q1 is turned off, Q2 is turned on. While this example shows a buck converter, with proper drive circuitry it is possible to replace diodes with MOSFET switches in all designs.

2.8 Charge Pumps

Charge pumps use a capacitor to either increase or invert the input voltage. An ideal voltage-doubling charge pump is shown in Figure 2.13. The charge pump capacitor is called a *flying* capacitor (presumably because the switches resemble flapping wings as they change state). During charging, the flying capacitor is charged by the switches. Then the capacitor is connected to the load in series with the input supply to provide a voltage above the input.

Figure 2.14 shows a different arrangement of the switches that allows a charge pump to provide a negative voltage nearly equal in magnitude to the input voltage.

Charge pumps are typically used in applications where a low current is necessary, such as in a bias supply for an IC or a FET amplifier. Charge pumps are not able to supply large amounts of current without using large value capacitors. The practical limit to output current is approximately 250 mA.

A voltage multiplier circuit is also a form of charge pump. Figure 2.15 illustrates a traditional voltage multiplier circuit driven by a totem-pole switch square wave generator.

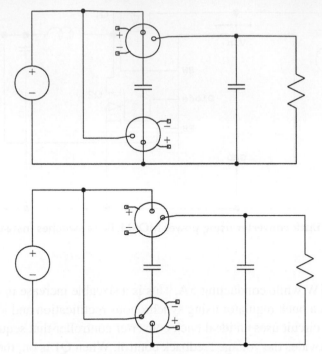

Figure 2.13: Idealized voltage-doubling charge pump

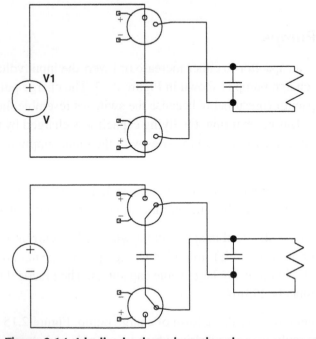

Figure 2.14: Idealized voltage inverting charge pump

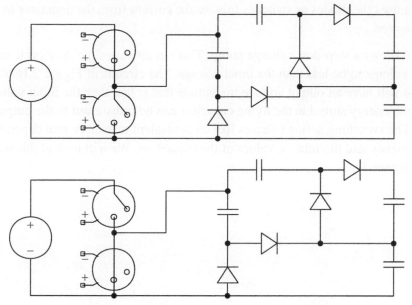

Figure 2.15: Square wave driven voltage multiplier

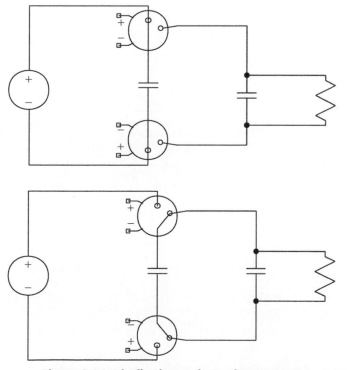

Figure 2.16: Idealized step-down charge pump

This circuit uses the diodes as switches to steer the current from the generator to the output capacitor.

Figure 2.16 shows a step-down charge pump. This circuit varies the duty cycle to allow the output voltage to be less than the input voltage. The circuits of Figure 2.16 and Figure 2.14 will have an output voltage magnitude that is less than the input voltage. Not all of the energy stored in the flying capacitor can be transferred to the output capacitor. The switching action behaves like an equivalent resistance that depends on the switch frequency and the relative values of the capacitors. We will look at this in detail in Chapter 4.

CHAPTER 3

DC-DC Converter Design and Magnetics

Sanjaya Maniktala

"Magnetics", a term that strikes fear in the heart of the average electronics engineer, along with terms such as feedback loop compensation, and electromagnetic interference. Before you throw up you hands in a nervous sweat, consider this, we all had to go through this before you and survived with a deepened understanding.

Magnetism can also be quite analogous to the flow of water through a pipe. B (flux density) can be seen as the number of water molecules flowing through a crossection of pipe. H (field strength) can be seen as the pressure causing the flow. Hysteresis loss is the amount of work done to reorient the magnetic domains within the core. Saturation is when there are no more magnetic domains remaining to reorient. Eddy current loss is the localized circulating flux disturbed by obstacles such as corners in the core just as water does in streams around rocks.

The terms, units and equations may be unusual. Be careful to not mix the units between the MKS and the CGS systems, or you will have results that will extraordinarily large.

Understanding the operation of the magnetics within a switchmode power supply and their design is fundamental to the design of a switchmode converter. The relationships between current, voltage and power are all controlled by the magnetic elements.

Their operation can sometimes seem counter-intuitive. The design of a transformer or inductor not only involves the electrical design but also its physical design. A poorly designed transformer, for instance will ring excessively, increase the radiated noise, and lower the overall efficiency of the switching power supply.

All the results from these calculations are "calculated estimates". That is, the results are midpoint results that get you in the area you wish to operate, but the resulting values can be increased or decreased to enhance a particular feature such as size, cost, etc. Changing the value from its midpoint, though, will also affect other operational factors, such as heat, and efficiency, which must also be considered.

Sanjaya Maniktala does a very good job in describing the various factors in designing the magnetic elements within a switchmode power supply.

—**Marty Brown**

The magnetic components of any switching power supply are an integral part of its topology. The design and/or selection of the magnetics can affect the selection and cost of all the other associated power components, besides dictating the overall performance and size of the converter itself. Therefore, we really should not try to design a converter without looking closely at its magnetics, and vice versa. With that in mind, in this chapter we will be introducing the basic concepts of magnetics, in parallel with a formal DC-DC converter design procedure.

Note that in the area of DC-DC converters, we have only a single magnetic component to consider—its inductor. Further, in this particular area of power conversion, it is customary to just pick an *off-the-shelf* inductor for most applications. Of course, there cannot possibly be enough "standard" inductors around to cover all possible application scenarios. But the good news is that, given a certain inductor and knowing its performance under a *stated set of conditions*, we can easily calculate how it will perform under our specific application conditions. Thereby, we can either validate or invalidate our initial selection. It may take more than one iteration or attempt, but by moving in this direction, we can almost always find a standard inductor that fits our application.

Later we will take up *off-line* power supply design. Such converters usually work off an AC (mains) input that ranges from 90 to 270 V. To protect users from the high voltage, these converters almost invariably use an isolating transformer, in addition to or in place of the inductor. But though these topologies are really just derivatives of standard DC-DC topologies, in terms of their magnetics, they are quite different. For example, we encounter significant (nonnegligible) high-frequency effects within the transformer—like skin depth and proximity effects—the analysis of which can be quite challenging. In addition, we find that there are definitely not enough general-purpose (off-the-shelf) parts around to meet all possible permutations and combinations of requirements, as can arise in off-line applications. So in these applications, we usually end up having to custom-design the magnetics. As mentioned, this is not a mean task, but by trying to first understand *DC-DC converter* design, and the selection of off-the-shelf inductors, we are in a much better position to tackle off-line power supplies. We can thereby build up basic concepts and skills, while garnering a much-needed "feel" for magnetics.

Off-line converters and DC-DC converters are also relatively different in terms of some rather implicit (often completely unstated) differences in basic design strategy—like the issue relating to the size of the magnetics vis-à-vis the current limit of the converter, as we will soon learn. With regard to their similarities, we should remember that both can have a *wide-input voltage range*, not a *single-value input voltage*, as is often assumed in related literature. Having a wide-input range raises the following question—what voltage point within the prescribed input range is the "worst-case" (or maximum) for a given stress parameter? Note that in selecting a power component we often need to consider the worst-case stress it is going to endure in our application. And then, provided that that particular stress parameter happens to be a relevant and decisive factor in its selection, we usually add an additional amount of safety margin, for the sake of reliability. However, the problem is that different stress parameters do not attain their worst-case values at the same input voltage point. We therefore realize that the design of a wide-input converter is necessarily going to be "tricky." For sure, designing a *functional* switching converter may be considered "easy," but designing it *well* certainly isn't.

Toward the end of this chapter, we will finally present the detailed DC-DC converter design procedure. But to account for a wide-input range, we will proceed in two distinct steps:

- A *general inductor design procedure*, for choosing and validating an off-the-shelf inductor for our application. We will see that, depending on the topology at hand, this is to be carried out at a certain, specified voltage end—one that we will identify as being the "worst-case" from the viewpoint of the inductor.

- Then we will consider the other power components. We will point out which particular stress parameters are important in each case, and also the input voltage at which they reach their maximum, and how to ultimately select the component.

Note that, although the design procedure may be seen to specifically address only the buck topology, the accompanying annotations clearly indicate how a particular step or equation may need to change if the procedure were being carried out for a boost or a buck-boost topology.

3.1 DC Transfer Functions

When the switch turns ON, the current ramps up in the inductor according to the inductor equation $V_{ON} = L \times \Delta I_{ON}/t_{ON}$. The current *increment* during the on-time is $\Delta I_{ON} = (V_{ON} \times t_{ON})/L$. When the switch turns OFF, the inductor equation $V_{OFF} = L \times \Delta I_{ON}/t_{OFF}$ leads to a current *decrement* $\Delta I_{OFF} = (V_{OFF} \times t_{OFF})/L$.

The current increment ΔI_{ON} must be equal to the decrement ΔI_{OFF}, so that the current at the end of the switching cycle returns to the *exact* value it had at the start of the cycle—otherwise we wouldn't be in a repeatable (steady) state. Using this argument, we can derive the input-output (DC) transfer functions of the three topologies, as shown in Table 3.1. It is interesting to note that the reason the transfer functions turn out different in each of the three cases can be traced back to the fact that the *expressions for* V_{ON} *and* V_{OFF} *are different.* Other than that, the derivation and its underlying principles remain the same for all topologies.

Table 3.1: Derivation of DC transfer functions of the three topologies

	Applying Voltseconds Law and $D = t_{ON}/(t_{ON} + t_{OFF})$		
Steps	$V_{ON} \times t_{ON} = V_{OFF} \times t_{OFF}$		
	$\dfrac{t_{ON}}{t_{OFF}} = \dfrac{V_{OFF}}{V_{ON}}$		
	$\dfrac{t_{ON}}{t_{ON} + t_{OFF}} = \dfrac{V_{OFF}}{V_{OFF} + V_{ON}}$		
	Therefore,		
	$D = \dfrac{V_{OFF}}{V_{ON} + V_{OFF}}$ (duty cycle equation for all topologies)		
	Buck	**Boost**	**Buck-Boost**
V_{ON}	$V_{IN} - V_O$	V_{IN}	V_{IN}
V_{OFF}	V_O	$V_O - V_{IN}$	V_O
DC Transfer Functions	$D = \dfrac{V_O}{V_{IN}}$	$D = \dfrac{V_O - V_{IN}}{V_O}$	$D = \dfrac{V_O}{V_{IN} + V_O}$

3.2 The DC Level and the "Swing" of the Inductor Current Waveform

From $V = L\, dI/dt$ we get $\Delta I = V\Delta t/L$. So the "swinging" component of the inductor current "ΔI" is completely determined by the applied voltseconds and the inductance. *Voltseconds* is the applied voltage multiplied by the time that it is applied for. To calculate it, we can either use V_{ON} times t_{ON} (where $t_{ON} = D/f$), or V_{OFF} times t_{OFF} (where $t_{OFF} = (1 - D)/f$), and we will get the same result (for that is how D gets defined in the first place!). But note also that if we apply 10V across a given inductor for 2 µs, we will get the same current swing ΔI as if we apply say, 20V for 1 µs, or 5V for 4 µs, and so on. So, *for a given inductor*, the voltseconds and ΔI are effectively one and the same thing.

What does voltseconds depend on? It depends on the input/output voltages (duty cycle) and the switching frequency. *Therefore, only by changing L, f, or D can we affect ΔI.* Nothing else! See Table 3.2. In particular, changing the load current I_O does nothing to ΔI. I_O is therefore, in effect, an altogether *independent* influence on the inductor current waveform. But what part of the inductor current does it specifically influence/determine? We will see that I_O is proportional to the average inductor current.

Table 3.2: How varying the inductance, frequency, load current, and duty cycle influences ΔI and I_{DC}

Action:													
		$L \uparrow$ (increasing)			$I_O \uparrow$ (increasing)			$D \uparrow$ (increasing)			$f \uparrow$ (increasing)		
		Buck	Boost	Buck-Boost	Buck	Boost	Buck-Boost	Buck	Boost	Buck-Boost	Buck	Boost	Buck-Boost
Response:	$\Delta I = ?$	$\downarrow$	$\downarrow$	$\downarrow$	$\times$	$\times$	$\times$	$\downarrow$	$\uparrow \downarrow^*$	$\downarrow$	$\downarrow$	$\downarrow$	$\downarrow$
	$I_{DC} = ?$	$\times$	$\times$	$\times$	$\uparrow(=)$	$\uparrow$	$\uparrow$	$\times$	$\uparrow$	$\uparrow$	$\times$	$\times$	$\times$

$\uparrow \downarrow$ indicates it increases and decreases over the range
* maximum at $D = 0.5$
"$\times$" indicates no change
$\uparrow(=)$ indicates—I_{DC} is increasing and is equal to I_O

The inductor current waveform is considered to have another (independent) component besides its swing ΔI: this is the *DC* (average) level I_{DC}, defined as the level *around* which the swing ΔI takes place symmetrically—that is, $\Delta I/2$ above it, and $\Delta I/2$ below it. See Figure 3.1. Geometrically speaking, this is the "center of the ramp." It is sometimes also called the *platform* or *pedestal* of the inductor current. The important point to note is that I_{DC} is based only on *energy flow requirements*—that is, the need to maintain an *average* rate of energy flow consistent with the input/output voltages and desired output power. So if the application conditions, that is, the output power and the input/output voltages, do not change, there is in fact *nothing* we can do to alter this *DC* level; in that sense, I_{DC} is rather "stubborn" (see Figure 3.1). In particular,

- Changing the inductance L doesn't affect I_{DC}.

- Changing the frequency f doesn't affect I_{DC}.

- Changing the duty cycle D *does* affect I_{DC}—for the boost and buck-boost.

To understand the last bullet above, we should note the following equations that we will derive a little later:

$$I_{DC} = I_O \text{ (buck)} \qquad (3\text{-}1)$$

$$I_{DC} = \frac{I_O}{1 - D} \text{ (boost and buck-boost)} \qquad (3\text{-}2)$$

The intuitive reason why the above relations are different is that in a buck, the output is in series with the inductor (from the standpoint of the DC currents—the output capacitor contributing nothing to the DC current distribution), and therefore the average inductor

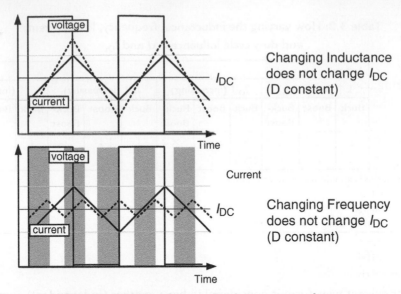

Figure 3.1: If *D* and *I_O* are fixed, *I_{DC}* cannot change

current must at all times be equal to the load current. In a boost and buck-boost, however, the output is likewise in series with the diode, and so the average diode current is equal to the load current.

Therefore, if we keep the load current constant, and change only the input/output voltages (duty cycle), we can affect I_{DC} *in all cases **except** for the buck*. In fact, the *only* way to change the DC inductor current level for a buck is to change the load current. Nothing else will work!

In the buck, I_{DC} and I_O are equal. But in the boost and buck-boost, I_{DC} depends also on the duty cycle. That makes the design/selection of magnetics for these two topologies rather different from a buck. For example, if the duty cycle is 0.5, their average inductor current is twice the load current. Therefore, using a 5 A inductor for a 5 A load current may be a recipe for disaster.

One thing we can be sure of is that, in the boost and buck-boost, I_{DC} is *always greater* than the load current. We may be able to cause this DC level to fall and even approach the load current value if we reduce the *duty cycle* close to 0 (i.e., a very small *difference between the input and output voltages*). But then, on increasing the duty cycle toward 1, the DC level of the inductor current will climb steeply. It is important we recognize this clearly and early on.

Another thing we can conclude with certainty is that in *all* the topologies, the DC level of the inductor current is *proportional* to the load current. So doubling the load current, for example (keeping everything else the same), doubles the DC level of the inductor current (whatever it was to start with). So in a boost with a duty cycle of 0.5, for example, if we have a 5A load, then the I_{DC} is 10A. And if I_O is increased to 10A, I_{DC} will become 20A.

Summarizing, changing the input/output voltages (duty cycle) does affect the DC level of the inductor current for the boost and the buck-boost. But changing D affects the swing ΔI in all three topologies, because it changes the duration of the applied voltage and thereby changes the voltseconds.

- Changing the duty cycle affects I_{DC} for the boost and the buck-boost.

- Changing the duty cycle affects ΔI for all topologies.

> **Note:** The off-line forward converter transformer is probably the only known exception to the logic presented above. We will learn that if we, for example, double the duty cycle (i.e., double t_{ON}), then almost *coincidentally*, V_{ON} halves, and therefore the voltseconds does not change (and nor does ΔI). In effect, ΔI is then independent of duty cycle.

Based on the discussions above, and also the detailed design equations, we have summarized these "variations" in Table 3.2. This table should hopefully help the reader eventually develop a more intuitive and analytical feel for converter and magnetics design, one which can come in handy at a later stage. We will continue to discuss certain aspects of this table, in more detail, a little later.

3.3 Defining the AC, DC, and Peak Currents

In Figure 3.2, we see how the AC, DC, peak-to-peak, and peak values of the inductor current waveform are defined. In particular we note that the AC value of the current waveform is defined as

$$I_{AC} = \frac{\Delta I}{2} \tag{3-3}$$

We should also note from Figure 3.2 that $I_L \equiv I_{DC}$. Therefore, sometimes in our discussions that follow, we may refer to the DC level of the inductor current as I_{DC}, and sometimes as the average inductor current I_L, but they are actually synonymous. In particular, we should not get confused by the subscript "L" in I_L. The "L" stands for *inductor*, not *load*. The load current is always designated as "I_O." Of course, we do realize that $I_L = I_O$ for a *buck*, but that is just happenstance.

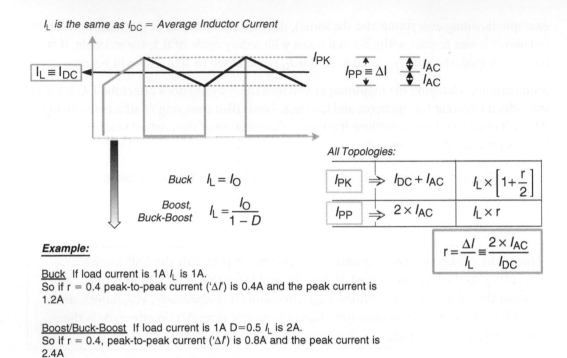

I_L is the same as I_{DC} = Average Inductor Current

All Topologies:

I_{PK}	$\Rightarrow I_{DC} + I_{AC}$	$I_L \times \left[1 + \dfrac{r}{2}\right]$
I_{PP}	$\Rightarrow 2 \times I_{AC}$	$I_L \times r$

$$r = \frac{\Delta I}{I_L} \equiv \frac{2 \times I_{AC}}{I_{DC}}$$

Buck $I_L = I_O$

Boost,
Buck-Boost $I_L = \dfrac{I_O}{1 - D}$

Example:

Buck If load current is 1A I_L is 1A.
So if r = 0.4 peak-to-peak current ('ΔI') is 0.4A and the peak current is 1.2A

Boost/Buck-Boost If load current is 1A D=0.5 I_L is 2A.
So if r = 0.4, peak-to-peak current ('ΔI') is 0.8A and the peak current is 2.4A

**Figure 3.2: The AC, DC, peak, and peak-to-peak currents,
and the current ripple ratio _r_ defined**

In Figure 3.2 we have also defined another key parameter called *r*, or the *current ripple ratio*. This connects the two independent current components I_{DC} and ΔI. We will explore this particular parameter in much greater detail a little later. Here, it suffices to mention that *r* needs to be set to an "optimum" value in any converter, usually around 0.3 to 0.5, regardless of the specific application conditions, the switching frequency, and even the topology itself. That therefore becomes a universal design *rule-of-thumb*. We will also learn that the choice of *r* affects the current stresses and dissipation in all the power components, and thereby impacts their selection. Therefore, setting *r* should be the *first step* when commencing any power converter design.

The DC level of the inductor current (largely) determines the I^2R losses in the copper windings (*copper loss*). However, the final temperature of the inductor is also affected by another term—the *core loss*—that occurs *inside* the magnetic material (core) of the inductor. Core loss is, to a first approximation, determined only by the AC (swinging) component of the inductor current (ΔI), and is therefore virtually independent of the DC level (I_{DC} or "DC bias").

We must pay the closest attention to the *peak current*. Note that, in any converter, the terms *peak inductor current*, *peak switch current* and *peak diode current* are all synonymous. Therefore, in general, we just refer to all of them as simply the *peak current* I_{PK} where

$$I_{PK} = I_{DC} + I_{AC} \tag{3-4}$$

The peak current is in fact the most critical current component of all, because it is not just a source of *long-term* heat buildup and consequent temperature rise, but a potential cause of *immediate* destruction of the switch. We will show later that the inductor current is instantaneously proportional to the magnetic field inside the core. So at the exact moment when the current reaches its peak value, so does this field. We also know that real-world inductors can *saturate* (start losing their inductance) if the field inside them exceeds a certain "safe" level, that value being dependent on the actual *material* used for the core (not on the geometry, or number of turns or even the air-gap, for example). Once saturation occurs, we may get an almost *uncontrolled* surge of current passing through the switch because the ability to limit current (which is one of the reasons the inductor is used in switching power supplies in the first place), depends on the inductor *behaving* like one. Therefore, losing inductance is certainly not going to help! In fact, we *usually* cannot afford to allow the inductor to saturate *even momentarily*. And for this reason, we need to monitor the peak current closely (usually on a cycle-by-cycle basis). As indicated, the peak is the likeliest point of the inductor current waveform where saturation can start to occur.

Note: A slight amount of core saturation may turn out to be acceptable on occasion, especially if it occurs only under temporary conditions, like power-up, for example. This will be discussed in more detail later.

3.4 Understanding the AC, DC and Peak Currents

We have seen that the AC component ($I_{AC} = \Delta I/2$) is derivable from the voltseconds law. From the basic inductor equation $V = L dI/dt$, we get

$$2 \times I_{AC} = \Delta I = \frac{\text{voltseconds}}{\text{inductance}} \tag{3-5}$$

So the current swing $I_{PP} \equiv \Delta I$ can be intuitively visualized as "voltseconds per unit inductance". If the applied voltseconds doubles, so does the current swing (and AC component). And if the inductance doubles, the swing (and AC component) is halved.

Let us now consider the DC level again. Note that *any capacitor has zero average (DC) current through it in steady-state*, so all capacitors can be considered to be missing altogether when calculating DC current distributions. Therefore, for a buck, since energy flows into the output during *both* the on-time and off-time, and via the inductor, the average *inductor* current must always be equal to the load current. So

$$I_{L} = I_{O} \quad \text{(buck)} \tag{3-6}$$

On the other hand, in both the boost and the buck-boost, energy flows into the output only during the off-time, and via the diode. Therefore, in this case, the average *diode* current must be equal to the load current. Note that the diode current has an average value equal to I_{L} *when it is conducting* (see the dashed line passing through the center of the down-ramp in the upper half of Figure 3.3). If we calculate the average of this diode current over the entire switching cycle, we need to weight it by *its* duty cycle, that is, $1 - D$. Therefore, calling I_{D} the average diode current, we get

$$I_{D} = I_{L} \times (1 - D) \equiv I_{O} \tag{3-7}$$

solving

$$I_{L} = \frac{I_{O}}{1 - D} \tag{3-8}$$

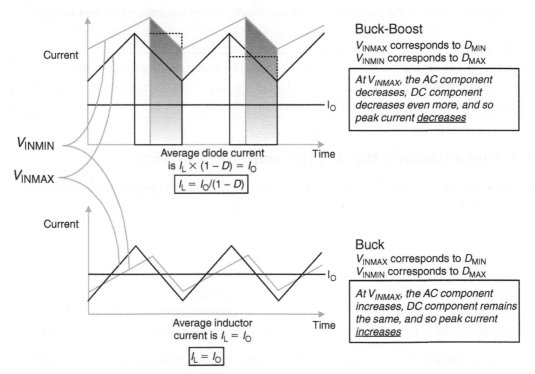

Buck-Boost

V_{INMAX} corresponds to D_{MIN}
V_{INMIN} corresponds to D_{MAX}

At V_{INMAX}, the AC component decreases, DC component decreases even more, and so peak current <u>decreases</u>

Average diode current is $I_{L} \times (1 - D) = I_{O}$

$I_{L} = I_{O}/(1 - D)$

Buck

V_{INMAX} corresponds to D_{MIN}
V_{INMIN} corresponds to D_{MAX}

At V_{INMAX}, the AC component increases, DC component remains the same, and so peak current <u>increases</u>

Average inductor current is $I_{L} = I_{O}$

$I_{L} = I_{O}$

Figure 3.3: Visualizing the AC and DC components of the inductor current as input voltage varies

Note also, that for any topology, a high duty cycle corresponds to a low input voltage, and a low duty cycle is equivalent to a high input. So increasing D amounts to decreasing the input voltage (its magnitude) in all cases. Therefore, in a boost or buck-boost, if the difference between the input and output voltages is large, we get the highest DC inductor current.

Finally, with the DC and AC components known, we can calculate the peak current using

$$I_{PK} = I_{AC} + I_{DC} \equiv \frac{\Delta I}{2} + I_L \tag{3-9}$$

3.5 Defining the "Worst-case" Input Voltage

So far, we have been implicitly assuming a *fixed* input voltage. In reality, in most practical applications, the input voltage is a certain *range*, say from V_{INMIN} to V_{INMAX}. We therefore also need to know how the AC, DC, and peak current components *change as we vary the input voltage*. Most importantly, we need to know at what specific voltage within this range we get the maximum *peak* current. As mentioned, *the peak is critical from the standpoint of ensuring there is no inductor saturation*. Therefore, defining the "worst-case" voltage (for inductor design) as the point of the input voltage range where the peak current is at its maximum, we need to design/select our inductor at this particular point always. This is in fact the underlying basis of the general inductor design procedure that we will be presenting soon.

We will now try to understand where and why we get the highest peak currents for each topology. In Figure 3.3, we have drawn various inductor current waveforms to help us better visualize what really happens as the input is varied. We have chosen two topologies here, the buck and the buck-boost, for which we display two waveforms each, corresponding to two different input voltages. Finally, in Figure 3.4 we have plotted out the AC, DC, and peak values. Note that these plots are based on the actual design equations, which are also presented within the same figure. While interpreting the plots, we should again keep in mind that for all topologies, a high D corresponds to a low input. The following analysis will also explain certain cells of the previously provided Table 3.2, where the variations of ΔI and I_{DC}, with respect to D, were summarized.

a)   *For the buck*, the situation can be analyzed as follows:

   • As the input *increases*, the duty cycle decreases in an effort to maintain regulation. But *the slope of the down-ramp $\Delta I/t_{OFF}$ cannot change*, because it is equal to V_{OFF}/L, that is, V_O/L, and we are assuming V_O is fixed. But now, since t_{OFF} has increased, but the slope $\Delta I/t_{OFF}$ has not changed, the only possibility is that ΔI must have increased (proportionally). So we conclude

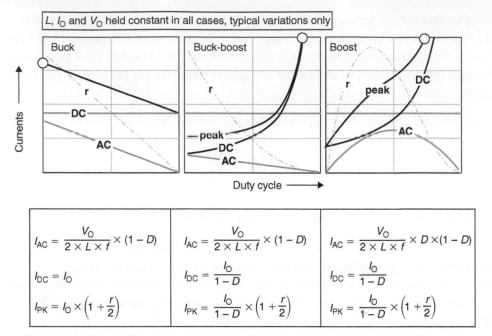

L, I_O and V_O held constant in all cases, typical variations only

$$I_{AC} = \frac{V_O}{2 \times L \times f} \times (1 - D)$$

$$I_{DC} = I_O$$

$$I_{PK} = I_O \times \left(1 + \frac{r}{2}\right)$$

$$I_{AC} = \frac{V_O}{2 \times L \times f} \times (1 - D)$$

$$I_{DC} = \frac{I_O}{1 - D}$$

$$I_{PK} = \frac{I_O}{1 - D} \times \left(1 + \frac{r}{2}\right)$$

$$I_{AC} = \frac{V_O}{2 \times L \times f} \times D \times (1 - D)$$

$$I_{DC} = \frac{I_O}{1 - D}$$

$$I_{PK} = \frac{I_O}{1 - D} \times \left(1 + \frac{r}{2}\right)$$

Figure 3.4: Plotting how the AC, DC, and peak currents change with duty cycle

that the AC component of the buck inductor current actually *increases as the input increases* (even though the duty cycle decreased in the process).

- On the other hand, the center of the ramp I_L is fixed at I_O, so we know the *DC level does not change.*

- So finally, since the peak current is the *sum* of the AC and DC components, we realize it also increases at high input voltages (see relevant plot in Figure 3.4).

Therefore, for a buck, it is always preferable to start the inductor design at V_{INMAX} (i.e., at D_{MIN}).

b) *For the buck-boost,* the situation can be analyzed as follows:

- As the input *increases*, the duty cycle decreases. But the slope of the down-ramp $\Delta I/t_{OFF}$ cannot change, because it is equal to V_{OFF}/L, that is, V_O/L, and V_O is fixed (same situation as for the buck). But since t_{OFF} has increased, ΔI must also increase to keep the slope $\Delta I/t_{OFF}$ unchanged. So we see that the *AC component ($\Delta I/2$) increases as the input increases* (duty cycle decreasing). Note that, up till this point, the analysis is the same as for the buck traced back to the fact that in both these topologies $V_{OFF} = V_O$.

- But now coming to the DC level I_L of the buck-boost, we will find it must change for this topology (though it remained fixed for the buck). Note that the shaded portion of the waveform in the upper half of Figure 3.3 represents the diode current. The average value of this during the off-time is the square dashed line passing through its center, that is, I_L. So the average diode current, calculated over the entire switching cycle, is $I_L \times (1 - D)$. And we know this must equal the load current I_O. So, as the input increases and duty cycle decreases, the term $(1 - D)$ increases. So the only way $I_L \times (1 - D)$ can remain constant at the value I_O is if I_L *decreases correspondingly*. We therefore realize that the DC level decreases as the input increases (duty cycle decreasing).

- Further, since the peak current is the sum of the AC and DC components, it also decreases at high input voltages (see relevant plot in Figure 3.4).

Therefore, for a buck-boost, we should always start the inductor design at V_{INMIN} (i.e., at D_{MAX}).

c) **For the boost**, the situation is a little trickier to understand. On the face of it, it is quite similar to the buck-boost, but there is a notable difference—and that is the reason why we did not even try to include it in Figure 3.3.

- Once again, as the input increases, the duty cycle decreases. But the difference here is that the slope of the down-ramp $\Delta I / t_{\text{OFF}}$ must decrease, because it is equal to V_{OFF}/L, that is, $(V_O - V_{\text{IN}})/L$ (magnitudes only), and we know that $V_O - V_{\text{IN}}$ is *decreasing*. Further, the required decrease in the slope $\Delta I / t_{\text{OFF}}$ can come about in two ways—either from an increase in t_{OFF} (which is already occurring as the duty cycle decreases), or from a decrease in ΔI. But in fact, ΔI may actually increase rather than decrease (as we increase the input). For example, if t_{OFF} is increasing more the increase in ΔI—then $\Delta I / t_{\text{OFF}}$ will still decrease as required. And in practice, that is what actually does happen in the case of the boost. With some detailed math, we can show that ΔI increases as *D approaches* 0.5, but decreases on either side (see Table 3.2 and Figure 3.4).

- It is therefore also clear that in either case above, the increase/decrease in the AC level does not dominate, and therefore, the peak current ends up being dictated only by the DC component. But we already know that the DC level of a boost changes in exactly the same way as for the buck-boost (discussed above) it decreases as the input increases (duty cycle decreasing).

- So we conclude that the peak current for the boost also decreases at high input voltages (see relevant plot in Figure 3.4).

Therefore, for a boost, we should always start the inductor design at V_{INMIN} (i.e., at D_{MAX}).

3.6 The Current Ripple Ratio *r*

In Figure 3.2 we first introduced the most basic yet far-reaching design parameter of the power supply itself—its current ripple ratio *r*. This is a geometrical ratio that compares and connects the AC value of the inductor current to its associated DC value. So

$$r - \frac{\Delta I}{I_{\text{L}}} \equiv 2 \times \frac{I_{\text{AC}}}{I_{\text{DC}}} \qquad (3\text{-}10)$$

Here we have used $\Delta I = 2 \times I_{\text{AC}}$, as defined earlier in Figure 3.2. *Once r is set by the designer (at maximum load current and worst-case input), almost everything else is pre-ordained*—like the currents in the input and output capacitors, the RMS (root mean square) current in the switch, and so on. Therefore, the choice of *r* affects component selection and cost, and it must be understood clearly, and picked carefully.

Note that the ratio *r* is defined for CCM (*continuous conduction mode*) operation only. Its valid range is from 0 to 2. When *r* is 0, ΔI must be 0, and the inductor equation then implies a very large (infinite) inductance. Clearly, *r* = 0 is not a practical value! If *r* equals 2, the converter is operating at the *boundary* of continuous and discontinuous conduction modes (boundary conduction mode or BCM). See Figure 3.5. In this so-called boundary (or "critical") conduction mode, $I_{\text{AC}} = I_{\text{DC}}$ by definition.

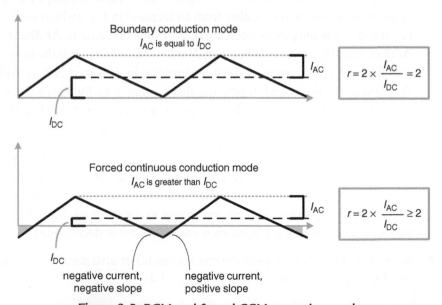

Figure 3.5: BCM and forced CCM operating modes

Note that an exception to the "valid" range of r from 0 to 2 occurs in *forced CCM* mode, discussed in more detail later.

3.7 Relating r to the Inductance

We know that current swing is voltseconds per unit inductance. So we can also write

$$\Delta I = Et/L_{\mu H} \quad \text{(any topology)} \tag{3-11}$$

Here Et is defined as the (magnitude of the) *voltµseconds* across the inductor (either during the on-time or off-time—both being necessarily equal in steady state), and $L\mu H$ is the inductance in µH. The reason for defining Et is that this number is simply easier to manipulate than voltseconds because of the very small time intervals involved in modern power conversion.

Therefore, the current ripple ratio is

$$r = \frac{\Delta I}{I_L} = \frac{Et}{L_{\mu H} I_L} \quad \text{(any topology)} \tag{3-12}$$

Note also that from now on, whenever L is paired up with Et in any given equation, we will drop the subscript of L, that is, "µH." It will then be "understood" that L is in µH.

Finally, we have the following key relationships between r and L:

$$r = \frac{Et}{(L \times I_L)} \equiv \frac{V_{ON} \times D}{(L \times I_L) \times f} \equiv \frac{V_{OFF} \times (1 - D)}{(L \times I_L) \times f} \quad \text{(any topology)} \tag{3-13}$$

Incidentally, the preceding equation, that is, the one involving V_{OFF}, assumes CCM, because it assumes that t_{OFF} (the time for which V_{OFF} is applied) is equal to the full available off-time $(1 - D)/f$.

Conversely, L as a function of r is

$$L = \frac{V_{ON} \times D}{r \times I_L \times f} \quad \text{(any topology)} \tag{3-14}$$

In subsequent sections we will often use the following easy-to-remember form of the previous equations. We are going to nickname this the **"$L \times I$" equation** (or rule)

$$L \times I_L = \frac{Et}{r} \quad \text{(any topology)} \tag{3-15}$$

But perhaps we are still wondering why we even need to talk in terms of r—why not talk *directly* in terms of L? We do realize from the above equations that L and r are related. However, the "desirable" value of inductance depends on the specific application conditions, the switching frequency, and even the topology. So it is just not possible to give a general design rule for picking L. But there is in fact such a general design rule-of-thumb for selecting r, one that applies almost universally. We mentioned that it should be around 0.3 to 0.5 in all cases. And that is why *it makes sense to calculate L by first setting the value of r*. Of course, once we pick r, L gets automatically determined, but only *for a given set of application conditions and switching frequency.*

3.8 The Optimum Value of r

It can be shown that, in terms of overall stresses in a converter and size, $r \approx 0.4$ represents an "optimum" of sorts. We will now try to understand why this is so, and later we will try to point out exceptions to this reasoning.

The size of an inductor can be thought of as being virtually proportional to its *energy-handling capability* (the effect of air-gap on size will be studied later). So, for example, we probably already know intuitively that we need bigger cores to handle higher powers. The energy-handling capability of the selected core must, at a bare minimum, match the energy we need to store in it in our application—that is, $1/2 \times L \times I_{PK}^2$. Otherwise the inductor will saturate.

In Figure 3.6, we have plotted the energy, $E = 1/2 \times L \times I_{PK}^2$, as a function of r. We see that it has a "knee" at around 0.4. This tells us that if we try to reduce r much lower

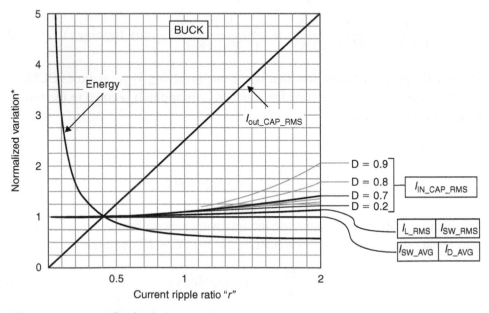

*All parameters normalized to their respective values at $r = 0.4$

Figure 3.6: How varying the current ripple ratio r affects all the components

than 0.4, we will certainly need a *very large inductor*. On the other hand, if we increase *r*, there isn't much greater reduction in the size of the inductor. In fact, we will see that beyond *r* ~ 0.4, we enter a region of diminishing returns.

In Figure 3.6, we have also plotted the capacitor RMS currents for a buck converter. We see that if *r* is increased beyond 0.4, the currents will increase significantly. This will lead to increased heat generation inside the capacitors (and other related components too). Eventually, we may be forced to pick a capacitor with a lower ESR and/or lower case-to-air thermal resistance (more expensive/bigger).

> **Note:** The RMS value of the current through any component is the current component responsible for the heat developed in it, via the equation $P = I_{RMS}^2 \times R$, where *P* is the dissipation, and *R* is the series resistance term associated with the particular component (e.g., the DCR of an inductor, or the ESR of a capacitor). However, it can be shown that the switch, diode, and inductor RMS current values are not very "shape-dependent." Therefore, the heat developed in them does not depend much on *r*, but mainly on the average value of the current. On the other hand, the RMS of the capacitor current waveforms can increase significantly, if *r* is increased. So capacitor currents are very "shape-dependent," and therefore depend strongly on *r*. The reason for that is fairly obvious—any capacitor in a steady state has zero average (DC) current through it. So since a capacitor effectively subtracts out the DC level of the accompanying current waveform, we are left with a capacitor current waveform that has a large "ramp portion" built-in into it. Therefore, changing *r* changes this ramp portion, thereby impacting the capacitor current greatly.

Note that in Figure 3.6, though we have used the buck topology as an example, the energy curve in particular is exactly the same for any topology. The capacitor current curves, though, may not be identical to those of the buck, but are *similar*, and so the conclusions above still apply.

Therefore, in general, a current ripple ratio of around 0.4 is a good design target for any topology, any application, and any switching frequency.

Later, we will discuss some reasons/considerations for *not* adhering to this *r* ~ 0.4 rule-of-thumb (under certain conditions).

3.9 Do We Mean Inductor? or Inductance?

Note that in the previous section, we said nothing explicitly about what the *inductance* was—we just talked about the *size* of the inductor. We know that, in theory, we can put almost any number of turns on a given core, and get almost any inductance. So *inductance* and *size* of inductor are not *necessarily* related. However, we will now see that in power conversion they often do turn out to be so, though rather indirectly.

Looking at Figure 3.6, we can see that a smaller r will require a higher energy-handling capability, and thus a larger inductor. Let us now formally go through *all* the possible ways of reducing r.

Since we are assuming our application conditions are fixed, the load current and input/output voltages are also fixed. Therefore, I_{DC} is fixed too. The only way we can cause r to decrease under these circumstances is to make ΔI smaller. However, ΔI is

$$\Delta I = \frac{\text{voltseconds}}{\text{inductance}} \ (\text{V-s/H}) \qquad (3\text{-}16)$$

But we know the applied voltseconds is fixed too (input and output voltages being fixed). So the only way to decrease r (for a given set of application conditions) is to increase the inductance. We can therefore conclude that if we choose a high inductance, we will invariably require a bigger inductor. It is therefore no surprise that when power supply designers instinctively ask for a "large inductance," they might well mean a "large inductor." Therefore the designer is cautioned against being too "ripple-phobic" in their designs. A certain amount of ripple is certainly "healthy."

However, we must not forget that if, for example, we increase the load current (i.e., a change in application conditions), we will clearly need to move to a larger inductor (with greater energy-handling capability). But simultaneously, we will need to *decrease* the inductance. That's because I_{DC} will increase, and so to keep to the "optimum" value of r, we will need to *increase* ΔI in the same proportion as the increase in I_{DC}. And to do this, we have to decrease, rather than increase, L.

Therefore, the general statement that a "large inductance is equivalent to a large inductor" only applies to a *given* application.

3.10 How Inductance and Inductor Size Depend on Frequency

The following discussion applies to *all* the topologies.

If keeping everything else fixed (including D) we double the frequency, the voltseconds will halve, because the durations t_{ON} and t_{OFF} have halved. But since ΔI is "voltseconds per unit inductance," this too will halve. Further, since I_{DC} has not changed, $r = \Delta I/I_{DC}$ will also halve. So if we started off with $r = 0.4$, we now have $r = 0.2$.

If we want to return the converter to the optimum value of $r = 0.4$, we will now need to somehow double the ΔI we were left with at the end of the last step. The way to do that is to halve the inductance.

- Therefore, we can generally state that inductance is inversely proportional to frequency.

Finally, having restored r to 0.4, the peak will still be 20% higher than the DC level. But the DC level has not changed. So the peak value is also unchanged (since r hasn't changed either, eventually). However, the energy-handling requirement (size of inductor) is $\frac{1}{2} \times L \times I_{PK}^2$. So since L has halved, and I_{PK} is unchanged, the required *size of the inductor has halved*.

- Therefore, we can generally state that the size of the inductor is inversely proportional to frequency.

- Note also that the required current rating of the inductor is independent of the frequency (since peak is unchanged).

3.11 How Inductance and Inductor Size Depend on Load Current

For all topologies, if we *double the load current* (keeping input/output voltages and D fixed), r will tend to halve since ΔI has not changed but I_{DC} has doubled. Therefore to restore r to its optimum value of 0.4, we need to get ΔI to double too. But we know that ΔI is simply "voltseconds per unit inductance," and in this case the voltseconds has not changed. So the only way to get ΔI to double is to halve the inductance.

- Therefore, we can generally state that inductance is inversely proportional to the load current.

What about the size? Since we doubled the load current, but still kept r at 0.4, the peak current $I_{DC}(1 + r/2)$ has also doubled. But the inductance has halved. So the energy-handling requirement (size of inductor), $\frac{1}{2} \times L \times I_{PK}^2$, will double.

- Therefore, we can generally state that the size of the inductor is proportional to the load current.

3.12 How Vendors Specify the Current Rating of an Off-the-shelf Inductor and How to Select it

The "energy-handling capability" of an inductor, $1/2 \times LI^2$ is one way of picking the size of the inductor. But most vendors do not provide this number upfront. However, they do provide one or more "current ratings." And if we interpret these current rating(s) correctly, that serves the purpose too.

The current rating may be expressed by the vendor either as a maximum rated I_{DC}, or a maximum rated I_{RMS}, or/and a maximum I_{SAT}. The first two are usually considered synonymous, since the RMS and DC values of a typical inductor current waveform are almost equal (we had indicated previously that the RMS of the inductor current is not very "shape-dependent"). So the DC/RMS rating of an inductor is by definition basically the direct current we can pass through it, such that we get a specified temperature rise (typically 40 to 55°C depending on the vendor). The last rating, that is, the I_{SAT}, is the maximum current we can pass, just before the core starts saturating. At that point, the inductor is considered close to the useful limit of its energy-storing capability.

We will also find that many, if not most, vendors have chosen the wire gauge in such a manner that the I_{DC} and I_{SAT} ratings of any inductor are also virtually the same. And by doing this, they can publish one (*single*) current rating—for example, "the inductor is rated for 5 A." Basically, having determined the I_{SAT} of the inductor, the vendor has then consciously tweaked the wire gauge (at the saturation current level), so as to also get the specified temperature rise too.

The rationale for wanting to set $I_{DC} = I_{SAT}$ is as follows—suppose the inductor had a DC rating of 3 A and an I_{SAT} of 5 A. The 5 A rating is then likely to be superfluous, because users would probably never select this inductor for an application that required more than 3 A anyway. Therefore, the excessive I_{SAT} rating in this case essentially amounts to an unnecessarily over-sized core. Of course, if we do find an inductor with different I_{DC} and I_{SAT} ratings, it is also possible the vendor may have (unsuccessfully) tried to exploit the larger size of the chosen core (by increasing the wire thickness), but the stumbling block was that the selected core geometry was somehow not conducive to doing so. Maybe it just did not have enough *window space* for accommodating the thicker windings.

In general, an inductor with a "single" current rating is usually the most optimum/cost-effective too.

However, in some rare off-the-shelf inductors, we may even find I_{SAT} stated to be less than I_{DC}. But what use is that? We can't operate beyond I_{SAT} in any case! So the only advantage, if any, that can be gleaned from such an inductor is that the temperature rise in a real application will be less than the maximum specified. Automotive applications?

In general, for most practical purposes, the current rating of the inductor that we need to consider is the **lowest** rating of all the published current ratings. We can usually simply ignore all the rest.

There are some subtle considerations and exceptions to the argument for always preferring an inductor with $I_{DC} \approx I_{SAT}$. For example, under transient/temporary conditions, the *momentary* current may exceed the normal steady operating current by a wide margin. So for example, suppose we are we using a switcher with an internally *fixed* current limit I_{CLIM} of 5 A— in *a 3 A application*. Then under startup (or sudden line/load steps), the current is very likely to hit the limiting value of 5 A for several cycles in succession as the control circuitry struggles to bring up the output rail into regulation. We will discuss this issue in greater detail below—in particular, whether this is even a concern to start with! However, assuming for now that it is, it then seems that it may actually make sense to use an inductor rated for 3 A continuous current, with an I_{SAT} rating of 5 A (provided such an inductor is freely available, and cheap). Of course, alternatively, we could just pick a standard "5 A inductor" (for the 3 A application), and thereby we would certainly avoid inductor saturation under all conditions (and the consequent likelihood of switch destruction). But we realize that, in doing so, our inductor may be considered slightly *over-designed* from the viewpoint of its copper/temperature-rise—the wire being unnecessarily thicker. However, we should keep in mind that larger cores certainly affect cost, but a little more copper rarely does!

3.13 What Is the Inductor Current Rating We Need to Consider for a Given Application?

Whenever we start up, or subject the converter to sudden line/load transients, the current no longer stays at the steady value it has under *normal* operation (i.e., when delivering the required maximum rated load current). For example, if we suddenly short the output, the control circuitry, in an effort to regulate the output, may momentarily expand the duty cycle to the highest permissible value (as set by the controller). We then are no longer in steady state, and so under the increased on-time voltseconds, the current ramps up progressively, and can reach the set current limit.

But then the inductor would probably be saturating! For example, if we are using a 5 A fixed current limit *buck* switcher IC for a 3 A application, we have probably picked an inductor rated for only around 3 A. But when we short the output, the current momentarily hits the current limit (which may be around 5.3 A for a "5 A buck switcher").

So the question is—should we select an inductor with a rating based on the current limit threshold (that it may encounter under severe transients), or simply on the basis of the maximum continuous normal operating current (under steady state operation in our application)? In fact, this question is not as philosophical as it may seem—it virtually separates standard industry off-line design procedures from those of DC-DC converters. To answer it effectively, a lot of factors may need to be considered, often on an individual or case-by-case basis. Let us address some of these concerns next.

Luckily, in most low-voltage applications, a certain amount of core saturation doesn't cause any problem. The reason for that is that if in the above example, the switch is rated for 5 A, and the current limiting circuit in the IC is known to act *fast enough* to prevent the current from ever rising beyond 5 A, then even if the inductor has started saturating as it gets to 5 A, there is no cause for concern—after all, if the switch doesn't break, we don't have a problem! And since the current doesn't exceed 5 A, the switch cannot break. So in this case, we could certainly pick a cost-effective "3 A inductor" for our application, knowing well in advance that it would saturate somewhat under various nonsteady conditions. Of course we don't want to operate a switching converter constantly (under its rated maximum load conditions) with a saturating inductor—we just "allow" it to do so under abnormal and temporary conditions, so long as we are sure that the switch can never be damaged.

However, this logic begs another key question to be answered—what exactly constitutes "fast enough"—that is, which factors affect our ability to turn the switch OFF fast enough to protect it from the consequences of a saturating inductor? Since this consideration may eventually end up dictating the size and cost of the inductor, it is important to understand this response-time issue well.

a) All current-limit circuitry takes some finite time to respond. There are inherent (internal) "propagation delays" as we move the overcurrent signal through the internal comparators of the IC, its op-amps, level-shifters, driver, and so on to the IC pin driving the switch.

b) If we are using a controller IC (as opposed to an integrated switcher, i.e., with an internal switch), the switch will necessarily be at a certain physical distance from its driver (which is usually inside the IC). In that case, the parasitic inductances of the intervening PCB traces (roughly 20 nH per inch of trace) will resist any sudden change in current, thereby creating an additional delay before the turn-off command issued by the IC actually reaches the gate/base of the switch.

c) Theoretically speaking, even if the current limiting circuitry had responded *immediately* to the overcurrent condition, *and* if the intervening traces had truly negligible inductance, the switch may still take a little time before it really turns itself OFF. During this delay, if the inductor is saturating, it will not be able to effectively prevent or limit the current spike that can be pushed through the transistor by the applied input DC source—well beyond the "safe" current limit threshold. Bipolar junction transistors (BJTs) are inherently slow, as compared to more modern devices like MOSFETs. But large MOSFETs (e.g. high-current, high-voltage devices) also produce delays because of their higher internal parasitic gate resistance and inductance and significant *inter-electrode parasitic capacitances* (that demand to be either discharged or charged as the case may be, before they allow the switch to change its state). Matters can get worse if we parallel several such MOSFETs together, as say for a very high-current application.

d) Many controllers and ICs incorporate an internal "blanking time" during which they deliberately "do not look" at the current waveform. The basic purpose is to avoid false triggering of the current limit circuitry by the noise generated at the turn-on transition. But this delay time could prove fatal to the switch, especially if the inductor has already started saturating, because the current limit circuitry won't even "know" if there is any overcurrent condition during this blanking interval. Further, in current-mode control ICs, the ramp to the PWM (pulse-width modulator) comparator stage is usually derived from the (noisy) switch current. So the blanking time is typically set even higher, typically about 100 ns for low-voltage applications and up to 300 ns for off-line applications.

e) Integrated high-frequency switchers (i.e., with the MOSFET or BJT switch contained in the same package as the control and driver) are usually the best-protected and most reliable, because the intervening inductances are minimized. Also, the blanking times can be set more accurately and optimally, since there is not going to be much variation in terms of different switches with widely varying characteristics. Therefore, integrated switchers can usually survive momentarily saturating inductors with almost no problem—unless the input voltage is very high (typically above 40–60 V), and *plus*, the inductor is sized very small.

f) If the input voltage is high, the rate of rise of the saturating inductor current ramp can become very large ("steep"). This follows from the basic equation $V = LdI/dt$. Here, if $L \to 0$, since V is fixed, the dI/dt must increase dramatically (see Figure 3.7). So now, even a small delay can prove fatal because a large ΔI can take place

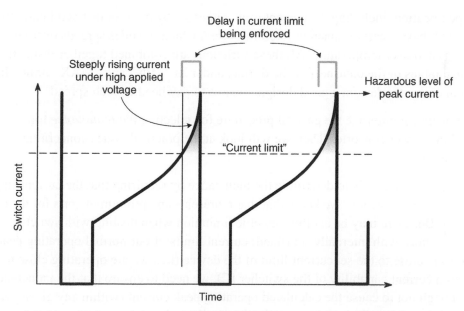

Figure 3.7: How higher voltages combined with inherent response-time delays can cause overstress in the switch when the inductor starts saturating

during a very small interval. The current can therefore overshoot the set current limit threshold by a very large amount, thereby endangering the switch. That is why, especially when we come to off-line applications, it is actually *customary* to select a core large enough to avoid saturation at the current limit threshold. And that usually gives enough time for the current limit circuitry to act—*before* the slope of the current has gone completely out of control.

Note, however, that the copper windings still only need to be proportioned to handle the continuous current (i.e., based on the maximum operating load).

In effect, what we are therefore always implicitly doing in off-line applications is *setting the I_{SAT} of the transformer higher than its I_{DC} rating.* That is clearly not what we usually do in low-voltage DC-DC converter design.

g) Generally speaking, in most low-voltage applications (i.e., V_{IN} typically less than about 40V), the inductors are selected based only on the maximum operating load current. The current limit is therefore, in effect, virtually ignored! This is the usual industry practice for DC-DC converter design, though it is probably not clearly spelled out in this way most of the time. But luckily, it seems to have worked!

3.14 The Spread and Tolerance of the Current Limit

Any specification, including the current limit, either set by the user or fixed internally in the IC, will have a certain inherent *tolerance band,* which includes spreads over process variations and over temperature. All these variations are combined together inside the electrical tables of the datasheet of the device, under its "MIN" and "MAX" limits. In a practical converter design, a good designer learns to pay heed to such spreads.

But let us first summarize the general procedure for selecting the *inductance* for a switching power converter. Then we will look at the practical issues concerning spreads/tolerance.

The normal procedure is to determine the inductance by requiring that the current ripple ratio is about 0.4, because we know that that represents an optimum of sorts for the entire converter. But there may be another possible limitation when dealing with switcher ICs, especially those with internally set (fixed) current limits: if our normal operating peak currents are close to the set current limit of the device (i.e., we are operating close to the maximum current capability of the switcher IC), we need to ensure that the inductance is large enough not to cause the calculated operating peak current (within any given cycle) to exceed the current limit. Otherwise foldback will obviously occur at the current limit threshold, and so the desired maximum output power cannot be guaranteed.

For example, if we have a "5 A buck switcher IC," being operated at 5 A load, with an r of 0.4, then the normal operating peak current is $5 \times (1 + 0.4/2) = 5 \times 1.2 = 6$A. So *ideally*, we would want the current limit of the device to be at least 6A. Unfortunately, when we come to such integrated switchers, that much "margin" is rarely available. Manufacturers always like to "bolster" the advertised ratings of their parts to be close to the maximum stress limits. So yes, if this particular part was declared to be a "4 A IC" instead of a "5 A IC," we would have been just fine. But as things stand, manufacturers usually pay scant regard as to what may constitute an *optimum* rating for the device, in relationship to its associated components and the overall design strategy. Therefore, for example, a certain commercial "5 A switcher IC" may have a published (set) current limit of only 5.3 A. But on analysis, we see that that allows only 0.3 A above, and 0.3 A below, the average level of 5 A. Therefore, the maximum allowed ΔI at 5 A load is only 0.6 A. And the maximum r is $0.6/5 = 0.12$ (when operated at a load current of 5 A). We can see that that is clearly much less than the optimum r of 0.4. And, no doubt, this lowered r will adversely impact the size of the inductor (and converter).

Now we take up the issue of the *spread* in current limit. So I_{CLIM} is actually two limits— I_{CLIM_MIN} and I_{CLIM_MAX} (i.e., the MIN and MAX of the current limit, respectively). The question is: which of these limits should we consider for designing the inductor?

- To guarantee output power, we need to look at the MIN of the current limit only. In most low-voltage DC-DC converter applications, the MIN limit is the only threshold that really counts—we can usually completely ignore the MAX (and of course the TYP value). The basic criterion for guaranteeing output power is—we must ensure that the calculated normal operating peak current in our application is always less than the MIN value of the current limit. Of course, if we are not operating close to the current limit of the device, this condition will be met without any struggle, and so we can then just focus on setting r to about 0.4.

- But like all components, inductors also have a typical tolerance—usually about $\pm 10\%$. So if we are operating very close to the limits of the device, and thereby r is being effectively dictated by the MIN of the current limit (rather than by its optimum or desirable value), then the (nominal) value of inductance we ultimately choose should be about *10% higher than the calculated value*. That will guarantee output power unconditionally, under all possible variations in current limit and inductance.

- Note that, ideally, we would also like to leave at least 20% additional margin (*headroom*) between the peak current of our application and the MIN of the current limit. This is usually necessary for getting a quick response (correction) to a sudden increase in load. So, in general, if we somehow manage to curtail the ability of the converter to respond quickly (for example, by not providing sufficient headroom in the current limit and/or maximum duty cycle), the inductor will not be able to ramp up current quickly enough to meet the sudden increase in energy demand. Therefore, the output will droop rather severely for several cycles before it eventually recovers.

But, unfortunately, once again, when dealing with fixed current limit (integrated) switchers, we will find that this "nice-to-have transient headroom" may be a luxury we just can't afford because, in most cases, the MIN current limit is set only slightly higher than the declared "rating" of the device. So in fact even a 20% headroom may not be available! And further, even assuming it is, this may demand a very *large* (impractical) inductance. And we know that that by itself is fairly counterproductive—a large inductance takes even more time for its current to ramp up, and that thereby effectively slows down the transient (loop) response—incidentally, just opposite to what we were hoping to derive here! Therefore, in general, we almost always end up ignoring this 20% or so step-response headroom/margin completely, especially when dealing with integrated switcher ICs.

As for the MAX of the current limit, whenever we deem that inductor saturation is of real concern to us (as in high-voltage applications), we must look at the MAX of the current limit to decide upon the size of the inductor—that being the worst-case in terms of peak current under overloads, inductor energy storage, and its possible saturation.

Therefore, in general, in high-voltage DC-DC (or off-line) applications, the MIN of the current limit may sometimes need to be considered when selecting inductance (as when operating close to current limit), but the MAX of the current limit will certainly always be used to determine the size of the inductor.

As a corollary, manufacturers of (low-voltage) DC-DC converter ICs actually need *not* (and probably justifiably do not) struggle too hard to minimize the spreads and tolerances of current limit (provided of course the MIN of the current limit is at least set high enough not to intrude on the declared power-handling capability of the IC). And for low-voltage DC-DC converter applications, the current limit is typically ignored altogether; the final selection of inductor current rating (and size) is simply based on the cycle-by-cycle peak inductor current under *normal* (steady) operation (i.e., the maximum load of the application, at the worst-case input voltage end).

On the other hand, manufacturers of off-line switcher ICs *do* need to maintain a tight tolerance on the current limit. In their case, the maximum power-handling capability of their particular device is in effect dependent only on the MIN (minimum limit) of the current limit specification, whereas the transformer size is determined entirely by the MAX of the current limit specification. So, in this case, a "loose" current limit specification effectively amounts to requiring *bigger components* (transformer) for the same maximum power-handling capability.

Note: Some makers of off-line integrated switcher ICs (e.g., the "Topswitch" from Power Integrations) often tout their "precise" current limit, thus suggesting that we get the best power-to-size ratio (i.e., converter power density) when using their products. However, we should remember that in most cases, their product families have a *discrete* set of fixed current limits. And that is a problem! For example, we may have devices available with current limits in steps of 2 A, 3 A, 4 A, and so on. So yes, we may indeed get a higher power density *when operating at the maximum rated output power* of a particular IC. But when operating at a power level *between* available current limits, we are not going to get an optimum solution. For example, in an application where the peak current is 2.2 A, then we would need to select the 3 A current limit part, and we will need to design our magnetics to avoid core saturation at 3 A. So in effect, we have a very imprecise current limit now! The best solution is to look for a part (integrated switcher or controller plus MOSFET solution) where we can precisely set the current limit *externally*, depending upon our application.

With all these subtle considerations in mind, a designer can hopefully pick a more appropriate inductor current rating for his or her application. Clearly, there are no hard and fast rules. Engineering judgment needs to be applied as usual, and perhaps some further bench-testing may also be needed to validate the final choice of inductor.

In the worked examples that follow, the general approach and design procedure will become clearer.

3.15 Worked Example (1)

A boost converter has an input range of 12 V to 15 V, a regulated output of 24 V, and a maximum load current of 2 A. What would be a reasonable goal for its inductance, if the switching frequency is (a) 100 kHz, (b) 200 kHz, and (c) 1 MHz? What is the peak current in each case? And what is the energy-handling requirement?

The first thing we have to remember is that, for this topology (as for the buck-boost), the *worst-case* is the lowest end of the input range, since that corresponds to the highest duty cycle and thus the highest average current $I_L = I_O/(1 - D)$. So for all practical purposes, we can completely disregard V_{INMAX} here—in fact it was a red herring to start with, for this particular analysis!

From Table 3.1, the duty cycle is

$$D = \frac{V_O - V_{IN}}{V_O} = \frac{24 - 12}{24} = 0.5$$

Therefore,

$$I_L = \frac{I_O}{1 - D} = \frac{2}{1 - 0.5} = 4 \text{ A}$$

Let us target a current ripple ratio of 0.4. So

$$I_{PK} = I_L \left(1 + \frac{r}{2}\right) = 4 \times \left(1 + \frac{0.4}{2}\right) = 4.8 \text{ A}$$

- We should remember that $r = 0.4$ always implies that the peak is 20% higher than the average. So we realize that, in effect, the peak current does *not* depend on the frequency. The inductor must be able to handle this above peak current without saturating. So in this example, we would be fine just picking an inductor rated for 4.8 A (or more), irrespective of frequency. In fact, we had previously learned in the section "How Inductance and Inductor Size Depend on Frequency" that the required current rating of an inductor is independent of the frequency (since the peak is unchanged). However, the size does change with frequency, because size is $\frac{1}{2} \times L \times I_{PK}^2$ and L changes as follows.

To calculate the inductance corresponding to the chosen value of r, we can use the following equation (presented previously). We also note from Table 3.1 that $V_{ON} = V_{IN}$ for the boost. Therefore for $f = 100\,\text{kHz}$,

$$L = \frac{V_{ON} \times D}{r \times I_L \times f} = \frac{12 \times 0.5}{0.4 \times 4 \times 100 \times 10^3} \Rightarrow 37.5\,\mu\text{H}$$

For $f = 200\,\text{kHz}$, we would get *half* of this, that is, $18.75\,\mu\text{H}$. And for $f = 1\,\text{MHz}$, we get $3.75\,\mu\text{H}$. We clearly see that high frequencies lead to smaller inductances.

We have previously observed that for a given application, small inductances invariably lead to small inductors. Therefore we conclude that on increasing the switching frequency, we will get smaller-sized inductors too. And that is the basic reason for hiking up switching frequencies in general.

The energy-handling requirement, if desired, can be explicitly calculated in each case, by using $E = \frac{1}{2} \times L \times I_{PK}^2$.

So far, we have been generally targeting $r = 0.4$ as an optimum value. Let us now understand all the reasons why this may not be a good choice on occasion.

3.15.1 Current Limit Considerations in Setting r

We had indicated previously that the current limit may be too low to allow r from being set to its optimum. Now we will also include the impact of the *spread* in the current limit.

So for example, in Table 3.3 we have the published specifications for the current limit of an integrated "5 A" switcher, the LM2679. To be able to guarantee the specified power output (or load current in this case) unconditionally, we need to guarantee that the peak current in our application never reaches even the *lower limit* ("MIN") of the published current limit specification. So in fact, in Table 3.3, we need to disregard all the numbers except for the MIN value—given as 5.3 A.

Now, if we are trying to get 5 A out of our converter with an r of 0.4, the estimated peak current will be $1.2 \times 5 = 6$ A. Clearly, as mentioned earlier, we are not going to get there with the LM2679! Unless we lower the value of r (increase inductance). Maximum value of r is

$$I_{PK} = I_O \times \left(1 + \frac{r}{2}\right) \le I_{CLIM\_MIN} \qquad (3\text{-}17)$$

Solving, with $I_O = 5\,$A, and $I_{CLIM\_MIN} = 5.3\,$A, we get

$$r \le 2\left(\frac{I_{CLIM\_MIN}}{I_O} - 1\right) = 2\left(\frac{5.3}{5} - 1\right) = 0.12$$

We can see from Figure 3.6 that this calls for an energy-handling capability (size of inductor) almost three times the optimum!

Actually, it turns out this part is just *specified* inappropriately. This part is in reality one with an *adjustable* current limit. And so we could have probably adjusted the current limit adjust resistor quoted in the electrical tables to allow for a "better" value of current limit, and thereby a better value of r (at maximum rated load). But, unfortunately, that is not clarified in the tables.

We should always remember that the minimum and maximum limits of the electrical tables are the only parts of a datasheet really guaranteed by any vendor (certainly not

Table 3.3: Published current limit specs for the LM2679

	Conditions		TYP	MIN	MAX	Units
Current Limit "I_{CLIM}"	$R_{CLIM} = 5.6\,k\Omega$	Room temperature	6.3	5.5	7.6	A
		Full operating temperature range		5.3	8.1	

the typical values!). So as a matter of fact, any other information in a datasheet just amounts to general design "guidance" and that includes any 'typical performance curves' provided. A prudent designer would never second-guess the vendor—in this case, as to whether the current limit resistor can indeed be adjusted to give us a smaller inductor, or not. Therefore, as it stands, if we are using the LM2679 for a 5A load current application, we do need an inductor three times larger than the optimum. Note that if the current limit can indeed be adjusted higher, the vendor should have picked the appropriate value for the current limit adjust resistor in the "conditions" column of the electrical table (and stated the limits accordingly).

Note also that when we talk of a "5A buck IC," that implies the part is supposed to deliver 5A *load current*. The current limit of course needs to be set (and stated) correctly for the rated load, as discussed above. However, we should be very clear that when we are talking of boost or buck-boost switcher ICs, a "5A" part, for example, does *not* give us a 5A load current. That is because the DC inductor current is not equal to I_O, but $I_O/(1 - D)$ for these topologies. So a "5A" rating in this case only refers to the *current limit* of the device. What load current we can derive from a "non-buck" IC depends on our specific application—in particular on the D_{MAX} (duty cycle at V_{INMIN}). For example, if the desired load current is 5A, and the (maximum) duty cycle in our application is 0.5, then the average inductor current is actually $I_O/(1 - D) = 10A$. Further, with an r of 0.4, the peak would be 20% higher, that is, $1.2 \times 10 = 12A$. So, for an optimum case, we would need to actually look for a device whose *minimum* current limit is 12A or more in this case. At the bare minimum, we need a device with a current limit higher than 10A, just to guarantee output power.

3.15.2 Continuous Conduction Mode Considerations in Fixing r

As discussed previously, under various conditions, we may enter discontinuous conduction mode (DCM). From Figure 3.5 we can see that, just as DCM starts to occur, the current ripple ratio is 2. However we can pose the question in the following manner—what if we have set the current ripple ratio to a certain value r' (i.e., the current ripple ratio *at the maximum load current*, I_{O_MAX}). And then we decrease the load current slowly—at what load does the converter enter DCM?

By simple geometry it can be shown that the transition to DCM will occur at $r/2$ times the maximum load. For example, suppose we set r' to 0.4 at 3A load; the converter will transition into DCM at $(0.4/2) \times 3 = 0.6A$.

But designers know that when DCM is entered, a lot of things within the converter change suddenly! The duty cycle, for one, will now starting pinching off toward zero as we decrease the load current further. In addition, the loop response of the converter (its ability to correct quickly for disturbances in line and load) also usually gets degraded in

DCM. The noise and EMI profile can change suddenly too, and so on. Of course there are some advantages of operating in DCM too, but let us for now assume that, for various reasons, the designer wishes to avoid DCM altogether, if possible.

Maintaining the converter in CCM, down to the minimum load of our application, enforces a certain *maximum* value for r. For example, if the minimum load is $I_{O\_MIN} = 0.5\,\text{A}$, then to maintain the converter in CCM at $0.5\,\text{A}$, the set current ripple ratio (r' at $3\,\text{A}$) needs to be lowered. Back calculating, we get the required condition for this

$$I_O \times \frac{r'}{2} = I_{O\_MIN} \tag{3-18}$$

So

$$r' = \frac{2 \times I_{O\_MIN}}{I_{O\_MAX}} \tag{3-19}$$

In our case we get

$$r' = \frac{2 \times 0.5}{3} = 0.333$$

We therefore need to set the current ripple ratio to less than 0.333 at maximum load, to ensure CCM at I_{O_MIN}.

Note that, generally speaking, we can make the converter operate in boundary conduction mode (BCM), or in full DCM, in three ways: a) by decreasing the load, b) by choosing a small inductance, or c) by increasing the input voltage.

We realize that decreasing the load will proportionally decrease I_{DC} to virtually any value, and so the condition $r \geq 2$ (BCM to DCM) will certainly occur sooner or later—below a certain load current. Similarly, decreasing L will necessarily increase ΔI, and so at some point we can expect the ratio $\Delta I/I_{DC}$ (i.e., r) to try to become greater than 2 (implying DCM).

However, as far as the third method of entering DCM mentioned above, we should realize that *solely* increasing the input voltage just might not do the trick! DCM or BCM can only happen under an input (line) variation, provided the load current is simultaneously below a certain value to start with (the value being dependent on L).

It is instructive to study the three topologies separately in this regard. Note that the general equation for r is

$$r = \frac{V_{ON} \times D}{I_L \times L \times f} \quad \text{(any topology, any mode)} \tag{3-20}$$

Applying the voltseconds law in CCM (or BCM), we also get

$$r = \frac{V_{OFF} \times (1-D)}{I_L \times L \times f} \quad \text{(any topology, CCM or BCM only)} \quad (3\text{-}21)$$

a) From the plots of r in Figure 3.4, we see that both the buck and the buck-boost have the highest value of r when D approaches zero—i.e., at maximum input voltage. For these topologies, the equation for r (derivable from the more general equation for r just given immediately above) is:

$$r = \frac{V_O}{I_O \times L \times f}(1-D) \text{ (buck)} \quad (3\text{-}22)$$

$$r = \frac{V_O}{I_O \times L \times f}(1-D)^2 \text{ (buck-boost)} \quad (3\text{-}23)$$

So putting $r = 2$ and $D = 0$ (i.e., highest input voltage plus BCM), we get the limiting condition

$$I_O = \frac{1}{2} \times \frac{V_O}{L \times f} \text{ (buck and buck-boost)} \quad (3\text{-}24)$$

Therefore, for these two topologies, if I_O is *greater* than the above limiting value, we will always remain in CCM, no matter how high we increase the input voltage.

b) Coming to the **boost**, the situation is not so obvious. From Figure 3.4, we see that r peaks at $D = 0.33$ (corresponding to the input being exactly two-thirds of the output). So the boost is most likely to enter DCM at $D = 0.33$—not, say, at $D = 0$ or $D = 1$. We can derive the following (exact) equation for r:

$$r = \frac{V_O}{I_O \times L \times f}D \times (1-D)^2 \text{ (boost)} \quad (3\text{-}25)$$

So putting $D = 0.33$, and $r = 2$ in this equation, we get the following limiting condition

$$I_O = \frac{2}{27} \times \frac{V_O}{Lf} \text{ (boost)} \quad (3\text{-}26)$$

Therefore, for the boost topology, if I_O is *greater* than this value, we will always remain in CCM, no matter how high we increase the input voltage.

Note that, if we do manage to enter DCM, the most likely input point for this to happen is an input of 0.67 times the output. In other words, if we are not in DCM at this particular input voltage, we can be sure we will be in CCM throughout the entire input range (whatever it may be).

3.15.3 Setting r to Values Higher Than 0.4 When Using Low-ESR Capacitors

Nowadays, with improvements in capacitor technology, we are seeing a new generation of very *low-ESR* capacitors—like monolithic multilayer ceramic capacitors (MLCs or MLCCs), polymer capacitors, and so on. Due to their extremely low ESRs, these capacitors usually have very high ripple (RMS) current ratings. Therefore, the required size of such capacitors in any application is no longer dictated by their ripple current handling capability. In addition, these capacitors also have almost no *aging* characteristics (or lifetime issues) that we need to account for beforehand in the design (as we customarily do for electrolytic capacitors, that can "dry out" over time). Further, due to their very high dielectric constant, these new capacitors have also become very *small* in size. So in fact, nowadays, increasing *r* may not necessarily cause a noticeable increase in the space occupied by the capacitors (or size of converter). On the other hand, increasing *r* may still lead to a relatively significant reduction in size of the inductor.

Summing up, with modern capacitors to the rescue, it may start making perfect sense to increase *r* from its traditional "optimum" of 0.4, to say around 0.6 to 1 on occasion (provided other considerations do not restrict this). If we do so, Figure 3.6 tells us, we can still get an additional 30 to 50% reduction in the size of the inductor. And that is certainly not insignificant, provided of course that that advantage is not offset by having to use larger capacitors in the bargain!

3.15.4 Setting r to Avoid Device "Eccentricities"

Surprisingly, device *eccentricities* may on occasion play a part in defining the limits of *r* too. For example in Figure 3.8 we have presented the current limit plot of an integrated high-voltage flyback switcher IC called the Topswitch®. On it we have superimposed a typical switch current waveform, just to make things a little clearer.

We see that, surprisingly, the current limit of this device is *time-dependent* for about 1.5 µs after the turn-on transition—something we don't intuitively ever expect. This "initial current limit" of the device occurs just as its internal current limit comparator starts to come out of its (valid) "leading edge blanking" time. As mentioned, during this blanking time the IC is just "not looking" at the current at all to avoid spurious triggering on the noise edge of the turn-on transition. But the problem is that once the current limit circuit gets down to monitoring the switch current again, it takes a certain time for the current limit threshold to settle down, and during this time it can be triggered at only about 75% of the supposed current limit!

Looking at the switch (or inductor) current waveform, we know that the current at the moment the switch turns ON is always *less* than the average value by the amount $\Delta I/2$. In other words this *trough* (*valley*) current I_{TR} is related to r according to the equation

$$I_{TR} = I_L \times \left(1 - \frac{r}{2}\right) \tag{3-27}$$

We realize that to avoid hitting the initial current limit of the device, we need to ensure that the trough falls below $0.75 \times I_{CLIM}$. So,

$$I_{TR} = I_L \times \left(1 - \frac{r}{2}\right) \le 0.75 \times I_{CLIM} \tag{3-28}$$

Now, we are assuming the power supply is at maximum load in this analysis. Therefore, the peak current is set equal to the current limit I_{CLIM}

$$I_{PK} = I_L \times \left(1 + \frac{r}{2}\right) = I_{CLIM} \tag{3-29}$$

Therefore, equating the two equations above, we get the limiting condition for r

$$\left(1 - \frac{r}{2}\right) \le 0.75 \times \left(1 + \frac{r}{2}\right) \tag{3-30}$$

or

$$r \ge 0.286 \tag{3-31}$$

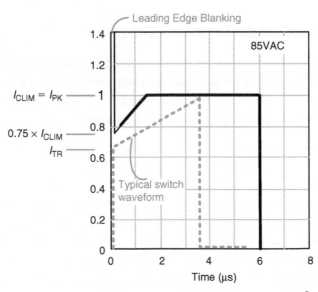

Figure 3.8: The initial current limit of the Topswitch®

Since *r* in any case is typically set to about 0.4, we should normally have no trouble with this "initial current limit" issue. However, note that on finer examination of the electrical tables of the datasheet, this $0.75 \times$ factor is specified only at 25°C. Unfortunately, very few power devices stay at 25°C for long! So, the bottom line is that, we, as designers, *do not really know* the value of the current limit as the device heats up. Yes, we can certainly make an educated guess, possibly leave an additional safety margin when fixing *r*, and certainly we may face no problem whatsoever. But the truth is we are on our own now—the vendor has *not* provided the requisite data (in the form of guaranteed limits within the electrical tables).

3.15.5 Setting r to Avoid Subharmonic Oscillations

Looking at Figure 3.9, we see that in any converter, the output voltage is first compared against an internal reference voltage. Then, the difference between the two (the "error") is filtered, amplified, and inverted by an error amplifier, the output of which (the *control voltage*) is fed to one of the two inputs of a pulse width modulator (PWM) comparator. On the other input of this PWM comparator, a *ramp* is applied, and this produces the switching pulses. So, for example, if the error at the output increases, the control voltage will decrease, and the duty cycle will thus decrease in an effort to reduce the output voltage. That is how regulation usually works.

In voltage mode control, the ramp applied to the PWM comparator is derived from an internal (fixed) clock. However in current mode control, it is derived from the *inductor current (or switch current)*. And the latter leads to a rather odd situation where even a slight disturbance in the inductor current waveform can become worse in the next cycle (see upper half of Figure 3.10).

Eventually, the converter may lapse into a strange "one pulse wide, one pulse narrow" switching waveform. This represents an operating mode that is definitely not "legitimate"

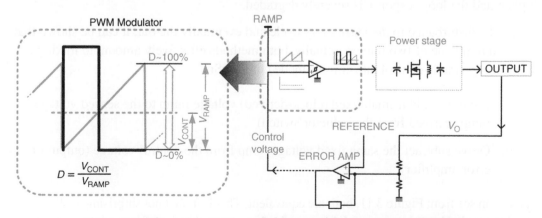

Figure 3.9: The pulse width modulator section of a power converter

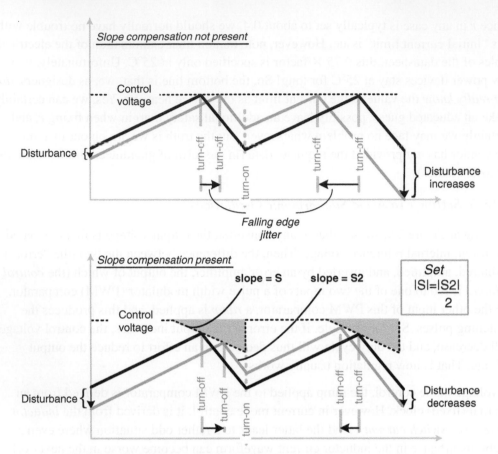

Figure 3.10: Subharmonic instability in current mode control, and avoiding it by slope compensation

or desirable for several reasons—in particular, the output voltage ripple is now much higher, and the loop response is severely degraded.

To get the disturbance to *decrease* every cycle and eventually die out, it can be shown that we need to do one of two things. Actually, both methods effectively amount to mixing a little voltage-mode control into current-mode control. So,

a) Either we add a small fixed (clock-derived) voltage ramp to the sensed voltage ramp (derived from the inductor/switch)

b) Or we subtract the same fixed voltage ramp from the control voltage (output of error amplifier).

As we can see from Figure 3.11, both are equivalent. That is in fact not surprising at all, considering that both the ramp and the control voltage go to the pins of a comparator. So if we

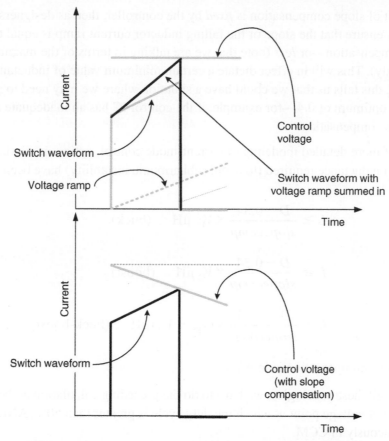

Figure 3.11: Adding a fixed ramp to the sensed signal, or modifying the control voltage, are equivalent methods of slope compensation in current-mode control

compare a signal A + B with a signal C, that is exactly equivalent to comparing A to C − B. And in both cases, equality at the input pins is established when A + B = C.

This technique is called *slope compensation*, and is the most recognized way of quenching the alternate wide and narrow pulsing (or subharmonic instability) associated with current-mode control (see lower half of Figure 3.10).

It can be shown that to avoid subharmonic instability, we need to ensure that the amount of slope compensation (expressed in A/s) is equal to *half* the slope of the falling inductor current ramp, *or more*. Note that, in principle, subharmonic instability can occur only if *D* is (close to or) greater than 50%. So slope compensation can be applied either over the full duty cycle range, or just for $D \geq 0.5$ as shown in Figure 3.10. Note that subharmonic instability can also occur only if we are operating in continuous conduction mode (CCM). So one way of avoiding it altogether is to operate in DCM.

If the amount of slope compensation is *fixed* by the controller, then as designers we need to personally ensure that the slope of the falling inductor current ramp is equal to *twice* the slope compensation—or *less* (note that we are talking in terms of the magnitudes of the slopes only). This will in effect dictate a certain minimum value of inductance. And in terms of *r*, this tells us that we could have a situation where we may need to set *r* to *less* than the optimum of 0.4—for example, if the control IC has an inadequate amount of built-in slope compensation.

As a result of more detailed modeling of current-mode control, optimum relationships for the minimum inductance required (to avoid subharmonic instability) have been generated as follows:

$$L \geq \frac{D - 0.34}{slopecomp} \times V_{IN} \, \mu H \quad \text{(buck)} \tag{3-32}$$

$$L \geq \frac{D - 0.34}{slopecomp} \times V_O \, \mu H \quad \text{(boost)} \tag{3-33}$$

$$L \geq \frac{D - 0.34}{slopecomp} \times (V_{IN} + V_O) \, \mu H \quad \text{(buck-boost)} \tag{3-34}$$

where the slope compensation is in A/μs.

Note that for all these topologies, we have to do the preceding calculation at the maximum input voltage point at which the duty cycle is greater than 50%, AND we are also simultaneously in CCM.

3.15.6 Quick-selection of Inductors Using "L × I" and "Load Scaling" Rules

Finally, having decided upon the value of *r* based on all the considerations outlined so far, we first present a *quick* method of picking an inductor for a given application. After that we will proceed to a more detailed analysis and worked example.

As mentioned previously, from the inductor equation $V = L \, dI/dt$, we can derive another useful relationship that we are calling the "$L \times I$" equation:

$$(L \times I_L) = \frac{Et}{r} \quad \text{(any topology)} \tag{3-35}$$

Symbolically

$$L \times I = \frac{voltseconds}{current \ ripple \ ratio} \quad \text{(any topology)} \tag{3-36}$$

So if we know the voltseconds (from our application conditions), and have a target value for *r*, we can calculate "$L \times I$." Then, knowing *I*, we can calculate *L*.

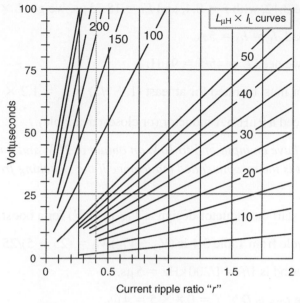

Figure 3.12: The "$L \times I$" curves for quick selection of inductance

Note that $L \times I$ can be visualized as a sort of inductance "per" ampere—except that the relationship is *inverse*. That is, if we *increase* the current, we need to *decrease* the inductance (by the same amount). So, for example, if we get an inductance of 100 μH for a 2 A application, then for a 1 A application, the inductance must be 200 μH, and for a 4 A application, the inductance would be 50 μH, and so on.

Note that because the $L \times I$ equation doesn't depend on topology, switching frequency, or on the specific input/output voltages, we can graph it out universally, as in Figure 3.12. That helps quickly pick an inductance for any application. Let us now exemplify the $L \times I$ graphical selection method for each topology.

3.16 Worked Examples (2, 3, and 4)

Buck: *Suppose we have an input of 15–20 V, an output of 5 V, and a maximum load current of 5 A. What is the recommended inductance if the switching frequency is 200 kHz?*

a) We need to start the inductor design at V_{INMAX} (20 V) for a buck.

b) The duty cycle from Table 3.1 is $V_O/V_{IN} = 5/20 = 0.25$.

c) The time period is $1/f = 1/200\,\text{kHz} = 5\,\mu\text{s}$.

d) The off-time t_{OFF} is $(1-D) \times T = (1-0.25) \times 5 = 3.75\,\mu\text{s}$.

e) The voltseconds (calculated using the *off-time*) is $V_O \times t_{OFF} = 5 \times 3.75 = 18.75\,\mu\text{s}$.

f) From Figure 3.12, with $r = 0.4$, and $Et = 18.75\,\mu s$, we get $L \times I = 45\,\mu HA$.

g) For a 5 A load, $I_L = I_O = 5\,A$.

h) Therefore, we need $L = 45/5 = 9\,\mu H$.

i) The inductor must be rated for at least $(1 + r/2) \times I_L = 1.2 \times 5 = 6\,A$.

Summarizing, we need a $9\,\mu H/6\,A$ inductor (or closest available).

Boost: *Suppose we have an input of 5 to 10 V, an output of 25 V, and a maximum load current of 2 A. What is the recommended inductance if the switching frequency is 200 kHz?*

a) We need to start the inductor design at V_{INMIN} (5 V) for a boost.

b) The duty cycle from Table 3.1 is $(V_O - V_{IN})/V_O = (25 - 5)/25 = 0.8$.

c) The time period is $1/f = 1/200\,kHz = 5\,\mu s$.

d) The on-time t_{ON}, is $D \times T = 0.8 \times 5 = 4\,\mu s$.

e) The voltseconds (calculated using the *on-time*) is $V_{IN} \times t_{ON} = 5 \times 4 = 20\,\mu s$.

f) From Figure 3.12 with $r = 0.4$, and $Et = 20\,\mu s$, we get $L \times I = 47\,\mu HA$.

g) For a 2 A load, $I_L = I_O/(1 - D) = 2/(1 - 0.8) = 10\,A$.

h) Therefore, we need $L = 47/10 = 4.7\,\mu H$.

i) The inductor must be rated for at least $(1 + r/2) \times I_L = 1.2 \times 10 = 12\,A$.

Summarizing, we need a $4.7\,\mu H/12A$ inductor (or closest available).

Buck-boost: *Suppose we have an input of 5 to 10 V, an output of −25 V, and a maximum load current of 2 A. What is the recommended inductance if the switching frequency is 200 kHz?*

a) We need to start the inductor design at V_{INMIN} (5 V) for a buck-boost.

b) The duty cycle from Table 3.1 is $V_O/(V_{IN} + V_O) = 25/(5 + 25) = 0.833$.

c) The time period is $1/f = 1/200\,kHz = 5\,\mu s$.

d) The on-time t_{ON} is $D \times T = 0.833 \times 5 = 4.17\,\mu s$.

e) The voltseconds (calculated using the *on-time*) is $V_{IN} \times t_{ON} = 5 \times 4.17 = 20.83\,\mu s$.

f) From Figure 3.12, with $r = 0.4$, and $Et = 20.83\,\mu s$, we get $L \times I = 52\,\mu HA$.

g) For a 2A load, $I_L = I_O/(1 - D) = 2/(1 - 0.833) = 12\,A$.

h) Therefore, we need $L = 52/12 = 4.3\,\mu H$.

i) The inductor must be rated for at least $(1 + r/2) \times I_L = 1.2 \times 12 = 14.4\,A$.

Summarizing, we need a $4.3\,\mu H/14.4\,A$ inductor (or closest available).

3.16.1 The Current Ripple Ratio r in Forced Continuous Conduction Mode (FCCM)

Finally, before we move on to magnetic fields, we make some closing remarks on designing with forced continuous conduction mode (FCCM).

We had said previously, that by definition, r is defined only for CCM, and therefore cannot exceed 2 (since that marks the boundary between CCM and DCM). However, in *synchronous regulators* (with diode replaced or supplanted by a low-drop MOSFET across it), we actually never enter DCM (unless the IC is deliberately designed to mimic that mode on demand). So now, on decreasing the load, we actually continue to remain in CCM. That is because for DCM to ever occur, the inductor current must be forced to *stay* at least for some part of the switching cycle at zero. And to get that to happen, we need to have a reverse-biased diode that prevents the inductor current from "going the other way." But in synchronous regulators, the MOSFET across the diode allows reverse-conduction even if the diode is reverse-biased, so we do not get DCM.

The CCM-type mode that replaces the DCM mode in synchronous regulators is distinguished from the usual (normal) CCM mode, by calling it the 'forced continuous conduction mode' (FCCM). The main switch is usually identified as the *top* (or "high-side") MOSFET, whereas the MOSFET across the diode is called the *bottom* (or "low-side") MOSFET. Further, in FCCM, r is legitimately allowed to exceed 2 (see Figure 3.5).

We can visualize FCCM as starting to occur when the load current is decreased sufficiently to cause part of the inductor current waveform to become "submerged" below ground— that is, with parts of it having a negative value (inductor current flowing momentarily away from the load). But note that as long as we are still drawing some load current *out* of the output terminals of the converter, the *average* value of the waveform, I_{DC} (center of ramp), is still positive—that is, going towards the load—on an average. Further, because I_{DC} is always proportional to the load current, it can be made to decrease all the way down to zero while still maintaining CCM. Since the *swing* in current, ΔI, depends *only* on the input and output voltages, which we have assumed have not changed, the ratio $r = \Delta I/I_L$ not only exceeds 2, but can in fact become extremely large.

All the basic design equations we can write for the RMS, DC, AC, or peak currents in the input/output capacitors and the switch, when operating in conventional CCM, apply to the converter in FCCM too (though there may be some additional losses, as for example when the current flows through the body diode of the top MOSFET). This, despite the fact that r can now exceed 2. In other words, the CCM equations do *not* get invalidated in FCCM. However, a specific computational problem can arise in some cases, because if r is infinite (zero load current), we can get a singularity—a "0" in the denominator. At first sight, that seems to make the CCM equations (presented the way we have been doing), unusable. But one trick we can employ to avoid the singularity is to *assume* a few milliamperes of minimum load, however small. Alternatively, we can substitute $r = \Delta I / I_{DC}$ back into the equations, and we will then see that I_{DC} cancels out (does not appear in the denominator anywhere). Either way, the equations of CCM apply to FCCM too.

3.16.2 Basic Magnetic Definitions

Having understood basic concepts like voltseconds, current components, worst-case voltage, and also how to do an initial (quick) selection of an off-the-shelf inductor, we will now try to go *inside* the magnetic component, so as to learn what happens in terms of the magnetic fields present inside its core. We will then use this information to do a more complete validation of a selected off-the-shelf inductor. Then we will find the remaining (worst-case) stresses of the converter.

At the outset, we should note that in magnetics, there are several different systems of units in use. This can become very confusing, since even the basic equations look different depending on the system in use. It is therefore a wise policy to stick to one system of units all the way through—converting to a different system, if required, only at the very end, that is, only at the level of numerical results (not at the level of the equations).

Further, unless otherwise stated, the reader can safely assume we are using the meter-kilogram-seconds system of units—that is, MKS system, also called the SI system (for System International).

Here are the basic definitions:

- *H*-field: Also called field strength, field intensity, magnetizing force, applied field, and so on. Its units are A/m.

- *B*-field: Also called flux density or magnetic induction. Units of B are tesla (T) or webers per square meter (Wb/m^2).

- Flux: This is the integral of B over a given surface area. That is,

$$\phi = \int_S B dS \ \text{Wb} \tag{3-37}$$

- If B is constant over the surface, we get the more common form $\phi = BA$ where A is the area of the surface.

> **Note:** The integral of *B* over a closed surface is zero since flux lines do not start or end at any given point but are continuous.

B is related to *H* at any given point by the equation $\mathrm{B} = \mu H$ where μ is the *permeability* of the material. Note that later we will use the symbol μ for relative permeability—that is, the *ratio* of the permeability of the material to that of air. So in MKS units we should actually preferably write $B = \mu_c H$, where μ_c is the permeability of the core (magnetic material). By definition, $\mu_c = \mu\mu_0$.

- The permeability of air, denoted by μ_0, is equal to $4\pi \times 10^{-7}$ henries/m in MKS units. In CGS units it is equal to 1. That is why in CGS units $\mu_c = \mu$, where μ is also automatically the relative permeability of the material (though units are different).

- Faraday's law of induction (also called Lenz's law) relates the induced voltage *V* that is developed across the ends of a coil (*N* turns), to the (time varying) *B*-field passing through it. So,

$$V = N\frac{d\phi}{dt} = NA\frac{dB}{dt} \tag{3-38}$$

- The "inertia" of a coil to a change in flux through it due to a time-varying current through it is its inductance *L*, defined as

$$L = \frac{N\phi}{I} \text{ henries} \tag{3-39}$$

- Since it can be shown that the flux is proportional to the number of turns *N*, the inductance *L* is proportional to the *square* of the number of turns. This proportionality constant is called the *inductance index* and is denoted by A_L. It is usually expressed as nH/turns^2 (though sometimes it is considered to be mH/1000 turns^2, both being *numerically* the same). So

$$L = A_L \times N^2 \times 10^{-9} \text{ henries} \tag{3-40}$$

- When *H* is integrated over a *closed loop*, we get the current enclosed by the loop

$$\oint Hdl = I \text{ amps} \tag{3-41}$$

where the integration symbol above reflects the fact that it is being performed over a closed loop. This is also called *Ampere's circuital law*.

- Combining Lenz's law with the inductor equation $V = L\, dI/dt$, we get

$$V = N\frac{d\phi}{dt} = NA\frac{dB}{dt} = L\frac{dI}{dt} \tag{3-42}$$

- From this we get the two key equations used in power conversion

$$\Delta B = \frac{L\Delta I}{NA} \quad \text{(voltage independent equation)} \tag{3-43}$$

$$\Delta B = \frac{V\Delta t}{NA} \quad \text{(voltage dependent equation)} \tag{3-44}$$

The first equation can be written symbolically as

$$B = \frac{LI}{NA} \quad \text{(voltage independent equation)} \tag{3-45}$$

And the latter equation can be written in a more "power-conversion-friendly" form as follows

$$B_{AC} = \frac{V_{ON}D}{2 \times NAf} \quad \text{(voltage dependent equation)} \tag{3-46}$$

For most inductors used in power conversion, if we reduce the current to zero, the field inside the core also goes to zero. Therefore, an implicit assumption of complete linearity is also usually made—that is, *B and I are considered proportional to each other* as shown in Figure 3.13 (unless of course the core starts saturating, at which point, all bets are off!). The voltage independent equation can then be expressed as any of the equations shown in the figure—in other words, this proportionality applies to the peak values of current and field, their average values, their AC values, their DC values, and so on. The constant of proportionality is equal to

$$\frac{L}{NA} \quad \text{(proportionality constant linking B and I)} \tag{3-47}$$

where N is the number of turns and A the actual geometrical cross-sectional area of the core (its center limb usually, or simply the effective area A_e given in the datasheet of the core).

3.17 Worked Example (5)—When Not to Increase the Number of Turns

Note that the voltage-independent equation is useful if, for example, we want to do a quick check to see if our core may be saturating. Suppose we are custom-designing our

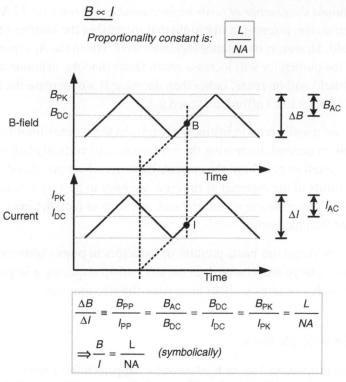

Figure 3.13: B and I can usually be considered proportional to each other

inductor. We have wound 40 turns on a core with an area of $A = 2\,\text{cm}^2$. Its measured inductance is $200\,\mu\text{H}$, and the peak inductor current in our given application is $10\,\text{A}$. Then the peak flux density can be calculated as follows:

$$B_{\text{PK}} = \frac{L}{NA}\,I_{\text{PK}} = \frac{200 \times 10^{-6}}{40 \times \dfrac{2}{10^4}} = 0.25 \text{ teslas}$$

Note that we have converted the area to m^2 in the above equation, because we are using the MKS version of the equation.

For most ferrites, an operating flux density of $0.25\,\text{T}$ is acceptable, since the saturation flux density is typically around $0.3\,\text{T}$.

Based on the B and I linearity, we can also linearly extrapolate and thus conclude that the peak current in our application should under no condition be allowed to exceed $(0.3/0.25) \times 10 = 12\,\text{A}$, because at $12\,\text{A}$, the field will be $0.3\,\text{T}$, and the core will then start to saturate.

But note: nor *should the number of turns be increased any further* (at 12 A). Looking at the B_{PK} equation above, it seems at first sight that increasing the number of turns will reduce the B-field. However, inductance increases as N^2 (from the A_L equation given previously), so the numerator will increase much faster than the denominator. Therefore, in reality, the B-field will *increase*, rather than decrease if we increase the number of turns, and we know we can't afford to exceed 0.3 T.

In other words, we usually tend to instinctively rely on the current-limiting properties of an inductor. And, in general, increasing the inductance will certainly help increase the inductance and therefore help limit the current. However, if we are already close to the energy-storage limits of the material of the core, we have to be very careful—a few extra turns could take us "over the edge" (saturation), and then in fact, the inductance will start collapsing rather than increasing.

We should also not forget our basic premise of inductors in power conversion—for a given application, a large inductance does usually end up requiring a large inductor! So, increasing the number of turns without increasing the size may naturally turn out to be a recipe for disaster.

3.17.1 The Field Ripple Ratio

Since I and B are proportional to each other, and r happens to be a ratio, we realize that r must apply equally to the field components as it does to the current components. So, in that sense, r can be looked at as a "field ripple ratio" too. We can therefore extend the definition of r as follows:

$$r = 2\frac{I_{AC}}{I_{DC}} = 2\frac{B_{AC}}{B_{DC}} \tag{3-48}$$

Therefore, r can also be used to relate the peak, AC, and DC values of both the current and field according to the equations

$$B_{DC} = \frac{2 \times B_{PK}}{r+2} \text{ or } I_{DC} = \frac{2 \times I_{PK}}{r+2} \tag{3-49}$$

$$B_{AC} = \frac{r \times B_{PK}}{r+2} \text{ or } I_{AC} = \frac{r \times I_{PK}}{r+2} \tag{3-50}$$

We can relate the peak to the swing too as follows:

$$B_{PK} = \frac{r+2}{2 \times r} \times \Delta B \text{ or } I_{PK} = \frac{r+2}{2 \times r} \times \Delta I \tag{3-51}$$

The latter form will in fact be used later by us in the worked example that will follow.

3.17.2 The Voltage Dependent Equation in Terms of Voltseconds (MKS units)

When discussing the current swing ΔI, we related it to the voltseconds. Now we can do the same for the B-field

$$\Delta B = \frac{L \times \Delta I}{N \times A} = \frac{Et}{N \times A} \text{ teslas}$$

(3-52)

So as for current, the voltseconds in our application also determines the *swing* of the magnetic field—though not its DC level.

3.17.3 CGS units

We may personally prefer to use the more broadly accepted MKS units, but we have to deal with the ground reality of the situation—that certain vendors (especially North American ones) still use CGS (centimeter-gram-seconds) units. Since we would certainly be evaluating and looking at their datasheets too, we will need to use the conversions in Table 3.4.

In particular, we should remember that the saturation flux density BSAT, which is around 0.3 T (300 mT) for most ferrites, is 3000 gauss (G) in CGS units. Also note that permeability of a material in MKS units needs to be divided by $4\pi \times 10^{-7}$ to get the permeability in CGS units. The reason for that is that permeability of air is set to 1 in CGS units, but in MKS units it is (numerically) equal to $4\pi \times 10^{-7}$.

3.17.4 The Voltage Dependent Equation in Terms of Voltseconds (CGS units)

It is also therefore helpful to know how to write the voltage dependent equation, (expressed in terms of Et), in CGS units instead.

So, converting A in m^2 to A in cm^2, we get from the previous equation:

$$\Delta B = \frac{100 \times E_t}{N \times A} \text{ gauss (A in } cm^2)$$

(3-53)

Table 3.4: Magnetic systems of units and their conversions

	CGS units	MKS Units	Conversions
Magnetic Flux	Line (or Maxwell)	Weber	1 Weber = 10^6 Lines
Flux Density (B)	Gauss	Tesla (or Wb/m²)	1 Tesla = 10^4 Gauss
Magnetomotive force	Gilbert	Ampere-turn	1 Gilbert = 0.796 Ampere-turn
Magnetizing Force Field (H)	Oersted	Ampere-turn/meter	1 Oersted = 1000/4π = 79.577 Ampere/meter
Permeability	Gauss/Oersted	Weber/m-Ampere-turn	$\mu_{MKS} = \mu_{CGS} \times (4\pi \times 10^{-7})$

3.17.5 Core Loss

The core loss depends on various factors: the flux swing ΔB, the (switching) frequency f, and the temperature (though we usually ignore this latter dependency for most estimates). Note however, that when vendors of magnetic materials express the dependency of core loss on a certain B, what they are really talking about is $\Delta B/2$, that is, B_{AC}. This happens to be the usual industry convention, but it is often quite confusing to power supply designers. In fact, there is more confusion caused by the fact that B may be expressed by the vendor, either in terms of gauss or in teslas. In fact, the dissipation also (due to the core loss) may be expressed either as mW or as W.

First let us look at the general form of core loss.

Core Loss = (Core Loss per unit volume) $\times$ Volume where "core loss per unit volume" is expressed generally as

$$\text{constant } t_1 \times B^{\text{constant } t_2} \times f^{\text{constant } t_3} \tag{3-54}$$

In Table 3.5 we have indicated the three main systems of units in use for describing the core loss per unit volume, and also provided the rules for converting between them. Note we are using V_e (effective volume) here—this can usually be considered to be simply the actual physical volume of the core, or we can just look it up in the datasheet of the core.

In Table 3.6 we have provided values for the constants in the core loss equation in one of these systems of units, besides some other operating limits. The reader is, however, advised to confirm these values from the respective vendors.

Table 3.5: The different systems in use for describing core loss (and their conversions)

	Constant	exponent of B	exponent of f	B	f	Ve	Units
System A	Cc	Cb	Cf	Tesla	Hz	Cm³	W/cm³
	$=\dfrac{C \times 10^{4 \times p}}{10^3}$	$=p$	$=d$				
System B	C	p	d	Gauss	Hz	cm³	mW/cm³
	$=\dfrac{Cc \times 10^3}{10^{4 \times Cb}}$	$=Cb$	$=Cf$				
System C	Kp	n	m	Gauss	Hz	cm³	W/cm³
	$=\dfrac{C}{10^3}$	$=p$	$=d$				

Table 3.6: Typical core loss coefficients of common materials

Material (Vendor)	Grade	C	$p^{(Bp)}$	$d^{(fd)}$	μ	$\approx B_{SAT}$ (gauss)	$\approx f_{MAX}$ (MHz)
Powdered Iron (Micrometals)	8	4.3E−10	2.41	1.13	35	12500	100
	18	6.4E−10	2.27	1.18	55	10300	10
	26	7E−10	2.03	1.36	75	13800	0.5
	52	9.1E−10	2.11	1.26	75	14000	1
Ferrite (Magnetics Inc)	F	1.8E−14	2.57	1.62	3000	3000	1.3
	K	2.2E−18	3.1	2	1500	3000	2
	P	2.9E−17	2.7	2.06	2500	3000	1.2
	R	1.1E−16	2.63	1.98	2300	3000	1.5
Ferrite (Ferroxcube)	3C81	6.8E−14	2.5	1.6	2700	3600	0.2
	3F3	1.3E−16	2.5	2	2000	3700	0.5
	3F4	1.4E−14	2.7	1.5	900	3500	2
Ferrite (TDK) Ferrite (Fair-Rite)	PC40	4.5E−14	2.5	1.55	2300	3900	1
	PC50	1.2E−17	3.1	1.9	1400	3800	2
	77	1.7E−12	2.3	1.5	2000	3700	1

Note: (a) E−(b) is the same as (a) $\times 10^{-(b)}$

3.18 Worked Example (6)—Characterizing an Off-the-shelf Inductor in a Specific Application

Now we will present the *general inductor design procedure* we have been talking about. We will be considering a wide-input voltage range here. The procedure is to be carried out at the "worst-case input voltage end"*with respect to the peak current*. The basic purpose is to ensure that we are avoiding inductor saturation under normal operation. So for the **buck**, we will work at V_{INMAX}, since that is the point at which the peak current is at its maximum. For a **boost or a buck-boost**, we need to conduct this procedure at V_{INMIN}, not V_{INMAX}, since that is the worst-case input voltage end with regard to the peak current, for these topologies.

The procedure will be illustrated by means of a step-by-step worked example. Though it is carried out for a buck, throughout the calculation, we will indicate precisely how the procedure and equations may need to change, were this a boost or a buck-boost. So for example, to the right of any equation presented below, we have indicated in brackets, which topology it is valid for.

A buck converter has an input of 18–24 V, an output of 12 V, and a maximum load of 1 A. We desire a current ripple ratio of 0.3 (at maximum load). We assume $V_{SW} = 1.5 V$, $V_D = 0.5 V$, and $f = 150,000 Hz$. An off-the-shelf inductor is to be selected and characterized for this application.

As mentioned, all the steps involved in the "general inductor design procedure" below are being carried out at a certain "V_{IN}"—which is the *maximum input voltage* for a **buck**, and *minimum input voltage* for a **boost or a buck-boost**.

3.18.1 Estimating Requirements

For a buck regulator, the duty cycle is (now including the switch and diode forward drops)

$$D = \frac{V_O + V_D}{V_{IN} - V_{SW} + V_D} \quad \text{(buck)} \tag{3-55}$$

So

$$D = \frac{12 + 0.5}{24 - 1.5 + 0.5} = 0.543$$

(For boost, use $D = \dfrac{V_O - V_{IN} + V_D}{V_O - V_{SW} + V_D}$, for buck-boost use $D = \dfrac{V_O + V_D}{V_{IN} + V_O - V_{SW} + V_D}$.)

The switch on-time is therefore

$$t_{ON} = \frac{D}{f} \Rightarrow \frac{0.543}{150000} \Rightarrow 3.62\,\mu s \quad \text{(any topology)}$$

$$t_{ON} = 3.62\,\mu s$$

The voltage across the inductor when the switch is ON is

$$V_{ON} = V_{IN} - V_{SW} - V_O = 24 - 1.5 - 12 = 10.5 \text{ V (buck)}$$

(For boost and buck-boost, use $V_{ON} = V_{IN} - V_{SW}$).

So the voltsμseconds is

$$Et = V_{ON} \times t_{ON} = 10.5 \times 3.62 = 38.0\,V\mu s \quad \text{(any topology)}$$

Using the "$L \times I$" equation

$$(L \times I_L) = \frac{Et}{r} \quad \text{(any topology)}$$

we get

$$(L \times I_L) = \frac{38}{0.3} = 127 \, \mu\text{H-A}$$

But the average inductor current is

$$I_L = I_O \quad \text{(buck)}$$

(For a boost and buck-boost, use $I_L = \frac{I_O}{1-D}$)

Therefore,

$$L = \frac{(L \times I_L)}{I_L} \equiv \frac{(L \times I_O)}{I_O} = \frac{127}{1} = 127 \, \mu\text{H} \quad \text{(any topology)}$$

The peak current will be 15% higher than I_L for $r = 0.3$. This follows from

$$I_{PK} = \left(1 + \frac{r}{2}\right) \times I_L = 1.15 \times 1 = 1.15 \, \text{A} \quad \text{(any topology)}$$

We now pick a promising off-the-shelf inductor—the PO150 from Pulse Engineering. Its inductance is 137 μH, which is close to our requirement of 127 μH, and it is rated for a continuous DC of 0.99 A, which is very close to our requirement of 1A. Its datasheet is reproduced in Table 3.7. Note that the other conditions mentioned by the vendor *do not match our application* (but that is not unexpected—what are the chances of an off-the-shelf inductor that precisely matches a given application?). Nevertheless, we can perform a full analysis, and thus either validate or invalidate our choice of component.

Table 3.7: Specifications of a selected inductor (the PO150)

I_{DC} (A)	L_{DC} (μH)	Et (Vμs)	DCR (mΩ)	Et_{100} (Vμs)
0.99	137	59.4	387	10.12
• The inductor is such that 380 mW dissipation corresponds to 50°C rise in temperature.				
• The core loss equation for the core is $6.11 \times 10^{-18} \times B^{2.7} \times f^{2.04}$ mW where f is in Hz and B is in gauss.				
• Et^{100} is the Vμ secs at which "B" is 100 gauss.				
• "B" is B_{AC}, i.e., $\Delta B / 2$.				
• Rated frequency of operation is 250 kHz.				

3.18.2 Current Ripple Ratio

We use the "$L \times I$" rule

$$(L \times I_L) = \frac{Et}{r} \quad \text{(any topology)} \tag{3-56}$$

So

$$r = \frac{Et}{L \times I_L} \quad \text{(any topology)} \tag{3-57}$$

The inductor has been designed by its vendor, for an r of

$$r = \frac{59.4}{137 \times 0.99} = 0.438$$

In our application we will get

$$r = \frac{38}{137 \times 1} = 0.277$$

This is very close to (and less than) our target of $r = 0.3$, and is therefore acceptable.

3.18.3 Peak Current

The inductor has been designed for a peak current of

$$I_{PK} = \left(1 + \frac{r}{2}\right) \times I_L = \left(1 + \frac{0.438}{2}\right) \times 0.99 = 1.21 \, \text{A (any topology)}$$

In our application we will get

$$I_{PK} = \left(1 + \frac{r}{2}\right) \times I_L = \left(1 + \frac{0.277}{2}\right) \times 1 = 1.14 \, \text{A} \quad \text{(any topology)}$$

The peak current in our application is considered "safe," being *less* than what the inductor was originally designed for. Therefore, we can safely assume that the peak B-field of our application also must be within the design limits of the inductor. However, it is instructive to confirm that directly, as we will do next.

Note that the frequency has not even entered the picture directly so far, since voltµseconds is all that really matters to an inductor. Different applications, with the same DC level of current and the same voltseconds, are essentially the same application from the viewpoint of the inductor. It just "doesn't care," for example, what topology this is, or what is the duty cycle. It doesn't even care about the frequency directly (though the exception to this is the core loss term, because that depends not only on the voltseconds, i.e., the current swing, but on the frequency too). However, we will also see that the core loss term is much smaller anyway, compared to the copper loss. So, for all practical purposes, if the rated voltseconds of a given inductor (current swing), and its DC current rating correspond to the voltseconds and DC current of our application, we are almost certainly going to be fine right off the bat. However, even if the rated voltseconds and DC level are quite different, as long as the peak flux density is close to or less than the rated value, we are OK from the saturation point of view. That's a good start, and we can then proceed to do a full validation analysis of the temperature rise and so on under our specific application conditions.

3.18.4 Flux Density

The vendor provides the following information (see Table 3.7):

$$Et_{100} = 10.12 \, V\mu s$$

This means that the voltµseconds that produces a B_{AC} of 100 gauss is 10.12. Since $B_{AC} = \Delta B/2$, the corresponding ΔB is 200 gauss (for every 10.12 V µs).

In Eq. (3-53) we presented the following relationship between ΔB and Et:

$$\Delta B = \frac{100 \times E_t}{N \times A} \quad \text{gauss} \quad \text{(any topology)}$$

Since ΔB and Et are proportional to each other (for a given inductor), we can conclude that the inductor has been designed for a flux density swing of

$$\Delta B = \frac{Et}{Et_{100}} \times 200 = \frac{59.4}{10.12} \times 200 = 1174 \text{ gauss (any topology)}$$

and a peak flux density of

$$B_{PK} = \frac{r+2}{2 \times r} \times \Delta B = \frac{0.438 + 2}{2 \times 0.438} \times 1174 = 3267 \text{ gauss (any topology)}$$

In our application this will give us a swing of

$$\Delta B = \frac{Et}{Et_{100}} \times 200 = \frac{38}{10.12} \times 200 = 751 \text{ gauss (any topology)}$$

and a peak of

$$B_{PK} = \frac{r+2}{2 \times r} \times \Delta B = \frac{0.277 + 2}{2 \times 0.277} \times 751 = 3087 \text{ gauss (any topology)}$$

We see that the peak field in our application is within the design limits of the inductor, as expected, so we need not worry about core saturation. This is a basic qualification the inductor must pass before we can proceed with the rest of the analysis.

$$\frac{L}{NA} = \frac{B_{PK}}{I_{PK}} = \frac{3087}{1.14} = 2708 \text{ gauss/A (any topology)}$$

> **Note:** If we break open the inductor and measure the number of turns, and also estimate/measure the cross-sectional area of the central limb of its core, we can verify this number.

3.18.5 Copper Loss

From the equations contained in Figure 3.14, we can calculate the RMS of the inductor current waveform. The inductor was designed for an RMS squared of

$$I_{RMS}^2 = \frac{\Delta I^2}{12} + I_{DC}^2 = I_{DC}^2\left(1+\frac{r^2}{12}\right) = 0.99^2\left(1+\frac{0.438^2}{12}\right)$$
$$= 0.996 \text{ A}^2 \text{ (any topology)}$$

and a copper loss of

$$P_{CU} = I_{RMS}^2 \times \text{DCR} = 0.996 \times 387 = 385 \text{ mW (any topology)}$$

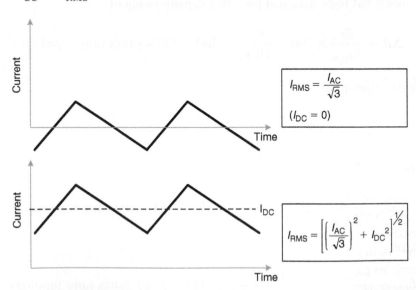

Figure 3.14: RMS value of an inductor current waveform

Whereas in our application we will get

$$I_{RMS}^2 = I_L^2 \left(1 + \frac{r^2}{12}\right) = 1^2 \left(1 + \frac{0.277^2}{12}\right) = 1.006 \, A^2 \text{ (any topology)}$$

and a copper loss of

$$P_{CU} = I_{RMS}^2 \times DCR = 1.006 \times 387 = 389 \, mW \text{ (any topology)}$$

3.18.6 Core Loss

Note that the vendor has already factored in the volume of the core and thus provided the following overall equation for the core loss of the inductor:

$$P_{CORE} = 6.11 \times 10^{-18} \times B^{2.7} \times f^{2.04} \, mW \quad \text{(any topology)}$$

where f is in Hz and B is in gauss. Note that B is $\Delta B/2$ here as per convention. So the core loss that the inductor was originally designed for is

$$P_{CORE} = 6.11 \times 10^{-18} \times \left(\frac{1174}{2}\right)^{2.7} \times (250 \times 10^3)^{2.04} = 18.8 \, mW$$

Whereas in our application

$$P_{CORE} = 6.11 \times 10^{-18} \times \left(\frac{751}{2}\right)^{2.7} \times (150 \times 10^3)^{2.04} = 2 \, mW$$

In general, we will find that in most ferrite-based off-the-shelf inductors, the designed core loss is only 5 to 10% of the total inductor loss (copper-plus-core loss). However, if the inductor uses a "powdered iron" core, this number may rise to about 20 to 30%.

> **Note:** Powdered iron cores tend to saturate more "softly" than ferrites, and that usually enhances their ability to withstand severe abnormal currents without leading to immediate switch destruction. On the other hand, powdered iron cores may have "lifetime" issues caused by slow degradation of the organic binder that holds their iron particles together. The vendor must be consulted about this possibility, and the steps necessary to avoid a premature end to our converter!

3.18.7 Temperature Rise

The vendor has stated that the inductor is such that 380 mW dissipation corresponds to 50°C rise in temperature. In effect this tells us that the thermal resistance of the core I is

$$Rth = \frac{\Delta T}{W} = \frac{50}{0.38} = 131.6°C/W \quad \text{(any topology)}$$

The inductor was originally designed for a total loss of

$$P = P_{CORE} + P_{CU} = 385 + 18.8 = 403.8 \text{ mW (any topology)}$$

This would have given a temperature rise of

$$\Delta T = Rth \times P = 131.6 \times 0.404 = 53°C \text{ (any topology)}$$

In our application

$$P = P_{CORE} + P_{CU} = 389 + 2 = 391 \text{ mW}$$

This will give a temperature rise of

$$\Delta T = Rth \times P = 131.6 \times 0.391 = 51°C$$

Provided we accept this temperature rise in our application (that will depend on our maximum operating ambient temperature), we can validate the chosen inductor. We have already confirmed it does not saturate in our application, and further, the current ripple ratio it provides is acceptable too.

This completes the general inductor design procedure.

3.19 Calculating the "Other" Worst-case Stresses

Having validated our choice of inductor, we can look a little more closely at the important issue of how the wide-input range impacts the *other* key parameters and stresses in our proposed converter. This also helps in correctly selecting the other power components.

3.19.1 Worst-case Core Loss

In the above so-called general inductor design procedure, we have actually been working at V_{INMAX} for a buck, and at V_{INMIN} for a boost or buck-boost. The reason was that the *inductor* sees the highest peak current, at this voltage end, so we have to "insure" the magnetics design at this particular point. But this point may not be the worst-case point for the other stresses in the power supply, and we need to start understanding that clearly now.

Let us first focus on the inductor itself. The point at which we are doing the inductor design will usually always give us the worst-case temperature rise too. But that is because the I_{DC} component of the inductor current is the dominant term. If for any reason, we are interested in knowing what the maximum *core loss* component of the total loss is, we should realize, looking back at Figure 3.4, that though the DC level may be going up, the AC component (on which the core loss term depends) may be decreasing (or even having an odd-shaped profile, as for the boost).

From Figure 3.4, we see that I_{AC} increases at high input voltages for both the buck and the buck-boost. For a *buck*, the general inductor design calculation above was carried out at V_{INMAX} and that just happens to be the point at which the core loss is a maximum too. Therefore, calculating the core loss at V_{INMAX} as we did in the previous example does coincidentally also give us the worst-case core loss.

However, if we were doing the calculation for a *buck-boost*, our general inductor design calculation would be being carried out at V_{INMIN}. But the core loss is a maximum at V_{INMAX}. Similarly, for a *boost*, we would also be carrying out the general inductor design calculation at V_{INMIN}. But the worst-case core loss for this topology occurs at $D = 0.5$ (see I_{AC} curve for boost in Figure 3.4). From the duty cycle equation of the boost, $D = 0.5$ corresponds to an input voltage equal to half the output.

> **Note:** If for the boost, the input range of the given application does *not* include the $D = 0.5$ point, we need to identify which voltage end of the range provides a duty cycle closest to $D = 0.5$. And we need to then do the worst-case core loss calculation at that end (if so desired).

Generally, the core loss term, being such a small component of the total loss, is of no great concern to us, so we won't even bother to do a numerical calculation here. But the general procedure to handle such cases will become apparent as we study the other worst-case loss terms of the converter below.

But let us now start *annotating (or subscripting)* some of the terms derived so far, just for gaining clarity in the discussion to follow. We should be clear that:

For a buck: The general inductor design procedure was carried out at V_{INMAX}, that is, D_{MIN}. So for example, the r we have set to 0.3–0.4 (and possibly recalculated with the selected inductor) is actually r_{DMIN} to be precise. Similarly, the voltseconds, Et, we have calculated so far is actually Et_{DMIN}.

For a boost and buck-boost: If a similar general inductor design procedure were carried for these topologies, it would be done at V_{INMIN}, that is, D_{MAX}. So, for example, the r we would have set to 0.3–0.4 (and possibly recalculated with the selected inductor) would actually be r_{DMAX}. Similarly, the voltseconds, Et, we would have calculated so far is actually Et_{DMAX}.

We need to keep these distinctions in mind; otherwise the following discussion can become confusing to no end!

3.19.2 Worst-case Diode Dissipation

The general equation for the average diode current is

$$I_D = I_L \times (1 - D) \qquad \text{(any topology)}$$

or equivalently

$$I_D = I_O \times (1 - D) \quad \text{(buck)} \tag{3-58}$$

$$I_D = I_O \quad \text{(boost and buck-boost)} \tag{3-59}$$

This leads to a diode dissipation of

$$P_D = V_D \times I_D = V_D \times I_O \times (1 - D) \text{ (buck)} \tag{3-60}$$

$$P_D = V_D \times I_D = V_D \times I_O \quad \text{(boost and buck-boost)} \tag{3-61}$$

For the buck, as the input voltage is raised, the duty cycle falls, and because the average inductor current I_L remains fixed at I_O, the average diode current *increases*. That means we get the worst-case diode current (and dissipation) at V_{INMAX} for a buck. So we can just use the numbers we already have derived from carrying out the general inductor design procedure (at V_{INMAX}).

For the boost and the buck-boost, as the input is raised, D decreases, but the average inductor current also falls, thereby keeping I_D always fixed at I_O. (We should remember that the boost and the buck-boost are unique in the sense that *all* the output current must pass through the diode when it conducts, so I_D must necessarily be equal to I_O at all times). That means the diode dissipation is independent of input voltage for these topologies. So we can, if we want, just use the numbers we already have derived from carrying out the general inductor design procedure (at V_{INMIN}).

Finally, for the ongoing buck converter design example, the calculation is as follows:

$$P_D = V_D \times I_O \times (1 - D_{MIN}) = 0.5 \times 1 \times (1 - 0.543) = 0.23 \text{ W (buck)}$$

Note that the general diode selection procedure is as follows:

The rule-of-thumb is to pick a diode with a current rating at least equal to, but preferably at least *twice* the worst-case average diode current given below (for low losses, since the diode forward drop decreases substantially if its current rating is increased):

- For a **buck**—maximum diode current is $I_O \times (1 - D_{MIN})$.

- For a **boost**—maximum diode current is I_O.

- For a **buck-boost**—maximum diode current is I_O.

- Its voltage rating is usually picked to be at least 20% higher ($\sim$ "80% derating"— i.e., safety margin) than the worst-case diode voltage given below:

- For a **buck**—maximum diode voltage is V_{INMAX}.

- For a **boost**—maximum diode voltage is V_O.

- For a **buck-boost**—maximum diode voltage is $V_O + V_{INMAX}$.

3.19.3 Worst-case Switch Dissipation

For all topologies the average input current (and therefore switch current) must increase as the input voltage decreases, so as to continue to satisfy the basic power requirement expressed by $P_{IN} = I_{IN} \times V_{IN} = P_O/\eta$ (where η is the efficiency, assumed fixed). Therefore, the switch RMS current is a maximum at V_{INMIN} (i.e., D_{MAX}) for all topologies.

For the boost and buck-boost, the general inductor design procedure is at D_{MAX} in any case. So we can directly use the numbers derived from that, to find the switch RMS current using Equation (3-62):

$$I_{RMS\_SW} = I_{L\_DMAX} \times \sqrt{D_{MAX} \times \left(1 + \frac{r_{DMAX}^2}{12}\right)} \quad \text{(any topology)} \qquad (3\text{-}62)$$

where I_{L_DMAX} and r_{DMAX} are, respectively, the average inductor current and current ripple ratio at D_{MAX} (i.e., at V_{INMIN}). D_{MAX} can be calculated using

$$D_{MAX} = \frac{V_O - V_{INMIN} + V_D}{V_O - V_{SW} + V_D} \quad \text{(boost)} \qquad (3\text{-}63)$$

$$D_{MAX} = \frac{V_O + V_D}{V_{INMIN} + V_O - V_{SW} + V_D} \quad \text{(buck-boost)} \qquad (3\text{-}64)$$

and we should remember that

$$I_{L\_DMAX} = \frac{I_O}{1 - D_{MAX}} \quad \text{(boost and buck-boost)} \qquad (3\text{-}65)$$

For the buck, the general inductor design procedure is at D_{MIN}. So we cannot directly use the numbers derived from that to find the switch RMS current (by the previously given equation). We need to calculate r_{DMAX}, but we only know r_{DMIN} so far. Let us proceed with the required steps.

$$r_{DMAX} = \frac{Et_{DMAX}}{L \times I_L} \quad \text{(any topology)} \qquad (3\text{-}66)$$

In other words, if we know the voltseconds at V_{INMIN}, we will know the corresponding current ripple ratio r_{DMAX} for the chosen inductor. But first we have to calculate D_{MAX}:

$$D_{MAX} = \frac{V_O + V_D}{V_{INMIN} + V_O - V_{SW} + V_D} = \frac{12 + 0.5}{18 - 1.5 + 0.5} = 0.735 \quad \text{(buck)}$$

The switch on-time is therefore

$$I_{ON\_DMAX} = \frac{D_{MAX}}{f} \Rightarrow \frac{0.735 \times 10^6}{150,000} = 4.9\,\mu s \quad \text{(any topology)}$$

The voltage across the inductor when the switch is ON is

$$V_{ON\_DMAX} = V_{INMIN} - V_{SW} - V_O = 18 - 1.5 - 12 = 4.5\,V \text{ (buck)}$$

So the voltsµseconds is

$$Et_{DMAX} = V_{ON\_DMAX} \times t_{ON\_DMAX} = 4.5 \times 4.9 = 22\,V\mu s \text{ (any topology)}$$

Therefore

$$r_{DMAX} = \frac{Et_{DMAX}}{L \times I_O} = \frac{22}{137 \times 1} = 0.16 \quad \text{(buck)}$$

Finally, we are in a position to calculate the switch dissipation

$$I_{RMS\_SW} = I_O \times \sqrt{D_{MAX} \times \left(1 + \frac{r_{DMAX}^2}{12}\right)} = 1 \times \sqrt{0.735 \times \left(1 + \frac{0.16^2}{12}\right)}$$

$$= 0.86\,A \quad \text{(buck)}$$

If for example, if the drain-to-source resistance is $0.5\,\Omega$, the dissipation in the MOSFET is

$$P_{SW} = I_{IRMS\_SW}{}^2 \times R_{DS} = 0.86^2 \times 0.5 = 0.37\,\text{W} \quad \text{(any topology)}$$

Note that the general switch selection procedure is as follows:

The rule-of-thumb is to pick a switch with a current rating at least equal to, but preferably at least *twice* the worst-case RMS switch current calculated above (for low losses, since the switch forward drop will decrease substantially if its current rating is increased)

Its voltage rating is usually picked to be at least 20% higher ($\sim$ "80% derating"—i.e. safety margin) than the worst-case switch voltage given below

- For a *buck*—maximum switch voltage is V_{INMAX}.

- For a *boost*—maximum switch voltage is V_O.

- For a *buck-boost*—maximum switch voltage is $V_O + V_{INMAX}$.

3.19.4 Worst-case Output Capacitor Dissipation

Coincidentally, the worst-case output capacitor RMS current for all three topologies occurs at the same point at which the general inductor design procedure for each of them is carried out. In other words, this point is V_{INMAX} for the buck, and V_{INMIN} for the boost and buck-boost. So we should have no trouble, directly using the numbers derived from the general inductor design procedure, to find the worst-case RMS current of the output capacitor, using the equations below.

For the buck, we get

$$I_{RMS\_OUT} = I_O \times \frac{r_{DMIN}}{\sqrt{12}} = 1 \times \frac{0.277}{\sqrt{12}} = 0.08\,\text{A} \quad \text{(buck)}$$

So for example, if the ESR of the output capacitor is $10\,\Omega$, we get the dissipation

$$P_{SW} = I_{RMS\_OUT}{}^2 \times \text{ESR} = 0.08^2 \times 10 = 0.064\,\text{W} \quad \text{(any topology)}$$

For the boost and the buck-boost, we need to use

$$I_{RMS\_OUT} = I_O \times \sqrt{\frac{D_{MAX} + \dfrac{r_{DMAX}{}^2}{12}}{1 - D_{DMAX}}} \quad \text{(boost and buck-boost)}$$

$$(3\text{-}67)$$

Note that the general output capacitor selection procedure is as follows:

The rule-of-thumb is to pick an output capacitor with a ripple current rating equal to or greater than the worst-case RMS capacitor current calculated above. Its voltage rating is usually picked to be at least 20 to 50% higher than what it will see in the application (i.e., V_O for all topologies). The output voltage ripple of the converter is also usually a concern. The total peak to peak output voltage ripple produced by the output capacitor is equal to its ESR multiplied by the worst-case peak to peak output current given below (ignoring the ESL of the capacitor):

- For a *buck*—peak to peak capacitor current is $I_O \times r_{DMIN}$. This is the same point at which the general inductor design procedure would have been carried out, and so r_{DMIN} is already known.

- For a *boost*—peak to peak capacitor current is $I_O \times (1 + r_{DMAX}/2)/(1 - D_{MAX})$. This is the same point at which the general inductor design procedure would have been carried out for this topology, so r_{DMAX} and D_{MAX} are already known.

- For a *buck-boost*—peak to peak capacitor current is $I_O \times (1 + r_{DMAX}/2)/$ $(1 - D_{MAX})$. This is the same point at which the general inductor design procedure would have been carried out for this topology, so r_{DMAX} and D_{MAX} are already known.

3.19.5 Worst-case Input Capacitor Dissipation

For the buck-boost, things are much simpler, since the worst-case input capacitor RMS current occurs at D_{MAX}, which is also the point at which we carry out the general inductor design procedure. So all the numbers available from that procedure can be used directly in Eq. (3-68):

$$I_{RMS\_IN} = I_{L\_DMAX} \times \sqrt{D_{MAX} \times \left(1 - D_{MAX} + \frac{r_{DMAX}^2}{12}\right)} \quad \text{(buck-boost)}$$

$$(3\text{-}68)$$

For the buck and the boost, the worst-case input RMS capacitor current occurs at $D = 0.5$. So we have to calculate r_{50}, that is, the current ripple ratio at $D = 50\%$ (or whatever voltage within the specified input range of our application range is closest to this point).

Let us do the numerical calculation for the buck, and the procedure will become clearer.

The input voltage at which $D = 50\%$ occurs for the buck is

$$V_{IN\_50} = 2 \times V_O + V_{SW} + V_D = 2 \times 12 + 1.5 + 0.5 = 26\,\text{V (buck)}$$

$$\text{(For the boost use } V_{IN\_50} = \frac{V_O + V_{SW} + V_D}{2} \approx \frac{V_O}{2}\text{)}$$

We see that our input range does not include this point. But the closest to it is V_{INMAX} and D_{MAX}. However, coincidentally, this is already the point at which the general inductor design procedure was carried out. So we can use all the numbers derived from that procedure to calculate the input capacitor RMS current, as follows:

$$I_{RMS\_IN} = I_O \times \sqrt{D \times \left(1 - D + \frac{r^2}{12}\right)} = 1 \times \sqrt{0.543 \times \left(1 - 0.543 + \frac{0.277^2}{12}\right)} \quad \text{(buck)}$$

$$\text{(For the boost use } I_{RMS\_IN} = \frac{I_O}{1 - D} \times \frac{r}{\sqrt{12}})$$

So finally

$$I_{RMS\_IN} = 0.502\,A$$

> **Note:** If for our worked buck example, the input range was not 18–24V but say 30–45V, then the general inductor design procedure would clearly be carried out at 45V. However, the input capacitor current would be a maximum at 30V. So we can use the above equation for the RMS current, but we would now need to use r_{DMIN} and D_{MAX}. Therefore, knowing only r_{DMAX} so far, we would need to calculate r_{DMIN} by the same procedure presented earlier—that is, by recalculating the voltseconds, and so on.

Note that the general input capacitor selection procedure is as follows:

The rule-of-thumb is to pick an output capacitor with a ripple current rating equal to or greater than the worst-case RMS capacitor current calculated above. Its voltage rating is usually picked to be at least 20 to 50% higher than what it will see in the application (i.e., *VIN_MAX* for all topologies). The input voltage ripple of the converter is also usually a concern because a small part of it does get transmitted to the output. There can also be EMI considerations involved. In addition, every control IC has a certain (usually unspecified) amount of input noise and ripple rejection, and it may misbehave if the ripple is too much. Typically, the input ripple needs to be kept down to less than $\pm 5\%$ to $\pm 10\%$ of the input voltage. The total peak to peak input voltage ripple produced by the input capacitor is equal to its ESR multiplied by the worst-case peak to peak input current given below (ignoring the ESL of the capacitor):

- For a *buck*—peak to peak capacitor current is I_O and $D_{MAX} \times (1 + r_{DMIN}/2)$. This is the same point at which the general inductor design procedure would have been carried out, and so r_{DMIN} is already known.

- For a *buck-boost*—peak to peak capacitor current is $I_O \times (1 + r_{DMAX}/2)/(1 - D_{MAX})$. This is the same point at which the general inductor design procedure would have been carried out for this topology, so r_{DMAX} and D_{MAX} are already known.

- For a **boost**—peak to peak capacitor current at the worst-case point for this parameter (i.e., $D = 0.5$) is equal to $2 \times I_O \times r_{50}$ where

$$r_{50} = \frac{V_{IN\_50}}{4 \times f \times L \times I_O} \text{ and } V_{IN\_50} - \frac{V_Q + V_{SW} + V_D}{2} \approx \frac{V_O}{2}$$

(3-69)

Note that if the input range does not include the $D = 0.5$ point, we need to look for the input voltage end closest to $D = 0.5$. Then we can use the general equation for the peak to peak input capacitor current

$$I_{PK\_PK} = \frac{I_O \times r}{1 - D}$$

(3-70)

where r and D correspond to this particular worst-case input voltage end. To find r we can use

$$r = \frac{V_O - V_{SW} + V_D}{I_O \times L \times f} \times D \times (1 - D)^2$$

(3-71)

where L is in H, and f is in Hz.

That completes the converter and magnetics design procedure.

CHAPTER 4

Control Circuits

Raymond Mack

> *The control section is important for the behavior of the switching power supply. It has a direct effect on output regulation, transient response, efficiency, and its reaction to adverse operating conditions. There are several methods of controlling a switching power supply, each enhancing certain aspects of the operation, but also possibly creating difficulties in another area. The selection of the best control method for your particular application should be based upon a good understanding of each of the needed operations.*
>
> *The feedback loop compensation (Chapter 9) is the last aspect of the design of the control section. The type of compensation used will determine the stability of the supply, the accuracy and the transient response of the output.*
>
> *Even after designing the perfect control section, other unseen factors can affect its proper operation. I once modified a supply design I had used before. I gave my schematic to the printed circuit board designer and later built the prototype. I simply could not get the supply to regulate and it oscillated badly. I then examined the PCB layout. It was electrically correct, but the control was on one side of a noisy ground node and the voltage sensing circuit was on the other side. Lots of noise was being injected into the sensitive input of the error amplifier. I then had to instruct the PCB layout person on the techniques of good layout practices. The next prototype operated flawlessly. In short, ignorant PCB layout designers and autoroute are your enemies.*
>
> *Ray Mack has a very good overview of the various methods of control, their strengths and shortcomings. There are optimum control methods for each power topology and Ray Mack makes these associations.*
>
> **—Marty Brown**

We will explore the various forms of controllers available from semiconductor manufacturers. A large variety of controllers are available, but each part is usually intended for a narrow application. I will refer to application notes from various manufacturers. These are available on each manufacturer's website or by contacting the manufacturer.

4.1 Basic Control Circuits

The simplest form of control circuit is variable frequency/constant on-time or pulse frequency modulation (PFM). In Figure 4.1, the oscillator has a constant on-time (basically a one-shot multivibrator similar to a 555 timer). As soon as the control voltage drops below the reference, the oscillator is triggered to turn on by the comparator. Under light loads, the frequency is low and the duty cycle is low. As the load increases, the frequency increases. The maximum frequency occurs at 50% duty cycle. The wide range of ripple frequency can cause problems for electromagnetic compatibility (EMC) and for ripple control on the output. The Texas Instruments TL-497 is a popular commercial example of this type of circuit.

EMC and ripple control are much more predictable and controllable if a constant frequency is used and the width of the pulse is varied. Pulse width modulation (PWM) uses a constant frequency and varies the on-time of the switch. Figure 4.2 illustrates the basics of a voltage mode PWM controller.

The voltage divider is used with the error amplifier and reference voltage to generate a scaled error signal. The oscillator is similar to a 555 oscillator and generates a constant frequency sawtooth wave. Typically, the timing resistor sets the charge current for the timing capacitor. Once the voltage on the timing capacitor reaches the trip point,

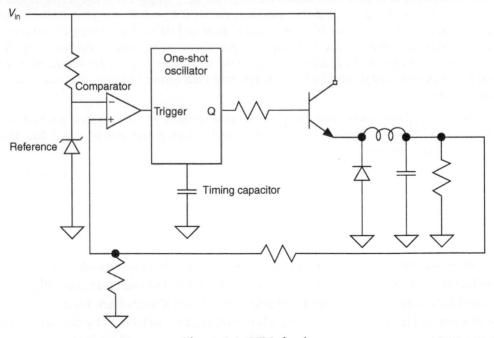

Figure 4.1: PFM circuit

a flip-flop in the oscillator turns on and rapidly discharges the timing capacitor to the lower trip point. The output switch is driven by comparing the error voltage and the oscillator voltage. Figure 4.3 shows how the switch signal is generated.

When the oscillator voltage is less than the error amplifier output voltage, the switch turns on. When the oscillator voltage goes above the error amplifier output voltage, the switch turns back off. If the error voltage is less than the lowest triangle voltage, the duty cycle will be 100%; if the error voltage is greater than the highest voltage of the triangle voltage, the duty cycle will be 0%.

Flyback and boost converters require a minimum amount of off-time so that energy stored in the inductor can be dumped to the output circuit. Some forward converter designs will also require a guaranteed amount of off-time. Modern voltage mode PWM controllers provide a mechanism to ensure a duty cycle less than 100%. This dead time is usually adjustable with an external resistor.

Current mode PWM control has inherent advantages over voltage mode control. These include improved transient response and a simpler control loop. Figure 4.4 illustrates the basics of a current mode PWM controller. In this circuit, the oscillator runs at a constant frequency. The pulse from the oscillator sets the flip-flop, which starts current flowing in the transistor switch. The current flow in the switch stops when the current as measured by R_{sense} creates a current sense voltage that equals the trip point set by the error amplifier.

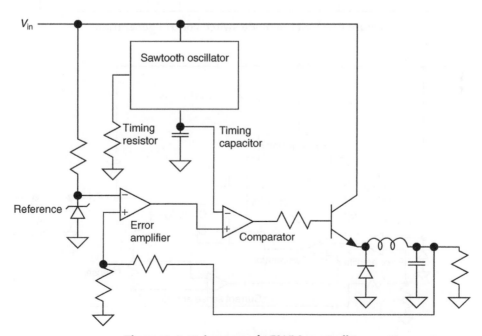

Figure 4.2: Voltage mode PWM controller

The comparator resets the flip-flop, which shuts off the switch. The error amplifier is used to adjust the trip point for the switch current so that the inductor current is the proper amount to maintain the output voltage. As the output voltage approaches the desired value, the error signal reduces the current trip point to maintain a constant average inductor current.

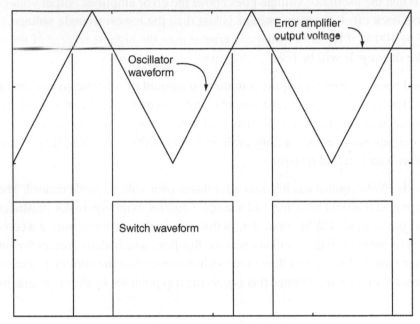

Figure 4.3: Voltage mode switch control generation

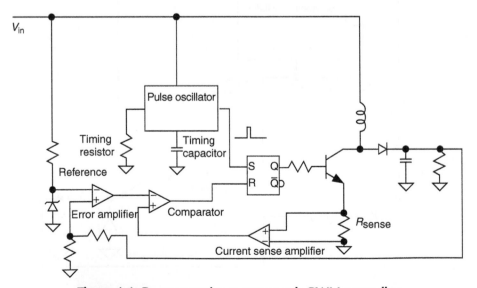

Figure 4.4: Representative current mode PWM controller

4.2 The Error Amplifier

Figure 4.5 shows the typical methods of setting up the error amplifier to control the output for a positive output supply and for a negative output supply. The negative output circuit uses a voltage divider connected to the reference to place the input to the amplifier above ground. PWM circuits are intended to operate from a single positive power supply. This means that all of the pins, especially the error amplifier and current sense pins, must not go more than one diode drop below ground.

You will also notice that there is a resistor (R_3) on the pin opposite the feedback pin. All bipolar transistor difference amplifiers (including op-amps and comparators) use the base of a transistor as the input. The input transistors require a small amount of bias current in order for amplification to occur. This bias current flows in R_1 and R_2 in addition to the normal voltage divider current, and it slightly changes the voltage at the feedback pin. The small amount of additional DC voltage due to the bias current will cause a small offset in the output voltage that depends on the closed loop gain of the amplifier and the values of R_1 and R_2. R_3 has a value equal to the parallel equivalent of R_1 and R_2. This ensures that both amplifier input pins are raised above ground by the same amount to balance the effects of the input bias current.

The output of the error amplifier is similar to a resistance-coupled DC circuit. Instead of a resistance, the load for the output transistor is a current source. The effect is that the current is split between the output transistor and the load. This is the equivalent of an open collector digital circuit, except that the transistor is operated in the linear region. Several "open collector" circuits can have their outputs connected together just like a wired-OR open collector digital circuit. The circuit that pulls the output to the lowest voltage is the one that controls the voltage at the input to the PWM comparator. The current source load for the output transistor makes it a transconductance amplifier rather than a voltage amplifier. The voltage gain is equal to the transconductance times the load resistance.

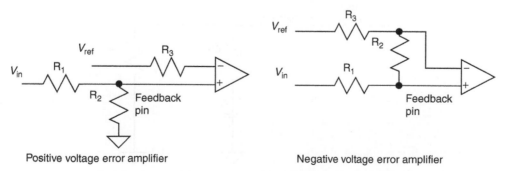

Positive voltage error amplifier   Negative voltage error amplifier

Figure 4.5: Positive and negative voltage error amplifiers

4.3 Error Amplifier Compensation

There is a broad class of electronic systems covered by classic feedback control theory. Closed loop op-amp circuits, electromechanical servos, phase-locked loops, linear power supplies, and switching power supplies can all be analyzed using control theory. A detailed description of feedback theory is beyond the scope of this book. Thomas Frederiksen gives a very good description of the effects of the transfer function in Chapter 4 of his book *Intuitive IC Op Amps* (National Semiconductor Technology Series, 1984). He describes how multiple poles and zeros can ensure stability or lead to oscillations in a closed loop system. There is also a condensed general description of frequency compensation of amplifier/power amplifier combinations at the end of Linear Technology Application Note 18. Consult a control theory textbook for a complete understanding of compensation.

The error amplifier in PWM controllers is not the equivalent of a 741 or 1458 op-amp. Op-amps have internal compensation that places a low-frequency pole somewhere below 100 Hz (usually under 5 Hz). This pole dominates the overall closed loop amplifier performance by rolling off the gain as frequency increases. The error amplifier in PWM controllers usually has no internal compensation. PWM controllers bring the output of the error amplifier out to a pin so that poles and zeros can be added to the closed loop system to provide frequency compensation to the system.

Numerous effects in a switching power supply tend to increase the phase delay around the loop. Two major contributors are the inductor and the filter capacitor, including its equivalent series resistance (ESR). The combination of the inductor and capacitor in the output circuit are the equivalent of a series resonant circuit and will cause two complex poles in the response. The transfer function changes with changes in the load current and power line voltage. The output capacitor and its ESR form a zero, and the load and output capacitor create a pole. Figure 4.6 shows the equivalent circuit of the output capacitor, ESR, and load resistance. You will notice that ESR is a contributor to both the pole and the zero.

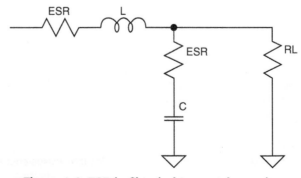

Figure 4.6: ESR in filter inductor and capacitor

The goal of compensation is to ensure that the final power supply will have a quick response to load and input transients, and will not oscillate. Compensation that is heavily damped will guarantee that the output voltage will not oscillate, but the output will likely have a large, long-lasting transient response to rapidly changing input or output. It is also likely to result in significant overshoot during recovery from short circuits. Response that is too rapid will result in oscillations in the control loop.

Figure 4.7 shows a typical compensation network for a buck or forward converter. The resistor and capacitor add a pole to the transfer function. This compensation network needs to be optimized for both gain and frequency. The resistor and capacitor act as a damper to lower the Q of the circuit.

Figure 4.8 shows a typical compensation circuit for a continuous mode boost or flyback converter. All continuous inductor current boost and flyback converters have a zero in the right-half plane. This requires the second pole added to the feedback response. This pole must roll off gain below the frequency of the right-half plane zero. Poles and zeroes in

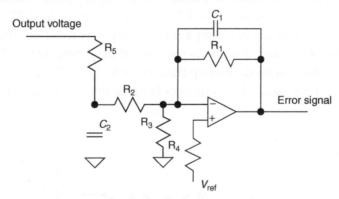

Figure 4.7: Typical compensation circuit for a buck or forward converter

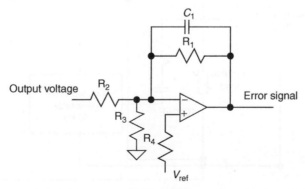

Figure 4.8: Typical compensation circuit for a continuous mode boost or flyback converter

the right-half plane are associated with responses that are steadily increasing in the time domain. The effect of this zero is obvious if you run a simulation of the startup of a boost converter without the second pole. The output voltage will have tremendous overshoot.

None of the application notes from IC manufacturers gives a rigorous method of evaluating the response of a switching supply using a mathematical approach. Application Note U-95 from Texas Instruments gives some guidance on math for linear power supply compensation that can be used for switching power supply analysis. However, if you understand the math involved, you probably don't need this book.

I prefer the empirical method described in Linear Technology Application Notes 19 and 25 for ensuring that the compensation circuit is optimal for the design. This approach uses time domain analysis rather than frequency domain analysis. The description in these application notes is specific to the LT1070 series of current mode controllers, but the technique is applicable to all switching power supplies that have transconductance error amplifiers.

Figure 4.9 shows a test setup based on Linear Technology application notes. There are three pieces of test equipment required. The first is a variable load. This can be an active load that is adjustable or simply a set of high power resistors. The second is an oscilloscope for observing the transient response of the power supply. The last is a function generator that will introduce step changes to the load. We are only interested in the step response, so we place a low-pass filter between the supply output and the input

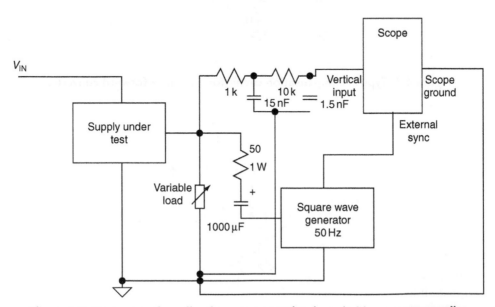

Figure 4.9: Test setup for adjusting compensation in switching power supplies

channel of the scope. Thus, we only see the DC value and not any switching frequency energy. We trigger the scope with the output of the function generator.

4.3.1 Test Sequence

1. Start compensation with a resistance of 1 kΩ and capacitance of 2 μF. This will load the error amplifier for high frequencies and create a dominant pole due to the capacitance and the load of the PWM circuit. There will be a zero in the response due to the resistance but it will have very little effect.

2. Verify that there are no ground loops by connecting the scope channel 1 probe to the ground connection. If channel 1 shows any response, you must isolate the scope or signal generator by breaking the safety ground connection. In order to maintain safety, you should use an isolation transformer between the test equipment and the power line.

3. Adjust the signal generator for a 5 V p-p square wave. This gives a 100-mA step input to the control loop. If the positive and negative step responses are not identical with light loading, reduce the signal generator voltage.

4. Verify that the response is a single pole "over-damped" response. If the response is not over-damped, increase the value of the resistor. The resistor should be increased first and then the capacitor value to ensure that we start with an over-damped condition.

5. Reduce the capacitor 2:1 per step until the response is slightly under-damped. This moves the pole frequency higher and increases the gain-bandwidth.

6. Start increasing the resistor value in 2:1 increments to decrease the response time and to increase damping. Stop when the response becomes over-damped again. Increasing this resistance moves the zero lower in frequency so that it starts to flatten out the gain response to mid-frequencies.

7. Continue to iterate, reducing both the capacitor and the resistor to give a fast damped response. The goal is to get the largest resistance and smallest capacitance that do not produce oscillations while giving rapid settling to the proper output voltage.

8. Now we have to verify that we have enough gain and phase margin over all conditions. One of the toughest problems is the value of the zero caused by the output capacitor and its ESR. ESR is very temperature-dependent. If the supply must operate at very low temperatures, the ESR will increase by orders of magnitude. The margin test will entail testing the response to ensure no oscillations for all combinations of temperature, load, and input voltage. A good

rule of thumb is to adjust for slight over-damping at the temperature extremes to ensure stable operation over the entire temperature range.

> Remember that breaking the electrical safety ground defeats the safety aspect of having a ground. You must use appropriate caution around the test equipment.

4.4 A Representative Voltage Mode PWM Controller

The 1526A family is representative of a second-generation, full-featured voltage mode PWM controller. This part is suitable for either DC–DC converter service or as an off-line controller at frequencies up to about 100 kHz. This part is especially suited to push-pull, half-bridge, and full-bridge circuits because it has two outputs. Figure 4.10 shows the internal block diagram of the controller.

The internal circuitry requires a stable, regulated voltage for proper operation. The reference regulator is a precision temperature-compensated linear regulator. It is capable of providing 20 mA to external circuits. The reference has a 2 V dropout, so the minimum supply voltage is 7 V. In the 1526A, the band gap reference is trimmed to make the final reference voltage accurate to ±1%.

The under-voltage lockout circuit compares the reference voltage to an internal band gap reference. The circuit pulls the reset pin low, disables the output drivers, and clamps the error amplifier output through the diode so there is no possibility of spurious output pulses until all of the circuitry has sufficient voltage for proper operation. The lockout continues until the reference voltage reaches 4.4 V. The lockout comparator has 200 mV of hysteresis. The circuit will not lock out once the reference reaches 4.4 V until the reference falls below 4.2 V. This prevents noise from causing spurious reset if the reference voltage is rising slowly.

Once the reset pin is released by the under-voltage lockout circuit, the normal soft start sequence begins. The soft start capacitor is connected to the error amplifier output through a clamp transistor that limits how high the error amplifier output voltage can rise during soft start. The clamp on the error voltage limits the maximum pulse width. As a consequence, the increase in inductor current and the rate of rise of output voltage while the system is starting is limited. The clamp is no longer active once the capacitor charges to 5 V. The soft start capacitor is charged with a constant current of 100 μA (typical), so we can use the capacitor definition and current definition to find the soft start time.

$$Q = CV \text{ and } I = \Delta Q/\Delta t \qquad (4\text{-}1)$$

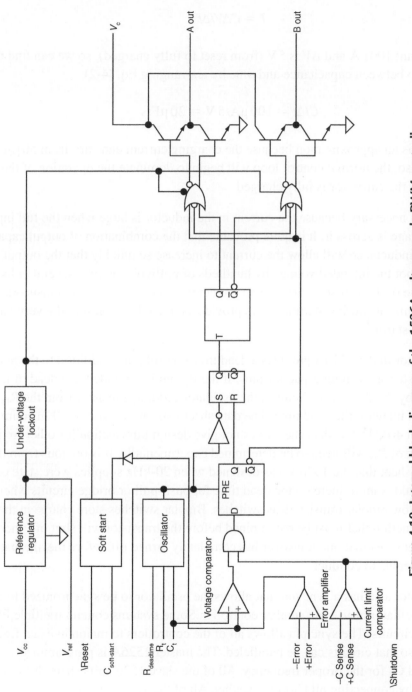

Figure 4.10: Internal block diagram of the 1526A voltage mode PWM controller

If we differentiate both sides of the capacitor equation we get:

$$I = C\Delta V/\Delta t \tag{4-2}$$

I is a constant $100\,\mu A$ and ΔV is $5\,V$ (from reset to fully charged), so we can find the relationship between capacitance and time by rearranging Eq. (4-2).

$$C/\Delta t = 100\,\mu A/5\,V = 20\,\mu F/s \tag{4-3}$$

This value is an approximation because the charging current can vary from $50\,\mu A$ to $150\,\mu A$. Also, the normal control loop will begin to dominate the operation of the system long before the capacitor is fully charged.

Soft start is necessary because the current in the inductor is large when the full input supply voltage is across it. It is quite probable that the combination of output capacitor and choke inductance will allow the current to increase so quickly that the output voltage can overshoot the intended voltage by hundreds of millivolts or even several volts. The purpose of the soft start circuit is to protect the diodes and switch transistors from excessive currents during startup and to provide a damped response to the very large transient at startup.

The oscillator in the 1526A provides a dead time control pin in addition to the normal timing resistor and timing capacitor pins. If the R_D pin is grounded, the dead time is controlled by the discharge circuit in the oscillator. Adding a resistor from the R_D pin to ground will increase the dead time. The data sheet lists an increase of $400\,ns/ohm$ when operating at $40\,kHz$. The data sheet does not give design information for other frequencies, so the value of R_D will need to be determined experimentally. It is obvious from this part of the data sheet that the 1526A was designed when 20-kHz supplies were state of the art. We would want to increase the dead time for push-pull or bridge circuits where we are using slow bipolar transistors as switches. Bipolar switches store charge in the base-collector junction that must be recombined before the transistor will shut off. Increasing dead time ensures that one transistor has completely turned off before the alternate transistor begins to conduct.

The oscillator also has a sync pin that allows the oscillator to be synchronized to an external oscillator or to sync another controller. Some systems contain multiple PWM controller circuits. The sync pin allows all of the controllers to maintain exact frequency and phase so that circuits can be paralleled. The master 1526A is programmed with R_T, R_D, and C_T for the proper frequency. All of the slave 1526 parts share the sawtooth waveform by connecting all C_T pins together. All of the sync pins must also be connected together. All of the slave R_T pins are left open.

The sync pin could also be used to sync the controller to an external logic clock if the system requires. To sync to an external logic signal, you must set the oscillator frequency approximately 10% below the desired frequency. The logic circuit should supply a short pulse (on the order of 500 ns) to the sync pin. This short pulse terminates the charge phase of the oscillator and restarts the cycle.

The sync pin, reset pin, and shutdown pin are all bidirectional, low active logic pins. Figure 4.11 shows how the internal circuits drive the pin as an open-collector output with internal pullup and as an input to the internal circuits. The shutdown pin can be used for fault conditions that require an immediate shutdown of the controller. The shutdown pin doubles as an output indicating that the current limit comparator is active. Pulling the shutdown pin low disables the outpu t drivers. The reset pin discharges the soft start capacitor and clamps the error amplifier output. Releasing the reset pin will initiate a soft start cycle. Each of these pins is compatible with TTL or CMOS logic.

The 1526A implements digital current limiting. The current sense comparator provides a logic output that terminates the output pulse. This allows the system to terminate each output pulse if the current limit is exceeded. Do not confuse this operation with current mode PWM control where the error signal controls the current trip point. This part has a fixed threshold for current limit action. The current sense amplifier has an internal 100 mV reference on the inverting pin, so the inverting pin can be grounded to provide for unipolar current sense input. This allows for a very low resistance current sense to minimize current sense power loss.

Other circuits, such as the SG2524, use a difference amplifier that subtracts voltage from the output of the error amplifier and reduces the pulse width output. The internal circuit of the SG2524 is shown in Figure 4.12.

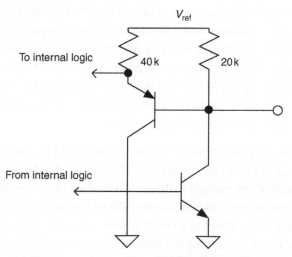

Figure 4.11: Internal circuitry of bidirectional pins in 1526A

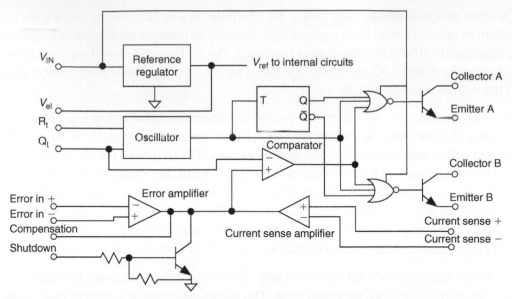

Figure 4.12: Internal circuitry of the SG2524

The 1526A PWM pulse generator uses digital logic to ensure that the comparator does not produce multiple pulses because of noise. The PWM comparator compares the oscillator ramp voltage to the error amplifier voltage and sends a pulse to set the flip-flop when the voltages are equal. The high signal from the PWM latch is sent to the output steering logic. The output pulse is terminated when the oscillator discharge pulse resets the PWM latch.

The output steering logic performs three functions. The first function is implemented by a toggle flip-flop that steers alternate output pulses to alternate output drivers. This allows the 1526A to be used in symmetric drive circuits such as push-pull or bridge circuits. The second function is output blanking. There is a minimum dead time on each output that is controlled by the length of the oscillator reset pulse. The output blanking is ANDed with the pulse command from the PWM latch so that it overrides the PWM latch signal. The third function of the steering logic disables the output drivers for fault conditions such as over-temperature and any time the reset pin is active.

The 1526A has two totem-pole outputs that can be connected to a power supply that is different from the control circuit power supply. This allows the drive to be tailored to the external switches. Each of the outputs is driven at one-half of the oscillator frequency. The pulses from the two outputs do not overlap. When the output is driven low, it saturates the lower transistor. There is a small amount of time where both transistors are on (cross-conduction time) because of the turn-off delay caused by saturation in the lower transistor. Because of the cross-conduction current, this device needs a small resistor in series with the V_C pin to limit the current. The 1526A is an improved version of the 1526 and limits the cross-conduction time to 50 ns. This length of time still requires the current limit resistor.

Figure 4.13 shows a typical drive circuit for FET switches. The 1526A output transistors can source or sink 100 mA. The capacitance of the FET can draw substantial current during charge and discharge. The resistor in series between the FET gate and the output pin protects the output transistors by limiting the peak current. Additionally, the drain to gate capacitance is usually quite large and can couple large inductive voltage transients from the drain circuit into the gate circuit. The Schottky diode ensures that the voltage on the output pin cannot be driven more than 0.3 V negative with respect to the IC ground pin.

4.5 Current Mode Control

Figure 4.14 shows the basic circuit of a current mode PWM controller in a boost converter. This circuit has two control loops. The outer loop measures the output voltage and provides an error signal to the inner loop. The inner loop compares the error signal and an analog of the inductor current to decide when to turn off the switch. The effect is to change the pulse width. The pulse width is a function of the inductor current rather than a function of the error signal.

The oscillator starts each cycle by setting the output latch to turn on the switch. The error amplifier generates the error signal that is used to compare against the inductor current signal. Once the peak inductor current signal is equal to the error signal, the comparator

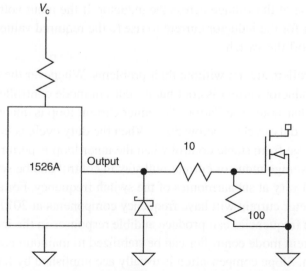

Figure 4.13: Typical drive circuit for FET switches

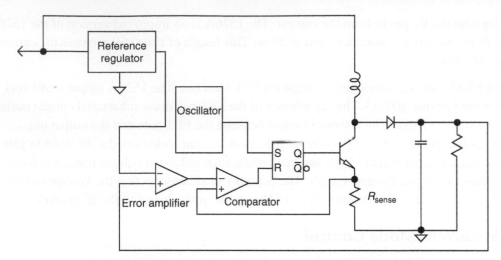

Figure 4.14: Basic circuit of a current mode PWM controller

resets the latch and turns off the switch. If the output voltage decreases, the error signal will increase and allow the peak current to increase with the next pulse.

The operation of the current mode controller has advantages over a voltage mode controller. The first is that the inductor current is a direct function of the error voltage, so for small signal analysis the inductor can be replaced by a voltage controlled current source. This removes one order from the transfer function. The control loop is easier to compensate than a voltage mode circuit. Another advantage is that input line voltage changes are removed from the compensation problem. The peak current through the inductor is a function of the voltage across the inductor. If the input voltage drops, it will just take longer for the inductor current to rise to the required value and for the comparator to shut off the switch.

Current mode controllers are not without their problems. Whenever the duty cycle exceeds 50% and inductor current is continuous, current mode controllers have a response called subharmonic oscillation. The inner current loop is unconditionally stable as long as the duty cycle is below 50%. When the duty cycle is larger than 50%, the output will diverge from stable control when the inner loop is perturbed by noise or transients. The average inductor current will stay in control and be set by the error amplifier, but it will vary at subharmonics of the switch frequency. For a 40-kHz switch frequency, the inductor current will have frequency components at 20 kHz, 10 kHz, etc. These subharmonic frequencies can produce audible responses in the inductor and other components. A current mode controller can be stabilized to maintain control by adding slope compensation. Slope compensation is usually accomplished by feeding some of the voltage from the oscillator capacitor into either the current sense amplifier or the error amplifier. Slope compensation changes the current trip from a constant voltage to a sawtooth waveform at the switch frequency. The trip current decreases as the duty cycle increases. There is a minimum compensation slope that will guarantee that the system is unconditionally stable. The inequality below describes this relationship:

$$S_{\text{COMPENSATION}} \geq S_{\text{CHARGE}}(2DC - 1)/(1 - DC) \tag{4-4}$$

$S_{\text{COMPENSATION}}$ is the slope of the compensation voltage and S_{CHARGE} is the slope of the inductor charging waveform. Fortunately, most modern current mode ICs provide internal slope compensation that can be used "as is" or modified if necessary. For older parts, such as the 1846A, either a manufacturer's application note or the data sheet will give the information necessary to calculate the appropriate amount of slope compensation. TI Application Note U-97 and Linear Technology Application Note 19 give detailed analyses of slope compensation.

4.6 A Representative Current Mode PWM Controller

The 1846A is representative of a third-generation controller. Figure 4.15 illustrates the internal circuitry of the 1846A. The oscillator and reference are basically the same circuit as used in the 1526A. The 1846A oscillator can be synchronized to another 1846A or to an external oscillator in the same way as performed on the 1526A. The under-voltage lockout circuit is different in that it uses the input voltage to make the lockout decision rather than the reference voltage. The under-voltage lockout holds the device in reset as long as the input voltage is below 8.0 V. The lockout circuit has 0.75 V of hysteresis to ensure that noise or slowly rising input voltage will not cause an unstable condition at turn on.

The error amplifier is a transconductance amplifier with an "open collector" output similar to that in the 1526A.

The current sense amplifier is a voltage difference amplifier with a gain of three. The diode and voltage source in series with the inverting input of the PWM comparator limit the voltage to about 3.5 V (4.6 V error signal max. minus 0.5 V minus one diode drop). This means that a current sense amplifier output above 3.5 V will not shut off the output pulse. This constrains the current sense voltage to be less than 1.1 V because of the gain of three in the current sense amplifier.

The inverting and noninverting inputs have a common mode range of ground to $V_{\text{IN}} - 3$ V. This allows the current sense amplifier to be used in boost, buck, forward, and flyback designs. Figure 4.16 shows three different methods of implementing current sense. The resistor and capacitor in Figure 4.16(a) are usually necessary to reduce the size of turn-on transients in the switch. In both bipolar and FET switches, there is coupling between

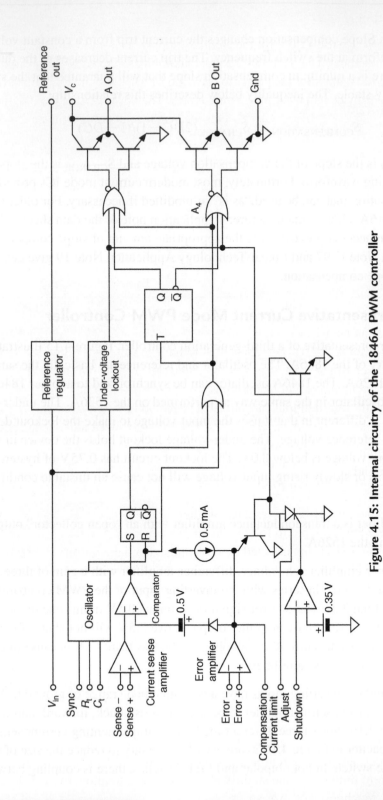

Figure 4.15: Internal circuitry of the 1846A PWM controller

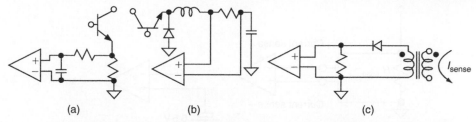

(a)   (b)   (c)

Figure 4.16: Three different methods of implementing current sensing: (a) grounded resistor; (b) floating resistor; and (c) with an isolating current transformer

the high voltage side of the switch (collector/drain) and the current sense resistor. The transient that couples to the current sense resistor can cause a false termination of the output pulse. The resistor and capacitor limit the rise time and reduce the transient so proper operation occurs. Buck designs will require that the input voltage is at least 3 V above the output voltage if a current sense resistor is used. In circuits where there is not sufficient common mode range or when total isolation is required (as in bridge circuits), the current limit amplifier can be driven by an isolating current transformer. A current transformer is also advantageous in very-high-current applications because it can reduce the voltage and, therefore, the power consumed by the current sense. The diode in Figure 4.16(c) is necessary so that the voltage at the noninverting amplifier input does not go more than one diode drop negative from ground.

The shutdown circuit, the under-voltage lockout circuit, and the current limit circuit clamp the output voltage of the error amplifier. The current limit pin is used to limit the maximum inductor current by clamping the error amplifier output below the 4.6 V maximum of the error amplifier. The error amplifier output is clamped to a voltage equal to one diode drop (the base-emitter voltage) of the current limit set transistor. Figure 4.17 shows a typical connection to the current limit pin. The current limit is not directly set by the voltage at the current limit pin, but, rather, it sets the current sense output voltage that will terminate a pulse. Since the diode drop in series with the inverting comparator input is roughly equal to the base-emitter voltage, the trip point is equal to the current limit voltage minus the 0.5 V offset. The equations below allow you to set the current limit:

$$V_{\text{CURRENT LIMIT}} = R_1 / (R_1 + R_2) V_{\text{REF}} \tag{4-5}$$

$$V_{\text{CURRENT SENSE}} = (V_{\text{CURRENT LIMIT}} - 0.5)/3 \tag{4-6}$$

$$I_{\text{CURRENT LIMIT}} = V_{\text{CURRENT SENSE}} / R_{\text{SENSE}} \tag{4-7}$$

R_2 has a secondary function of supplying the holding current for the shutdown latch. If you desire the shutdown to latch, R_2 must be below 2.5 kΩ to supply at least 1.5 mA of

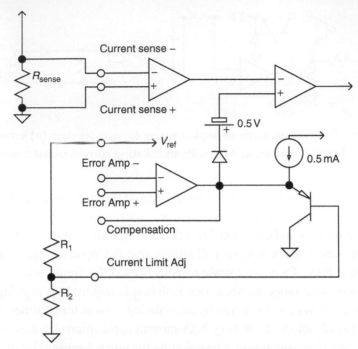

Figure 4.17: 1846 current limit pin implementation

current to hold the latch. When the shutdown signal goes below 350 mV, the shutdown circuit will shut off the PWM latch and hold the IC in reset until a power cycle occurs. Selecting R_2 greater than 5 kΩ will allow the shutdown circuit to reset the PWM latch and discharge any capacitance on the current limit set pin, but when the shutdown signal is removed, a new start sequence will begin.

This IC does not provide soft start circuitry. Soft start is accomplished by adding a capacitor to ground on the current limit pin. The current limit pin sets the peak current trip point, so raising the voltage slowly on the current sense pin will provide the soft start function.

You will notice that it is possible for the comparator to fail to set the flip-flop before a new oscillator cycle occurs if the inductor current is quite low and the error signal commands a large inductor current. This would cause the duty cycle to be greater than 100%. The signal presented to the output logic is the OR of the oscillator pulse and the flip-flop output. The short pulse from the oscillator will guarantee a short dead time on the output equal to the discharge time of the timing capacitor. You can adjust the length of the dead time by changes to the relative values of the timing resistor and capacitor. The data sheet gives a nomograph for setting the dead time.

The output logic and totem-pole outputs of the 1846A are similar to that of the 1526A. You must follow the same precautions of limiting the current into the collector supply during the crossover in the output transistors. You must similarly limit the output current when driving FET switches by using series resistors.

4.7 Charge Pump Circuits

IC manufacturers continue to improve the output capability of charge pump converters. Switch frequency and switch on resistance are the two parameters that affect power dissipation and indirectly affect efficiency and maximum output current. Charge pump circuits have an equivalent series resistance that is given by:

$$R_{EQ} = 1/F_{SWITCH}C_{FLYING} \tag{4-8}$$

This equivalent resistance is a property of the switched capacitor circuit and is not an actual physical resistance. You can see that we can improve performance (lowering R_{EQ}) by raising the frequency or increasing the flying capacitor. Performance is only increased until the internal physical resistance of the switches approaches the equivalent resistance of the switching circuit. In general, charge pump ICs can be used in parallel to achieve greater output current.

Figure 4.18 shows the internal circuit of the LTC3200, which is a representative voltage doubling charge pump that provides a regulated output. The circuit contains a 2-MHz fixed-frequency oscillator that drives the switch circuit with a two-phase nonoverlapping clock. The error amplifier compares the voltage at the feedback pin to the internal 1.268 V zener voltage reference. The output of the error amplifier controls the amount of current that can flow into the flying capacitor during phase one of the clock. Phase two of the clock connects the flying capacitor in series with the input voltage to provide current to the load and output capacitor.

This charge pump controller has a soft start and switch control circuit to limit the current draw from the input supply. The switch control circuit shuts down the system if the IC gets hotter than 160°C and reenables the circuit around 150°C. This circuit also limits output current to 225 mA in the case of a short circuit.

The LTC3200 will produce a regulated output between 1.268 V and 5.5 V, up to 100 mA. The input voltage range is 2.7 V to 4.5 V. This can accommodate a single lithium cell, three alkaline cells, three NiCad cells, or three NiMH cells. The current control circuit allows the IC to regulate the output to a voltage above or below the input voltage. Efficiency suffers for output below the input, however. The output voltage is set with a voltage divider between the output pin and the feedback pin. The equation for the output voltage is:

$$V_{OUT} = 1.268(1 + (R_1/R_2)) \tag{4-9}$$

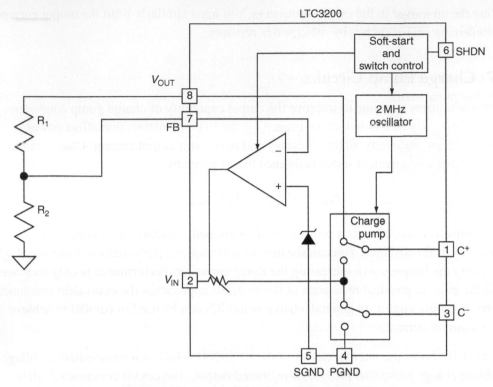

Figure 4.18: Internal circuitry of the LTC3200

The resistors can range from several kilohms to 1 MΩ. If the output voltage will be below the input voltage, it is necessary to put a 1-mA load on the output to ensure that the voltage does not creep up at very light loads.

The input capacitor, output capacitor, and flying capacitor are all required to have low ESR. These capacitors need to be greater than 0.5 µF, but values up to only 1 µF will be adequate for good output current and low ripple. Electrolytic and tantalum capacitors will not have low enough ESR to work properly. Ceramic capacitors are the preferred type. Ceramic capacitors have a significant temperature coefficient depending on the type of dielectric. X5R and X7R capacitors have the smallest change in value over temperature. Another consideration is the change in capacitance with applied voltage. Z5U and Y5V capacitors have significant changes in capacitance with applied voltage. The output capacitor ESR must be below 0.3 Ω in order for the error amplifier to remain stable. If the ESR is higher, the amplifier response is no longer a single pole rolloff and can become unstable.

The LTC3200 uses a variable resistance to modulate the charge current, so there is some amount of power dissipated in the IC in order to maintain a regulated output. The LT1516 is an example of a charge pump that uses burst mode to maintain a regulated 5.0 V output.

This circuit trades off higher ripple voltage (100 mV at full load) and a second-order filter problem for increased efficiency.

Figure 4.19 shows the internal circuit of the LT1516. Comparator 2 compares the divided output voltage against the internal reference. If the voltage is below the threshold, the charge pump switches are enabled and charge is transferred from the input to the output until the output rises above the upper comparator 2 trip point. This burst mode causes a low-frequency ripple in the output equal to the hysteresis in comparator 2. There is also high-frequency ripple in the output due to the switching of the charge pump while charging the output capacitor.

The LT1516 uses two flying capacitors to implement either a voltage tripling or voltage doubling configuration. Whenever V_{IN} is less than 2.55 V, comparator 1 forces the control logic to put the device in voltage tripler mode. During the charge phase, the switches place both flying capacitors from input to ground. During the discharge phase, the flying capacitor C1 is placed in series with C2 and the series combination is placed in series with the input. Once V_{IN} is greater than 2.55 V, the IC switches to voltage doubler mode and just uses C2 as the flying capacitor. Comparator 3 has a 50 mV offset from the feedback voltage at comparator 2. If the voltage droops by 50 mV or more, comparator 3 puts the IC back in tripler mode until the voltage rises above the upper trip point of comparator 3.

The input and output capacitors can be tantalum or electrolytic with the LT1516 because we have comparator control (bang-bang) instead of an error amplifier (proportional control), so there is no control loop to have oscillations. ESR is no longer a consideration

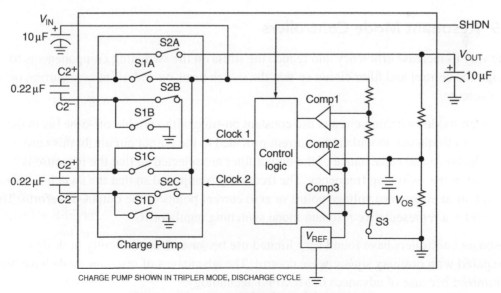

CHARGE PUMP SHOWN IN TRIPLER MODE, DISCHARGE CYCLE

Figure 4.19: Internal circuitry of the LT1516

for control stability. The only effect ESR will have is on ripple voltage. A good solution is to parallel a low-ESR ceramic capacitor (about 1 µF) with the higher-capacitance electrolytic or tantalum (about 10 µF). The ceramic capacitor reduces 600 kHz ripple from the bursts of charging and the electrolytic reduces ripple at the control frequency.

4.8 Multiple Phase PWM Controllers

The demands of power supplies for Pentium® class CPUs are significantly different from low power and traditional CPU needs. A Pentium, Athalon, or Opteron CPU requires very low voltage at tens of amps. A typical Pentium 4 power supply must supply 1.4 V at 65A.

All of the controllers mentioned so far are single-phase controllers. Current mode PWM controller-based regulators can be operated in parallel to increase current capacity. Many IC manufacturers produce control ICs that operate multiple power supplies in parallel with nonoverlapping phases. The LT3730 is a representative buck controller IC designed for Intel portable computer applications. It can operate up to 600 kHz per phase. Since the phases do not overlap, this gives a ripple frequency of 1.8 MHz. The inductor for each phase can be one-third as large as would be required in a single-phase design. The output capacitor can also be one-third the size of a comparable single-phase design. The input and output capacitor ripple current decreases as you add additional phases. Increasing the number of phases increases the efficiency of the power supply by reducing the losses due to ripple current in the capacitors. Polyphase operation is also possible for boost converters.

Polyphase operation of charge pump controllers can also provide similar improvements in efficiency by effectively increasing the frequency of operation.

4.9 Resonant Mode Controllers

One way to increase efficiency and reduce the stress on the switching components is to design the control and filter circuit so that the switch turns on and off at zero current or zero voltage.

Resonant mode switching circuits use constant on-time with variable off-time (as in the TL497) to frequency modulate the current provided to the output circuit. In this case, the inductance and capacitance of the output filter are selected so that the response is resonant at the switching frequency. The frequency is adjusted so that the switch is turned on and off at either zero voltage points or zero current points of the output waveform. The UC1860 is a representative resonant mode switching regulator IC.

Resonant controllers have found very limited use because of the difficulty of design compared with ordinary square wave control. The advantages of resonant mode have been minimized because of advances in MOSFET technology.

CHAPTER 5

Non-isolated Circuits

Raymond Mack

> *Non-isolated switching power supplies are those supplies where the input and output share a common ground and other signals. They can also sometimes propagate failures to other portions of the system, if not designed properly.*
>
> *They can only be used when the input power source is considered "safe" to the user. This is where the input voltage is less than 42 VDC. Any input voltage greater than 42 VDC is considered "hazardous" to the user and requires dielectric isolation between the input and output (see Chapter 6).*
>
> *One finds this type of switching power supply in portable equipment and low voltage printed circuit boards. There are many design aids available from the semiconductor suppliers in the market today.*
>
> *Ray Mack uses very straightforward design flows to illustrate the design of several non-isolated switching converters, based upon ICs from Linear Technology.*
>
> *—Marty Brown*

In this chapter, we will look at detailed designs of non-isolated converters. The applications include remote regulation for line-operated systems or power management in battery-operated systems. I find at least one new application or new device suitable for a non-isolated circuit every week in the trade journals. The applications have exploded during the past five years and show no signs of slowing. The trend is toward ever smaller, more efficient, and more specialized controllers. The designs we will look at here will give general design methodology as part of the specific designs.

Engineers seem enamored of creating new jargon and acronyms or initials. The new term for remote regulation is point of load (POL). Point of load regulators are almost always non-isolated circuits.

All of the designs shown here use current mode PWM control because of its inherent advantages in loop stability and current control. One of the problems with current mode control is subharmonic oscillation at duty cycles greater than 50%. Older ICs required external means to provide slope compensation to eliminate subharmonic oscillations. The modern ICs described in this chapter all contain internal slope compensation, so there is one less design task to complete.

5.1 General Design Method

There are many design variables for power supply designs, so each design will be different from the last one. The sequence below will help beginners with a starting framework. A complete design will usually require iterations of several steps.

1. Choose a converter type based on the input voltage range and the output voltage. Input always above the output indicates a buck converter. Output voltage always above the input indicates a boost design.

2. Choose an IC based on the output power, physical size, etc. This is probably the most daunting task for a beginner because there are so many parts from so many vendors. The complexity of the circuit is usually dictated by the output power. The more power, the larger and more complex the circuit. The requirements will normally force the selection of switching frequency at this step. We usually decide on diode or synchronous rectification here.

3. Choose the ripple current in the inductor based on the output ripple voltage requirement. This decision affects the choice of the input and output capacitor because of the interaction of capacitor ESR and ripple current.

4. Calculate the inductor value based on the ripple and average current.

5. Calculate the required current sense resistor based on the IC data sheet.

6. Select the switch transistor and diode based on the inductor current.

7. Calculate the input and output capacitor values based on the ripple current and ripple voltage requirements.

8. Select a first try at the loop compensation circuit.

9. Select the soft start components, if required.

5.2 Buck Converter Designs

The LT1765 is a full function current mode PWM IC with an integral NPN transistor switch, current sense resistor, and slope compensation. The switch frequency is fixed at

1.25 MHz. Figure 5.1 shows a representative buck converter design using the LT1765. This part is available in either an SO8 or 16-pin TSSOP package. The SO8 package uses the lead frame connected to the ground pin for heat dissipation. The TSSOP package has an integral heat sink pad under the package to conduct heat to the ground plane. This part is designed for small size and low bill of materials.

Any buck converter that uses an NPN transistor or NMOS switch will require a voltage above the input voltage in order to fully turn on the switch. A bipolar switch will only need a control voltage that is 0.7 V larger than the input voltage. The control voltage for an NMOS switch will be higher than for a bipolar switch. If an NMOS switch is used, the best choice for a buck converter is a logic-level switch that will only require about 2 V above the input voltage.

Figure 5.1 shows a charge pump implementation that supplies the necessary switch control voltage. This concept will work for both bipolar and NMOS designs. When the switch is closed, the voltage of the boost capacitor will add to the switch voltage so that the switch can saturate. When the switch opens, the boost capacitor will be connected across the output and charged to the output voltage minus the two diode drops from D1 and D2 (about 1 V less than the output voltage). The diode voltage drops and the internal supply circuit voltage drop limit the output voltage for full efficiency to about 3.3 V. If a lower output voltage is required, the switch will no longer saturate and power dissipation will increase dramatically. The data sheet for the LT1765 recommends a 0.18 μF boost capacitor for most applications. This value is calculated based on 700-ns on time (87% duty cycle), 90-mA boost current, and 0.7 V of ripple on the boost voltage. A ceramic capacitor with ESR below 1 Ω will be required to fully charge the capacitor during the shortest off time.

This circuit will bootstrap the boost voltage during powerup. When the circuit starts, the output voltage and the voltage at the switch pin will be zero. The control circuit will turn

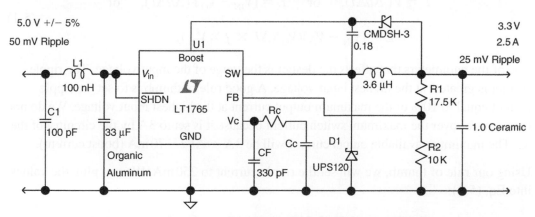

Figure 5.1: Representative buck converter using the LT1765

on the switch and the switch pin voltage will be 0.6 V below the input voltage because of V_{BE}. The transistor will not be saturated, but it will begin supplying inductor current and begin charging the output capacitor. As the output voltage rises above 1.0 V, the boost diode will conduct when the switch is off and begin charging the boost capacitor. The power dissipation in the switch will rapidly decrease as the boost voltage increases.

Current mode PWM controllers provide inherent output current limiting in a buck converter. The output current will be limited to the peak inductor current. For PWM ICs that have a shutdown pin, you can use external circuitry to detect a fault and shut down the power supply.

The size of the inductor determines the amount of ripple current. We use the inductor equation and the duty cycle equation to determine the relationship between the inductor and the ripple current.

Equation (2-6) from Chapter 2 gives the duty cycle in terms of the voltages:

$$V_O = V_{IN}DC, \quad \text{or} \quad DC = V_O/V_{IN}$$

Equation (2-1) from Chapter 2 gives the inductor voltage in terms of inductance and change in current:

$$V = L(\Delta I/\Delta t)$$

The amount of time for the current to go from minimum to maximum is:

$$\Delta t = T \times DC, \quad \text{or} \quad \Delta t = (1/f)\, DC \quad \text{or} \quad \Delta t = (1/f)\,(V_O/V_{IN})$$

where T is the period of the switching frequency f.

We can rearrange the inductor equation to yield:

$$L = V(\Delta t/\Delta I), \quad \text{or} \quad L = (V_{IN} - V_O)(\Delta t/\Delta I), \quad \text{or}$$

(5-1)

$$L = (V_{IN} - V_O)(V_O/(\Delta I \times f \times V_{IN}))$$

One of the parameters that affects the design is the range of the input voltage. The ripple current is greatest at the highest input voltage. A good rule of thumb is to set the ripple current equal to 10% of the maximum output current at the highest input voltage. We do not have control over the maximum switch current because it is set to 3 A by the circuitry of the IC. The maximum available output current will be 3 A $-$ $\Delta I/2$ -70 mA (boost current).

Using our rule of thumb, we will set the ripple current to 250 mA. We can plug the values into Eq. (5-1):

$$L = (5.0 - 3.3)\,(3.3/(0.25 \times 1.25 \times 10^6 \times 5.0)) = 3.6\,\mu H$$

Transient response and ripple current are related. A large ripple current will allow faster response to load changes. However, large ripple current combined with the ESR of the output capacitor will increase the output ripple voltage. Figure 5.2(a) shows the equivalent AC circuit for the output when the output capacitor is infinite. If (10 × ESR) is less than the value of R_L, then we can make the simplifying assumption that all of the ripple current flows in the ESR of the capacitor. If we consider the capacitor leg to be ESR in series with the capacitive reactance, as in Figure 5.2(b), then we can use this impedance to set the output ripple voltage.

Ripple voltage is usually set as a design parameter, so we can use it to select the size of the capacitor and its ESR.

The peak-to-peak ripple voltage is found by:

$$\Delta V = \Delta I (\text{ESR} + X_C)$$

Substituting and rearranging, we get:

$$\text{ESR} + X_C = \Delta V (L \times f \times V_{IN}) / (V_O (V_{IN} - V_O))$$

A good rule of thumb is to allocate two-thirds of the total impedance to the capacitor ESR and the remaining one-third to the capacitor. We can use the formula for capacitive reactance to decide on a capacitance value:

$$C = 1 / (2\pi f \times X_C)$$

This value will give a slightly larger value than necessary because the waveform is triangular rather than a sine wave and the higher harmonics will be attenuated to a greater extent. The consequence of the decision to allocate one-third of the impedance to the capacitor is that as ESR goes down, we can use smaller capacitor values. It is possible

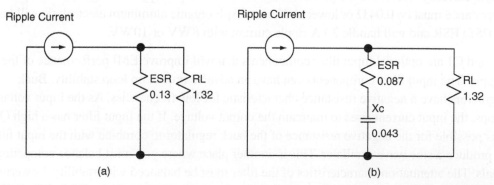

(a)   (b)

Figure 5.2: (a) Equivalent AC circuit for the output when the output capacitor is infinite; (b) equivalent AC circuit with ESR in series with the capacitive reactance

that a capacitor with the required capacitance will have ESR that is larger than the target ESR, especially for aluminum electrolytic capacitors. If this is the case, then it will be necessary to increase the capacitance or allocate more of the ripple budget to the ESR of the capacitor. Obtaining reasonable transient response, ripple voltage, and loop stability may require several iterations to obtain a design that meets all the criteria.

The specification in Figure 5.1 calls for 25 mV of ripple. Using the equation above,

$$\text{ESR} + X_C = 0.025 \, \text{V}(3.6 \, \mu\text{H} \times 1.25 \, \text{MHz} \times 5.0 \, \text{V})/(3.3 \times (5.0 - 3.3))$$
$$= 0.100 \, \Omega$$

We are looking for a capacitor with $0.07 \, \Omega$ ESR and $0.03 \, \Omega$ reactance. This calculates to $4.3 \, \mu\text{F}$. A multilayer ceramic capacitor is a reasonable choice for this capacitor. Since ceramic capacitors have almost no ESR, a capacitor between $1.4 \, \mu\text{F}$ and $4.3 \, \mu\text{F}$ will be likely to satisfy our output ripple requirement.

Buck regulators present two problems to the input power supply. The first is that the input current is a square wave with a peak value equal to the output current of the supply. The current draw while the switch is off is zero. This very large square wave is reflected back into the input supply. L1, C1, and the $33 \, \mu\text{F}$ of Figure 5.1 provide filtering to average out the current supplied from the input. Another problem is that any stray inductance combined with the $33 \, \mu\text{F}$ will act as a high frequency resonant circuit that is excited by the fast rise and fall times of the current. This can cause EMI problems at harmonics of the switch frequency. See the section on Layout Considerations later in this chapter for more details.

The RMS ripple current in the input capacitor is determined by:

$$I_{\text{RMS}} = I_{\text{OUT}}(\text{DC} - \text{DC}^2)^{1/2} \tag{5-2}$$

It is important to choose a capacitor that is rated for this ripple current. The RMS input ripple current is 1.2 A. We are given a budget of 50 mV input ripple voltage, so the capacitor impedance must be $0.04 \, \Omega$ or lower. A Kemet $33 \, \mu\text{F}$ organic aluminum electrolytic will have $0.028 \, \Omega$ ESR and will handle 2.1 A ripple current with 8 WV or 10 WV.

L1 and C1 are optional input filter components that will improve EMI performance of the supply. The input filter components can have an adverse impact on loop stability. Buck regulators have a negative resistance characteristic for low frequencies. As the input voltage drops, the input current rises to maintain the output voltage. If the input filter has a high Q, it is possible for the negative resistance of the buck regulator to combine with the input filter to produce a sine wave oscillator. This is another place where you must balance competing goals. The attenuation characteristics of the filter must be balanced with stability. Lowering the resonant frequency will increase attenuation, but can lead to loop instability. This is a place where iteration in the lab is likely to be necessary to obtain a stable power supply.

The data sheet gives us guidance on compensating the feedback loop. We will start with 330 pF for C_C, and 0 for R_C and C_F. If we were to build this design, we would adjust the values of these three components in the lab to account for second-order effects of the components and the effects of the circuit layout, using the compensation method described in Chapter 2.

The data sheet also gives us guidance on selecting R1 and R2. Linear Technology suggests 10k for R2 to minimize the offset voltage due to bias current of the feedback pin. The formula for R1 is given as:

$$R1 = \frac{R2 \times (V_{OUT} - 1.2)}{1.2 - (R2 \times 0.25\,\mu A)} = 17.5\,K$$

Figure 5.3 is a circuit from the data sheet that gives a soft start circuit using external components connected to the compensation pin. This circuit can be used for any current mode PWM controller that does not provide an internal soft start circuit. The soft start works by limiting the voltage rise on the compensation pin. The circuit effectively adds the soft start capacitor (C_{SS}) to the compensation capacitor to create a very heavily damped response. As the output approaches the final value, the extra damping gradually decreases so that only the 330 pF controls compensation.

Diode D1 of Figure 5.1 will have a forward voltage drop of 0.4 V at 3 A current. The equations up to this point have made the simplifying assumption that the diode forward drop is so small that it can be ignored. In the case of Figure 5.1, this is arguably not valid. As long as the input voltage is well regulated, the errors do not affect the final outcome;

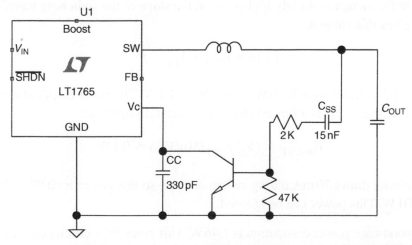

Figure 5.3: Soft start circuit using external components connected to the compensation pin

the circuit will still be able to maintain control. However, if the input voltage has a larger range, the circuit may have more trouble maintaining control. We need to add the diode drop to the output voltage in each of the equations where V_O appears as the voltage across the inductor in order to get accurate results.

DC becomes:

$$DC = (V_O + V_D)/V_{IN}$$
$$\text{so } DC = (3.3 + 0.4)/5.0 = 0.74 \text{ instead of } 0.66$$

This change in duty cycle will affect the value of ripple voltage, ripple current, and the value of the inductor. Eq. (5-3) gives a more accurate equation for the inductor value:

$$L = (V_{IN} - V_O)(V_O + V_D)/(\Delta I \times f - V_{IN}) \qquad (5\text{-}3)$$

For Figure 5.1, this changes the inductor from 3.6 µH to 4.0 µH.

The average output diode current can be found from:

$$I_{AVG} = I_{OUT} \times (1 - DC)$$

The power dissipation in the diode at full load in Figure 5.1 is $(2.5 (1 - 0.74) \times 0.4) = 0.26$ W. We also need to account for losses in the switch in the IC. The worst-case saturation voltage is 0.43 V. The average switch current is:

$$I_{AVG} = I_{OUT} \times DC$$

The switch power in Figure 5.1 is $(2.5 \times 0.74 \times 0.43) = 0.80$ W. The actual power dissipated in the switch is slightly higher due to the slope of the switching waveform. The data sheet gives this value as:

$$17 \text{ ns} \times I_{OUT} \times V_{IN} \times f$$

This gives total switch loss of 0.80 W $+ 0.27$ W $= 1.1$ W. The boost circuit also dissipates power. The data sheet gives a formula for boost circuit dissipation:

$$P_{BOOST} = (V_O^2(I_{OUT}/50)/V_{IN}) = 0.1 \text{ W}$$

The boost circuit draws 70 mA during switch on time, so this power is $(0.07 \times 0.74) \times 0.3$ V $= 0.01$ W. This power can be ignored.

The total worst-case power dissipation is 1.46 W. This gives 86% efficiency for this circuit.

If we rerun the analysis for an input voltage of 12.0 V, we will see that the output diode power becomes a more significant portion of the power loss.

$$DC = (3.3 + 0.4)/12 = 0.31$$

$$P_{SWITCH} = (2.5 \times 0.31 \times 0.43) + (17\,ns \times 2.5 \times 12 \times 1.25\,MHz) = 0.97\,W$$

$$P_{BOOST} = (V_O^2 \times (I_{OUT}/50)/V_{IN}) = 0.05\,W$$

$$P_{DIODE} = (2.5 \times (1 - 0.31) \times 0.4) = 0.69\,W$$

The total worst-case power dissipation is 1.71 W. The efficiency only drops to 84% because the total loss in the switch is less due to the shorter duty cycle. Both efficiency numbers are worst-case and will be better when the IC has characteristics listed as typical in the data sheet. Also, the saturation voltage of a bipolar transistor decreases as the temperature increases, as would be expected at full output power.

Efficiency in the 85% range is adequate for systems that are powered from off-line sources, such as a desktop PC or consumer entertainment equipment. But for battery-operated equipment such as mobile phones that operate from a battery pack composed of a few cells, every extra percentage point of efficiency increases battery life. Figure 5.4 shows a buck regulator using the LT1773 synchronous controller to implement a high efficiency buck converter. The LT1773 is representative of complementary symmetry synchronous controllers available from a number of IC manufacturers.

Synchronous rectification using an NMOS transistor instead of a diode cuts losses significantly. Likewise, using a PMOS high side transistor eliminates the need for a boost supply. The top driver pulls the PMOS gate to ground to turn it on and to V_{IN} to turn it off. The bottom driver pulls the NMOS gate to V_{IN} to turn it on and to ground to turn

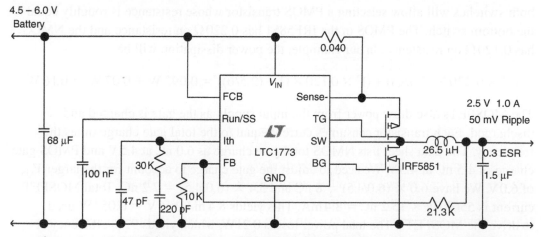

Figure 5.4: High efficiency buck converter using the LT1773 synchronous controller

it off. Current can flow in either direction through a MOSFET switch when it is turned on. The current in the NMOS switch actually flows from source to drain during normal operation. At low output current, it is possible for the inductor current to go to zero. When you use a diode, the inductor current stops as soon as the diode becomes reverse biased. With an NMOS switch, the inductor current can decrease to zero and begin to draw current from the output capacitor. The LT1773 uses the SW connection to detect when the current changes direction. When inductor current goes negative, the IC shuts off the lower switch.

The maximum input supply forces our choice of MOSFETs. The gate-source voltage will be equal to the input voltage for both MOSFETs. There are basically three classes of MOSFETs: low input, logic input, and normal input. The low input voltage MOSFETs will turn on around 1 V, but the maximum gate-source voltage is only around 8 to 10 V. Logic level devices usually have a maximum gate-source around 15 V and turn on around 3 V. Normal level devices have gate-source ratings around 20 V, but turn on around 4 to 5 V.

Synchronous rectifier controllers must ensure a minimum amount of time between turning off the top switch and turning on the bottom switch. If both transistors are on at the same time, you get a destructive short circuit from V_{IN} to ground. The inductor current must continue to flow during this dead time. The body-drain diode of the NMOS switch provides the path for the current during the dead time. This current will store charge in the diode junction until the switch turns on and then the charge will be dissipated in the switch. A small increase in efficiency is possible if the NMOS switch is paralleled with a Schottky diode. Schottky diodes do not store charge in the junction.

Figure 5.4 shows a design that is optimized for small size and low bill of materials by using a single package containing a PMOS and NMOS transistor. PMOS has roughly twice the on resistance of NMOS for the same geometry. Using individual MOSFETs for both switches will allow selecting a PMOS transistor whose resistance is roughly equal to the bottom switch. The PMOS in the IRF5851 has $0.220\,\Omega$ on resistance and the NMOS has $0.120\,\Omega$ on resistance. In our example, the power dissipation will be:

$$(1^2 \times 0.220 \times (2.5/6)) + (1^2 \times 0.120 \times (1 - (2.5/6))) = 0.092\,\text{W} + 0.07\,\text{W} = 0.16\,\text{W}$$

The MOSFETs also draw power from the input supply as the gate is charged and discharged. Each transistor consumes current equal to the total gate charge times the frequency. The data sheet lists NMOS total gate charge as 6.0 nC at 4.5 V and PMOS gate charge of 4.5 nC at 4.5 V. We need to adjust the gate charge to account for the larger V_{GS} of 6.0 V. We have $6.0 \times (6.0/4.5) = 8$ nC and $5.4 \times (6.0/4.5) = 7.2$ nC. Total MOSFET current is 550 kHz $\times$ 15.2 nC = 8.4 mA. This yields 8.4 mA $\times$ 6 V = 0.054 W used to drive the MOSFETs. The total power lost is 0.21 W, which yields 92% efficiency at maximum output. The efficiency will improve slightly as the battery discharges since there will be less power consumed driving the MOSFETs. If you use typical values for on resistance, you get $0.106\,W + 0.054\,W$ for 94% efficiency.

High-efficiency controllers frequently implement burst mode for low power output situations. As the output power declines, the controller will produce a burst of pulses to charge the output and then shut off the controller while the output slowly drops to the low trip voltage where a new burst is output. This operation is very similar to how pulse frequency modulation controllers operate. Instead of a single long pulse that changes the frequency of the control, it produces one or more pulses at the fixed frequency followed by periods of no pulses. This improves EMI control because the filters only have to deal with the frequency of the oscillator.

5.3 Boost Converter Designs

Figure 5.5 shows a boost converter based on the LT1680 current mode PWM controller IC. This controller is intended for high power applications using large external NMOS switches. It includes adjustable frequency, selectable maximum duty cycle, high switch drive current, soft start, and 60 V common mode range on the current sense amplifier. The data sheet for this IC walks you through selecting all of the components needed for the design. The first selection is the operating frequency and duty cycle limit. Ours is a typical design and uses 100 kHz and 90% maximum duty cycle.

Boost converters cannot implement output short circuit protection using the control IC and the PWM circuitry. The diode provides a path from the input supply to the output independent of the switch, so the controller IC cannot turn off current flow. The only way to implement current limiting for a boost converter is to provide a linear current limit on either the output or the input of the supply. This is a serious consideration if current limiting is a design requirement. A transformer isolated design is usually a better choice if short circuit current limiting is a consideration.

The size of the inductor, the value for the current sense, the MOSFET, and the output capacitor are affected by the decision to use continuous mode or discontinuous mode operation. Discontinuous operation will preclude using the average current limit function available in this IC. However, discontinuous mode allows using a smaller inductor than continuous mode.

Discontinuous mode operation has advantages in transient response, slope compensation, and switch losses. Discontinuous mode allows faster transient response. This is especially true for a rapid decrease in output current. Since the inductor current goes to zero for each cycle, a sudden drop in output current demand can be adjusted on the very next cycle by shortening the duty cycle. This is called load dump. The only thing necessary to accommodate a rapid decrease in load current is to draw down the current of the last

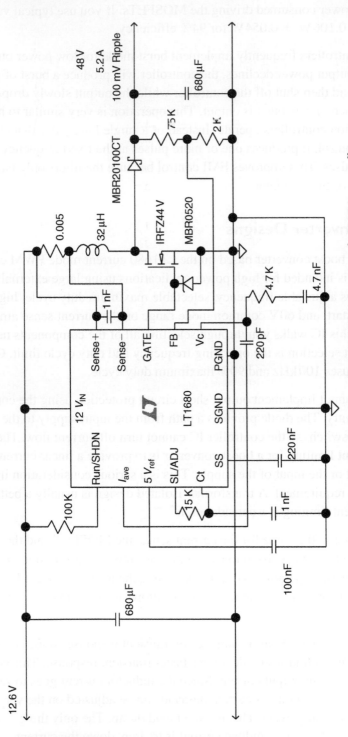

Figure 5.5: Boost converter based on the LT1680 current mode PWM controller

pulse stored in the capacitor. There is no inductor current to consume. Likewise, a rapid increase in output current can be accommodated quickly because a large amount of the new current can come from increasing both the duty cycle and peak current. Another advantage of discontinuous mode is that the circuit will not be affected by subharmonic oscillations and will not require slope compensation. Since the inductor current is zero and the switch node voltage is zero when the switch turns on, there is no switching power consumed as the switch turns on. Turning on the switch with zero current flow is the best case for switching loss.

The down side to discontinuous operation is that the peak inductor current, peak switch current, and ripple current are very large. The large ripple current requires a larger value output capacitor with a small ESR. Also, the switch will have a very large ratio of peak to average current, so it must have a very large peak current rating. The total output power is limited by the peak inductor current, and the peak inductor current is limited by the saturation characteristic of the inductor. Once the inductor saturates, it can no longer store additional energy. The inductor current is no longer controlled by the applied voltage when saturated, so the switch current can increase very quickly. Damage to the switch is likely if the inductor saturates. The ripple voltage is a function of load current for discontinuous operation. Larger output current translates directly to larger ripple voltage.

Continuous mode operation has advantages in ripple current, peak inductor current, peak switch current, and maximum output power. Load dump for a rapid decrease in output current is problematic because all of the energy stored in the inductor must be dumped into the load. Even though the switch is turned off for multiple cycles, it is possible for the output voltage to increase quickly because of the energy stored in the inductor. The slow transient response in continuous mode makes soft start even more important. The slow response makes it very likely for large output voltage overshoot without soft start. In essence, soft start mixes a very slow transient response during startup with a faster transient response for normal operation. The lower ripple current in continuous mode allows using output capacitors with lower capacitance and higher ESR for reasonable ripple voltage. The ripple voltage in continuous mode is constant.

The switch must have a larger power rating in continuous mode service because the switch will turn on with full output voltage applied while carrying full inductor current. This is the worst case for switching losses in a switch. Continuous operation requires slope compensation for duty cycles greater than 50%. Slope compensation also requires that the inductor have a minimum value to ensure that the slope compensation stays in control. The larger inductor allows larger output power but at the expense of transient response.

Our example will use continuous mode operation since the application is a 48 V telecom application with relatively constant output power. Again, we will choose ripple current equal to 10% of the full inductor current. For the buck converter, the peak inductor current was equal to the output current plus one-half of the ripple current. This is not the case in the boost converter. We can start from the recognition that the energy stored in the inductor while the switch is closed is equal to the energy delivered to the load:

$$V_{IN} \times I_{L\text{-}AVG} \times DC = (V_{OUT} - V_{IN}) \times I_{OUT} \tag{5-4}$$

$$DC = (V_{OUT} - V_{IN})/V_{OUT} \text{ for a boost converter.}$$

Rearranging gives average inductor current:

$$I_{L\text{-}AVG} = (V_{OUT} \times I_{OUT})/V_{IN}$$

Substituting for the maximum load condition gives:

$$I_{L\text{-}AVG} = (5.2 \times 48.0)/12 = 20.8 \text{ A}$$

The peak inductor current will be 20.8 A + one-half of the ripple current = 20.8 + 2.1 = 22.9 A.

Now we are ready to determine the size of the current sense. From the data sheet we find that:

$$R_{SENSE} = 120 \text{ mV}/I_{LIMIT}, \text{ so } R_{SENSE} = 0.12/22.9 = 0.005 \,\Omega$$

Notice that R_{SENSE} is placed between the inductor and the input supply. It could also be placed between the inductor and the switch, but this would cause a problem for the current sense amplifier. Placing the resistor at the power input keeps the common mode voltage for the current sense amplifier stable and close to the supply voltage. Placing the sense resistor on the switch side of the inductor will cause the common mode voltage to change from ground to full output voltage in every cycle. The additional AC voltage caused by limited common mode rejection would disrupt proper operation of the current sense amplifier.

We can use the inductor equation to derive the boost converter equivalent of Eq. (5-1):

$$L = V_{IN} \times (V_{OUT} - V_{IN})/(\Delta I \times f \times V_{OUT}) \tag{5-5}$$

Substituting the values from Figure 5.5 gives:

$$L = 12.0 \times (48.0 - 12.0)/(2.8 \times 100 \text{ kHz} \times 48.0) = 32 \,\mu\text{H}$$

The next component to pick is the switch transistor. At 100 kHz, a MOSFET is the only reasonable choice. The breakdown voltage must be higher than the output voltage.

The switch needs a small amount of margin for safety. The IRFZ44 V has a minimum breakdown of 60 V, which gives a 25% margin. We also have to be sure that there is adequate current capability and power dissipation.

The peak current for our design is 22.9 A, so we are well below the 39 A (100°C) continuous rating of the part. The last consideration is adequate power dissipation. The worst-case duty cycle is 90% and maximum current is 22.9 A. The device worst-case on resistance is 0.0165 Ω, so maximum power is 22.9 × 22.9 × 0.0165, or 8.7 W. The LT1680 data sheet shows that the rise and fall times will be 50 ns with the 1800 pF of the IRFZ44 V. It is safe to assume the same rise and fall times from the data in the IC data sheet. The switching losses are:

$$50\,\text{ns} \times I_{\text{PK}} \times V_{\text{OUT}} \times f = 50\,\text{ns} \times 22.9 \times 48\,\text{V} \times 100\,\text{kHz} = 5.5\,\text{W}$$

The total dissipation of 14.2 W is well within the capability of the switch with a proper heat sink.

The peak diode current is equal to the peak inductor current, so we need a diode with a peak current rating of 23 A and breakdown voltage at least as large as the output voltage. The duty cycle of the diode current is much smaller than the switch duty cycle, so the average power is much less than the peak power. The MBR20100CT dual Schottky diode will provide more than enough margin with 100 V breakdown, 20 A current per device, and 0.9 V forward drop. Worst-case power dissipation occurs during the short time of a load dump where the full peak inductor current flows continually. This will give 0.9 V × 22.9 A, or 20.6 W. The diode duty cycle of 25% will give an average power of 5.1 W. This power will require a heat sink for the diode.

The input and output noise problems are opposite to the buck converter case. The input current for a boost converter is constant (with continuous operation) with ripple current equal to the ripple current of the inductor. This makes the filter requirements relatively easy to implement. The filter job is also easier since the waveform is a triangle wave instead of a square wave. We can approximate the RMS ripple current as 0.707 × (P-P ripple current/2). This is not exact, but we don't need an exact value. We will need margin anyway, so a minor error will just be factored into the margin.

The output current applied to the output capacitor is essentially a sawtooth wave with a peak value equal to the peak inductor current. The output capacitor ESR is very important because of the very large value of the ripple current.

The RMS ripple current can be determined from:

$$I_{\text{RMS}} = I_{\text{PK}}(\text{DC} - \text{DC}^2) \tag{5-6}$$

We can use the same process as in the buck converter design to assign one-third of the ripple voltage to the capacitor impedance and two-thirds of the ripple voltage to the ESR of the output capacitor. As in the buck converter case, we can end up needing more capacitance in order to meet the voltage ripple and capacitor dissipation requirements because of large ESR.

The boost design does not lend itself to synchronous rectification. It is possible to implement synchronous rectification, but it must use discrete components. I have found only one boost IC that implements synchronous rectification. Boost converters use a diode as the rectifier, which reduces the best possible efficiency. In applications where the duty cycle is around 50%, the diode dissipation can be much larger than the switch dissipation. The efficiency of this design is approximately 89%.

Figure 5.6 shows a battery application for a boost converter where the input is supplied by one lithium cell or multiple NiMH cells. The MAX1896 is a 6-pin IC that is designed for minimum bill of materials and extremely small size. It is a current mode PWM controller that implements all current mode functions such as slope compensation, feedback compensation, switch frequency, and current sense inside the IC. Operation at 1.4 MHz contributes to its small size because the inductor can be very small and the filter capacitors can be either ceramic or tantalum. The control circuit also implements pulse skipping to allow operation at low output current.

This IC takes advantage of the control of parameters that is possible in monolithic circuits. Since the FET switch on resistance is well controlled, it can be used as the current sense for the PWM circuit. The voltage at the L_X pin is directly proportional to the inductor current when the switch is on. The current sense across the $0.7\,\Omega$ on resistance sets a current trip between 550–800 mA. The current limit is a function of both on resistance and slope compensation (and, indirectly, of duty cycle).

This circuit is a little different from the previous example where we had a fairly stable input voltage. The battery voltage will change significantly during use. The voltage will

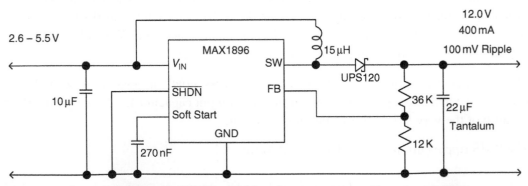

Figure 5.6: Example NiMH or lithium battery-operated boost converter

decrease rather rapidly near end of charge for NiMH cells. We have to design the circuit assuming the lowest input voltage to ensure enough duty cycle to store adequate energy in the inductor. Again, we must choose the ripple current. We can choose a larger ripple current of 100 mA since the high frequency allows use of relatively small capacitors while still having small ripple voltage. We use the values from Figure 5.7 and Eq. (5-5):

$$L = 2.6 \times (12.0 - 2.6)/(0.10 \times 1.4\,\text{MHz} \times 12.0) = 15\,\mu\text{H}$$

The 1.4-MHz switching frequency allows very small filter capacitors that can be either tantalum or ceramic. The internal compensation circuit relies on a low frequency zero provided by a tantalum capacitor and its ESR. If a ceramic capacitor is used, the very low ESR will place the zero at a much higher frequency. The other problem with ceramic capacitors is that the equivalent inductance is significant. The inductance and small ESR complicates the loop equation. The data sheet provides the data necessary to calculate a feed forward capacitor that will externally compensate the feedback loop when using a ceramic capacitor.

This circuit has an internal soft start circuit that requires only a capacitor to set the soft start time. The circuit limits the switch current until the soft start pin reaches 1.5 V. We can use the comparator voltage and the 4 µA soft start current to calculate the capacitor value from the time needed. Our example needs 100 ms of soft start time, so we can use the definitions for capacitance, charge, and current:

$$\text{Total charge} = \text{current} \times \text{time} = 4\,\mu\text{A} \times 100\ \text{ms} = 400\,\text{nC}$$

$$C = Q/V = 400\ \text{nC}/1.5\,\text{V} = 266\,\text{nF}$$

So we pick a standard value of 270 nF for the soft start capacitor.

5.4 Inverting Designs

Figure 5.7 shows an inverting design using the MAX1846 inverting controller. This controller IC is intended for a full function design where maximum control of parameters is balanced with small size. This design is another in which we will create an output from a number of NiMH cells. The 3.0 V minimum input range of the controller IC precludes use in a single lithium cell application.

The first parameter to choose is the switching frequency. The controller frequency can be set from 100–500 kHz. The efficiency of the design will depend on the switch frequency because of the requirement for a P-channel FET. P-channel FETs have larger switching losses than N-channel devices because they are minority carrier devices. We need to balance the dynamic losses in the switch with the improved performance that comes with smaller components and higher frequency. The other parameter that constrains the switching frequency is the maximum duty cycle versus switching frequency. The

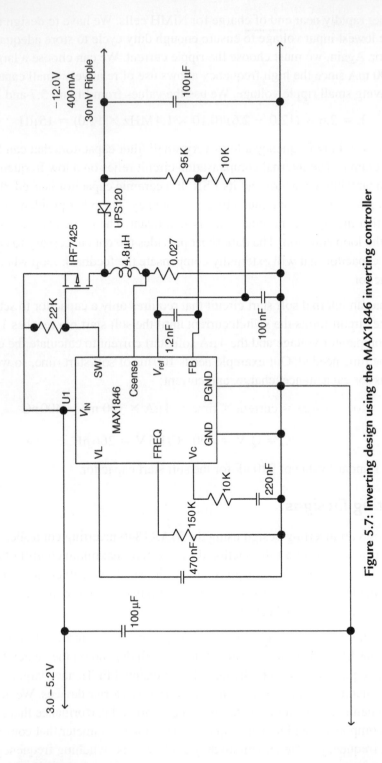

Figure 5.7: Inverting design using the MAX1846 inverting controller

controller has a minimum off time of 400 ns that lowers the maximum duty cycle as frequency increases. As the ratio of absolute output voltage to input voltage increases, the maximum duty cycle increases. We can rearrange Eq. (2-10) from Chapter 2 to get duty cycle as a function of input and output voltage:

$$\text{Duty Cycle} = V_{OUT}/(V_{OUT} - V_{IN})$$

Substituting our values, we see that the maximum duty cycle is:

$$\text{Duty Cycle} = (-12)/(-12 - 3.0) = 80\%$$

The data sheet shows that the typical value for a maximum duty cycle of 80% occurs for a 500 kHz switching frequency. There will not be enough margin because the worst-case conditions for the IC will not give an 80% duty cycle. We can select a switching frequency of 400 kHz to provide the needed margin.

This example power supply is intended for an analog system that may have significant current transients, so we need to design this supply for rapid transient response. The same relationship expressed in Eq. (5-4) applies for an inverting design.

$$V_{IN} \times I_{L\text{-}AVG} \times DC = V_{OUT} \times I_{OUT}$$

Substituting duty cycle and rearranging gives:

$$I_{L\text{-}AVG} = (V_{OUT} - V_{IN}) \times I_{OUT}/V_{IN} = (-12 - 3.0) \times -0.5/3.0 = 2.5 \text{ A} \qquad (5\text{-}7)$$

Notice that it is important to use proper signs for the output current and output voltage! We can select ripple current equal to 50% of the average inductor current at maximum load. This gives a peak inductor current of 3.13A. Setting such a high ripple current at maximum load and minimum input voltage is likely to cause discontinuous operation at very light loads and at maximum input voltage. The ripple voltage will be constant while the supply operates in continuous mode. The ripple voltage will be even less once the supply changes to discontinuous operation at low output current. It is important to verify loop stability in the lab for both continuous and discontinuous operation because the loop gain will be different depending on the mode.

We again use the inductor equation and rearrange for an inverting converter:

$$L = (V_{IN} \times V_{OUT})/((\Delta I \times f) \times (V_{OUT} - V_{IN})) \qquad (5\text{-}8)$$

Substituting the values from Figure 5.8 gives:

$$L = (3.0 \times (-12.0))/((1.25 \times 400 \text{ kHz}) \times (-12 - 3.0)) = 4.8 \,\mu\text{H}$$

Figure 5.8 gives the operating parameters for both the input and output. We used the minimum input voltage and maximum output current to determine the size of the inductor, assuming continuous mode operation. We can use the maximum input voltage and minimum output current to see the effects of discontinuous operation.

The duty cycle for fully charged cells will be:

$$\text{Duty Cycle} = (-12)/(-12 - 4.2) = 0.74$$

We can rearrange Eq. (5-8) to solve for ΔI:

$$\Delta I = \frac{(V_{IN} \times V_{OUT})}{Lf(V_{OUT} - V_{IN})} = \frac{4.2(-12)}{(4.8\,\mu H \times 400\,kHz)(-12 - 4.2)} = 1.62\,A$$

$I_{L\text{-AVG}}$ is equal to $\Delta I/2$ at the point where the mode changes from continuous to discontinuous. We can rearrange Eq. (5-7) to solve for I_{OUT}:

$$I_{OUT} = (I_{L\text{-AVG}} \times V_{IN})/(V_{OUT} - V_{IN}) = (0.81 \times 4.2)/(-12 - 4.2) = 210\,mA$$

This result indicates that our design will be in discontinuous mode when the cells are fully charged and the load current is less than 210 mA.

An inverting supply implemented with a current mode IC will give inherent short circuit current limiting because the switch disconnects the inductor from the input supply. The output short circuit current will be limited to the peak inductor current.

The current sense resistor is determined from the peak inductor current at maximum output using the formula from the data sheet:

$$R_{CS} = 0.085V/I_L = 0.085/3.13A = 0.027\,\Omega$$

The switch gate voltage is equal to the input voltage. This means we must start our search for a device that will fully turn on at 3.0 V. The drain-source voltage will be equal to the input voltage plus the output voltage, so the breakdown voltage must be greater than 16.2 V. The last parameter to determine is peak drain current. Our supply has a peak inductor current of 3.13 A. The IRF7425 is a reasonable device that meets these requirements.

Both the input capacitor current and the output capacitor current are discontinuous for an inverting design. The waveform on the input and output are sawtooth waves with peak amplitude equal to the peak inductor current. Both ESR and the ripple current rating are important considerations for the input and output capacitors. The input filter considerations of Q and negative input impedance are the same as we saw with the buck converter. The RMS input capacitor current is shown below:

$$I_{RMS} = I_{OUT}(DC/(1 - DC))^{1/2}$$

5.5 Step Up/Step Down (Buck/Boost) Designs

Figure 5.8 shows an implementation of a step up/step down design based on the MAX641 buck/boost converter.

The MAX641 is intended as a step-up regulating converter for fixed output use. It lends itself to the buck/boost design because it has complementary drive pins. The L_X pin is driven from an internal MOSFET and the Ext pin is intended to drive an external MOSFET switch in higher power designs. The V_{OUT} pin is actually used to supply power to the IC internal circuits. In normal boost mode, the inductor will supply current to bootstrap the system. In this design, we need to use the V_{OUT} pin tied directly to V_{IN} to supply IC current.

The design method is identical to the inverting design. The formula for the duty cycle for continuous mode is:

$$\text{Duty Cycle} = V_{OUT}/(V_{IN} + V_{OUT})$$

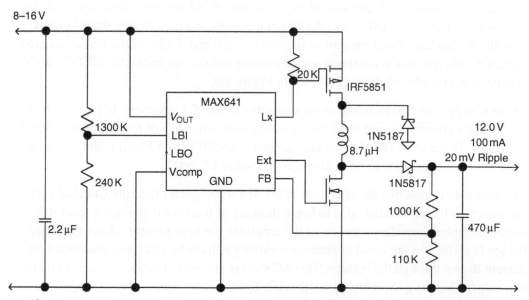

Figure 5.8: Step up/step down design based on the MAX641 buck/boost converter

The maximum inductor current will occur when the input voltage is below the output voltage and the system is acting as a boost converter. The same relationship expressed in Eq. (5-4) applies for this design.

$$V_{IN} \times I_{L\text{-}AVG} \times DC = V_{OUT} \times I_{OUT}$$

Substituting duty cycle and rearranging gives:

$$I_{L\text{-}AVG} = (V_{OUT} + V_{IN}) \times I_{OUT}/V_{IN} = (4.0 + 6.0) \times 1.0/4.0 = 2.5\,A$$

We can select ripple current equal to 20% of the average inductor current at maximum load. This gives a peak inductor current of 2.75 A. We will need to use the minimum input voltage to select an inductor small enough to allow adequate current flow.

We again use the inductor equation and rearrange for this converter:

$$L = (V_{IN} \times V_{OUT})/((\Delta I \times f) \times (V_{OUT} + V_{IN}))$$

Substituting the values from Figure 5.9 gives:

$$L = (4.0 \times 6.0)/((0.5 \times 550\text{ kHz}) \times (4.0 + 6.0)) = 8.7\,\mu H$$

The selection of the MOSFETs follows the criteria we have used before. First, we select parts based on the gate voltage and then based on the drain voltage. Finally, we ensure that the current rating is adequate.

This design is quite costly because of the extra switch and the extra diode required. These additional components add cost to the bill of materials and they reduce the efficiency. Another design that trades the cost of the extra switch and diode for the cost of an extra inductor and capacitor is the single-ended primary inductance converter (SEPIC) design. A representative SEPIC design is shown in Figure 5.9.

There is very little design information available on SEPIC converters. In my search of IC vendor websites, I found three barely adequate descriptions of the design of a SEPIC converter. These were DN48 from TI (Unitrode), AN1051 from Maxim, and an article by National Semiconductor in *EDN Magazine*, October 17, 2002.

The best way to describe the operation of the SEPIC converter (and the Cuk and Zeta converters, which are variants) is to begin thinking of the circuit as similar to an RC voltage coupled amplifier stage. In an RC amplifier, the load resistor allows the active device (a switch, in our case) to produce a varying voltage by changing the amount of current drawn through the resistor. This AC voltage is coupled to the load circuit by the capacitor, which is a short circuit for the AC. The capacitor blocks the DC applied to the amplifier from the load. At RF, the resistor can be replaced with a choke so that the

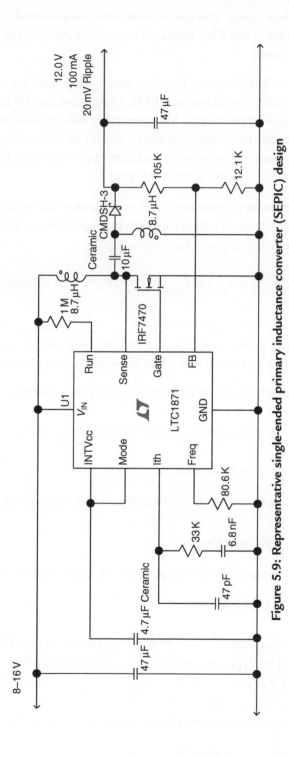

Figure 5.9: Representative single-ended primary inductance converter (SEPIC) design

amplifier dissipates less power. The square waveform is transferred to the load circuit where the diode, inductor, and filter capacitor convert the AC into the DC output voltage, just as in a buck converter.

The two inductors in a SEPIC circuit have the same voltage and the same current as long as the inductor values are identical. SEPIC circuits are usually designed with equal value inductors to simplify the design, but equal value inductors are not required. If the inductor values are equal, both inductors can be wound on the same core. DN 48 gives a reasonable design sequence. The inductor and duty cycle equations for a SEPIC circuit are the same as for the buck/boost design in Figure 5.8.

First, we select the switching frequency. Next, we calculate the highest inductor current. The maximum inductor current will occur when the input voltage is below the output voltage and the system is acting as a boost converter.

$$I_{\text{L-AVG}} = (V_{\text{OUT}} + V_{\text{IN}}) \times I_{\text{OUT}} / V_{\text{IN}} = (4.0 + 6.0) \times 1.0 / 4.0 = 2.5\,\text{A}$$

We can select ripple current equal to 20% of the average inductor current at maximum load. This gives a peak inductor current of 2.75 A. We will need to use the minimum input voltage to select an inductor small enough to allow adequate current flow.

We again use the inductor equation and rearrange for this converter:

$$L = (V_{\text{IN}} \times V_{\text{OUT}}) / ((\Delta I \times f) \times (V_{\text{OUT}} + V_{\text{IN}}))$$

Substituting the values from Figure 5.9 gives:

$$L = (4.0 \times 6.0) / ((0.5 \times 550\,\text{kHz}) \times (4.0 + 6.0)) = 8.7\,\mu\text{H}$$

This is the value for both inductors.

The next step is to determine the RMS ripple current in the coupling capacitor.

DN48 gives the following equation:

$$\begin{aligned} I_{\text{C\_RMS}} &= (I_{\text{OUT}}(\text{max})^2 \times \text{DC}(\text{max}) \times I_{\text{IN}}(\text{max})^2 \times (1 - \text{DC}(\text{max})))^{1/2} \\ &= (1 \times (4/(4+6)) \times 2.5^2 \times (1 - (4/(4+6))))^{1/2} \\ &= 1.22\,\text{A} \end{aligned}$$

We must select a coupling capacitor that has the power handling ability for this ripple current.

We pick the output capacitor to give the desired output ripple, again allocating two-thirds of the ripple voltage to ESR and one-third to X_C. The output ripple current is given by:

$$I_{RMS} = I_{OUT}(DC/(1-DC))^{1/2}$$

We use the equations shown in the buck converter section to calculate the capacitor value and determine the required ESR.

The gate voltage of the MOSFET is set by the internal voltage regulator to 5.2 V. This will require a logic level MOSFET. The drain voltage is equal to $V_{IN} + V_{OUT}$. The peak MOSFET current is $I_{IN} + I_{OUT}$. The peak diode current and peak reverse voltage are equal to the MOSFET voltage and MOSFET current.

5.6 Charge Pump Designs

The design sequence below should serve as a starting point for those new to charge pump design:

1. Choose an IC based on the output power, physical size, input voltage, etc. The ratio of input voltage to output voltage will determine if a step-up, step-down, or inverting converter is selected. You also select based on whether output voltage regulation is required.

2. Choose the switch frequency (if adjustable) and flying capacitor value.

3. Choose the output ripple voltage. Choose the output capacitor based on the output ripple voltage.

Figure 5.10 shows a step up charge pump with output regulation. The LTC3200-5 will produce a regulated 5.0 V over the input range of a single lithium cell. This circuit is

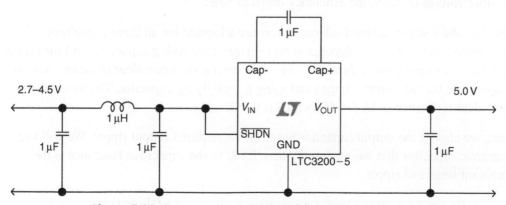

Figure 5.10: Step-up charge pump with output regulation

typical of charge pump ICs where the number of external components is very small. This IC requires just three capacitors to convert an unregulated low voltage into regulated 5.0 V. The IC provides an internal voltage divider for the feedback circuit to set the output voltage to 5.0 V.

Small ceramic capacitors are indicated for all three capacitors because of the 2-MHz switch frequency.

As mentioned in Chapter 2, the charge pump has an equivalent resistance of:

$$R_{EQ} = 1/(f \times C_{FLYING})$$

as long as the equivalent resistance is much larger than the internal switch resistance. This equivalent resistance is one source of power dissipation. It is possible to reduce the dissipation by using a larger value flying capacitor until the ESR dominates.

The data sheet warns against using tantalum or aluminum electrolytic flying capacitors because the flying capacitor can have negative voltage during the startup of the supply. These capacitors would not be a good choice for this supply in any case, because they will have significant ESR at the 2-MHz switch frequency.

The efficiency of the supply depends on the ratio of output and input voltage. The switching nature of the charge pump would yield nearly 100% efficiency for an output voltage twice the input voltage and very light load. As the load increases, the losses in the ESR of the capacitor and the internal resistances increase. The losses due to the equivalent switching resistance also increase as load current increases. These losses limit the output current at very low input voltages, as shown on the data sheet.

In order to regulate to a voltage less than twice the input, the IC must dissipate power for input voltages from 2.7–5.0 V. The IC operates much like a linear regulator in this case. At input voltage of 5.0 V, the efficiency drops to 50%.

The data sheet suggests that 1-μF capacitors are adequate for all three capacitors. Maximum output current is dependent on the size of the flying capacitor until the internal switch resistance begins to dominate. The data sheet gives equivalent resistance versus temperature for two input voltages and using a 1-μF flying capacitor. The switching equivalent resistance is $1/(f \times C)$, which is $0.5 \, \Omega$.

Next, we choose the output capacitor based on the required output ripple. We will use a ceramic capacitor that has essentially zero ESR, so the capacitive reactance is the dominant source of ripple.

$$V_{p\text{-}p} = I_{OUT}/(2 \times \pi \times f \times C) = 40 \text{ mA}/(6.28 \times 2 \text{ MHz} \times 1 \, \mu\text{F}) = 3 \text{ mV}$$

We use the value of I_{OUT} because the duty cycle is 50% and the current supplied to the output is essentially a square wave.

The value of the input capacitor has less effect on the ripple because the input current is essentially equal while the flying capacitor is charging and while current is transferred to the output. There is a very short period of time where the nonoverlapping clocks that drive the switches are all off. This time is approximately 25 ns for the LTC3200-5. The RMS value of such a short pulse is very small. However, the rise and fall times are still rather fast, so the capacitor must be very close to the IC to prevent the inductance of the input trace from creating a resonant circuit that will ring. Figure 5.10 shows an inductor and additional capacitor forming a pi filter to provide additional filtering of noise reflected to the input supply.

Figure 5.11 shows an inverting charge pump with output regulation. Again, this IC provides a regulated output voltage with a small component count (four capacitors and two resistors). This IC works by charging the two flying capacitors in parallel across the input supply during the charge phase. During the discharge phase, the switches are reconfigured to stack the two flying capacitors in series to produce a negative voltage equal to twice V_{IN}.

Maxim describes the regulation mechanism as PFM control, but the control mechanism is actually pulse dropping. The clock runs at a constant 450 kHz and the control circuit drops pulses as required to keep the output in control.

You will notice that one feedback resistor is tied to the input voltage. This is required because the feedback pin must be above ground. The design in Figure 5.11 requires the input voltage to be regulated because the feedback circuit uses the input voltage as the

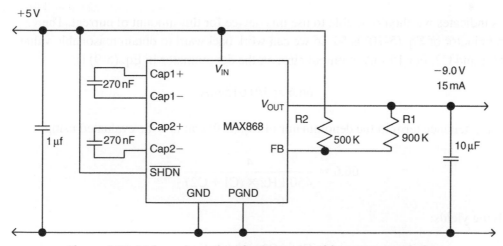

Figure 5.11: Voltage inverting charge pump with output regulation

reference. An alternative is to provide a second input voltage that is a reference for the regulator. The data sheet suggests setting R2 to a value between 100–500 K to limit the current draw from the divider. Then use the formula from the data sheet for R1:

$$R1 = R2 \times \left(\left| V_{OUT} \right| / V_{REF} \right)$$

The data sheet provides equations that allow us to calculate the required values for C1, C2, and C_{OUT}.

For a given value for C1 and C2, we can verify that the maximum output current available meets the design goal. Maxim chooses to give this equation rather than a formula to calculate the capacitors:

$$I_{OUT(MAX)} = \frac{2V_{IN} - \left| V_{OUT} \right|}{\dfrac{4}{f_{max} \times (C1 + C2)} + R_{OUT} \times \dfrac{10\,V}{V_{IN} + \left| V_{OUT} \right|}} \tag{5-9}$$

The data sheet gives 450 kHz as the maximum frequency and 70 Ω as the equivalent output resistance. Eq. (5-10) is the result of Eq. (5-9), based on setting C1 and C2 to infinity. This will give the absolute maximum current based only on the voltages and the IC characteristics. We substitute 5.0 V input and −9.0 V output to determine if we will be able to obtain 15 mA current out.

$$I_{OUT(MAX)} = \frac{(2 \times 5.0) - \left| -9.0 \right|}{70 \times \dfrac{10\,V}{5.0 + \left| -9.0 \right|}} \tag{5-10}$$

This indicates we should be able to use this device for this amount of current. The denominator of Eq. (5-10) is 50, so we can work backward to obtain reasonable values for C1 and C2. For 15 mA of output current, the denominator in Eq. (5-9) is:

$$66.6 = 1/0.015\,mA$$

The capacitance term in the denominator of Eq. (5-9) can now be evaluated using:

$$66.6 = \frac{4}{450\ kHz \times (C1 + C2)} + 50$$

Solving yields:

$$C1 = C2 = 0.27\mu F$$

The data sheet also gives an equation in terms of X_C, R_{OUT}, and ESR:

$$V_{\text{RIPPLE(P-P)}} = ((2 \times V_{IN}) - |V_{OUT}|) \times \left(\frac{1}{1 + \dfrac{4C_{OUT}}{C1 + C2}} + \frac{\text{ESR}}{R_{OUT}} \right)$$

Since R_{OUT} is 70 Ω, ESR of a ceramic capacitor will not contribute to that term. 10 µF is a reasonable start for C_{OUT}. This yields ripple voltage of:

$$V_{\text{RIPPLE(P-P)}} = ((2 \times 5.0) - |9.0|) \times \left(\frac{1}{1 + \dfrac{4 \times 10\,\mu\text{F}}{0.27\,\mu\text{F} + 0.27\,\mu\text{F}}} \right) = 13\,\text{mV}$$

Tantalum capacitors will have ESR on the order of 0.5–3 Ω in this capacitance range and voltage range (depending on manufacturer and technology), so the ripple voltage will be significantly larger with tantalum capacitors.

The input capacitor ESR is much more important for an inverting supply because the IC draws current only while charging the flying capacitors. The peak input current is double the output current. The input ripple is even more important if V_{IN} is used as the reference. Once again, a large value ceramic capacitor with low value ESR is appropriate.

5.7 Layout Considerations

The basic white protoboard you used in your beginner EE classes will work for a *small* power supply up to perhaps 20-kHz switching frequency. Not many useful power supplies run at such a low frequency any more. A modern switching regulator will run from 100 kHz up to several MHz. The harmonics of the switching waveform extend up to the VHF frequency range. Failure to use a PC board that uses good high-frequency layout will guarantee disappointing results (and, likely, lots of smoke).

There are two issues that we have to consider. The first is to design the layout of the power supply circuit so it does not interfere with its own operation. The second is to consider how the voltages and potentially huge current densities can interfere with the rest of the system if the power supply is placed too close to sensitive circuits.

Pentium CPUs can draw 40 A. Even 10 mΩ will produce a voltage drop of 0.4 V. In such a power supply, it is very important to keep low-level signals isolated from the high current paths of the rectifiers and switches. It is easy to overlook the magnetic consequences of such currents. Each loop where this current flows is a single turn inductor that we tend to ignore. Our example would create as much as 10 A-Turns of AC magnetic field that can easily couple into adjacent traces and loops in the power supply and other close circuits. Pentium applications are rather extreme, but they illustrate how easy it is to have otherwise inconsequential layout choices become important in switching supplies.

Figure 5.12 shows a representative PCB layout and schematic from the LT1871 data sheet. This gives a good example of the considerations in layout of a circuit. The figure does not show the bottom of the PCB. The layout needs a large continuous ground plane on the bottom of the board that extends from the right side of the board to the area of the via at the IC ground pin. The ground plane should narrow at this point and then expand to connect to the vias for the timing and measurement circuits. This is indicated in the

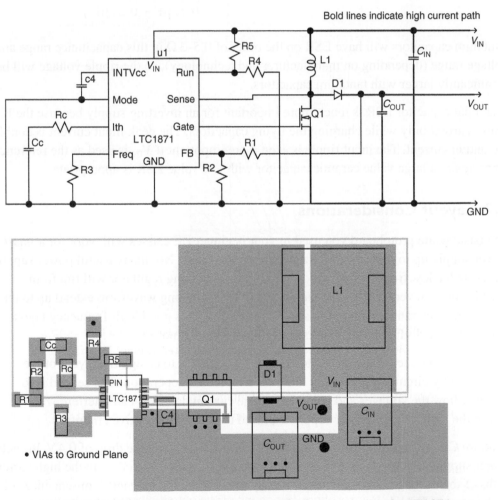

Figure 5.12: Representative PCB layout and schematic using the LT1871

schematic by the narrower ground connection between the GND pin and the components on the left in the schematic.

The first consideration of layout is to realize that the ground current of the input supply flows directly to the output circuit. Notice that the schematic has been drawn to roughly show how components will be physically placed on the PC board. All of the switch components, as well as C_{IN} and C_{OUT}, are placed near each other and away from the signal ground connection of the LT1871. The ground connection of the IC is part of the signal measuring circuit, so any voltage changes due to switching currents flowing from the input capacitor to the output capacitor can change the voltage applied to the sense circuits inside the IC. The ground current out of the IC can also be fairly large during the times that the MOSFET is switching. The peak gate current can be on the order of hundreds of mA at turn-on and turn-off of the switch. This indicates that a fairly large trace is needed between the GND pin of the IC and the common connection between C_{IN} and C_{OUT}. Notice that the top ground area is large and the IC ground pin is at one corner of the ground area to limit the voltage change due to AC current flowing in the ground area from C_{IN} and C_{OUT}. The majority of the DC current flow in this design flows on the ground plane on the bottom of the board (not shown in Figure 5.12). The figure shows the connections for V_{IN}, V_{OUT}, and GND. The input and output ground connections should be made between C_{IN} and C_{OUT} so that current flow is concentrated near the vias for the switching components.

The connections to the feedback resistors and the current sense input should be routed as far away as possible from the lines that drive the switch gate and the lines that connect the switches and the inductor. Again, there are large AC currents flowing in these traces and even small closed circuits nearby will be one-turn inductors that can produce sizeable voltages and disturb linear parts of the circuit. There are two major magnetic loops. The first is composed of L1, C_{IN}, and Q1. The second is C_{OUT}, D1, and Q1. We minimize magnetic pickup by the measuring circuits by keeping the traces small and as close together as possible. This minimizes the loop area and the induced voltage.

These same considerations apply to charge pump circuits where switching currents can be quite large. You will want to keep the common connection for the IC, C_{IN}, and C_{OUT} close and keep loops away from the feedback input if the converter is regulated.

It is important to use wide traces as much as possible at the frequencies of modern switching supplies. Even one-half inch of a narrow trace can have inductance of many tens of nH. All of the design rules in this chapter presume reasonable circuits with minimal parasitic elements. If you inadvertently design in parasitic inductances on the PC board, it is possible to create unintended additional voltage stresses on components when elements switch on or off. Where possible, it makes sense to use surface mount rather than through-hole components to help minimize the parasitic inductances in component leads.

CHAPTER 6

Transformer-isolated Circuits

Raymond Mack

Transformer-isolated switching power supply topologies offer a great many advantages over the non-isolated topologies. Three of the major factors are: it is very easy to create multiple output voltages from the same transformer, transformers create a natural dielectric boundary from the input to output circuits, and the input voltage levels are independent from the output voltages. Unfortunately, multiwinding magnetics are more expensive to manufacture than simple inductors, but this disadvantage may be attractive in the context of the overall system cost.

Transformer-isolated topologies can be used to create multiple output non-isolated converters or to dielectrically isolate a section of the system such as a floating sensor circuit. They are very flexible.

Transformer-isolated topologies are also used as the basis of all of the 48 VDC (telecom input bus voltages) and the AC/DC converters. Here, the advantage provided by isolation due to the construction of the transformer plays a very important role in the many "hazardous" applications that can present several thousands of volts between the input and output(s). This topic is covered in Chapter 11.

When testing these types of supplies, it is good to isolate the ground connection of the oscilloscope and the other instruments with an earth ground isolation plug. Otherwise you will be connecting the various isolated circuits together, which may result in pyrotechnic light shows on your workbench.

Ray Mack delves into the circuits needed for these topologies in an understandable way.

—Marty Brown

In this chapter, we will look at detailed designs of transformer isolated converters. The primary application is off-line power supplies, but these designs are also useful in applications where safety isolation is required or where the input voltage can vary above and below the output voltage. All of the designs shown here use current mode PWM control, just like the designs in Chapter 5, because of its inherent advantages in loop stability and current control.

6.1 Feedback Mechanisms

The following section applies to transformer circuit applications where the transformer is used for isolation, such as in off-line supplies. The output can be connected directly to the control IC in applications where isolation is not required.

Most transformer circuits use the magnetic circuit of the transformer to provide electrical isolation of the secondary circuit from the primary circuit. Putting the control IC on the input side of the supply requires that feedback of the output voltage to the control IC has to cross an isolation barrier. If the IC is powered from an isolated supply, then the switch control must cross the isolation barrier.

Using an optocoupler is the "easiest" way to transfer output voltage information across the isolation barrier to a control IC on the primary side. Optocouplers, in general, provide isolation of 2500 V or more between the LED and the photo transistor. There are several characteristics of optocouplers that make them less than ideal in this application. However, they are still a reasonable choice for this application because they are small and inexpensive when compared to transformers. The first problem is the large variation in transfer function from unit to unit. This change in current transfer ratio causes a large variation in the loop equation from unit to unit. The control loop must be designed conservatively in order to account for the worst-case optocoupler. This results in a nominal system being damped more heavily than necessary.

Another problem is the low corner frequency of the transfer function. Optocoupler phototransistors are built with a rather large base region to improve the conversion of light to current. The large base region creates a larger input capacitance and reverse transfer capacitance than in regular transistors. Although it is only a few picofarads, the Miller effect will amplify the capacitance to a much larger value. The phototransistor is used in a manner identical to an RC-coupled amplifier. The Miller capacitance creates a pole at a fairly low frequency. Just as in an RC amplifier, the frequency response can be improved by using a low collector resistance. This lowers the voltage gain of the optocoupler. Agilent, Clairex, and other manufacturers produce optocouplers with better frequency response, but they are significantly more expensive than ordinary devices such as the 4N27.

The usual method of compensating for the low optocoupler gain and the capacitance of the optocoupler is to use an amplifier and voltage reference on the isolated side of the supply. National Application Note AN-1095 gives a detailed design method with rigorous analysis of the control loop for an optocoupler isolated system. Figure 6.1 shows a common drive circuit using the TL431 shunt regulator. Resistors R1 and R2 divide the output voltage down to 2.5 V for the control pin of the TL431. There are two optional compensation circuits in Figure 6.1. These can be used to add a pole or a zero to the loop response. The TL431 and its variations provide the voltage reference, comparator,

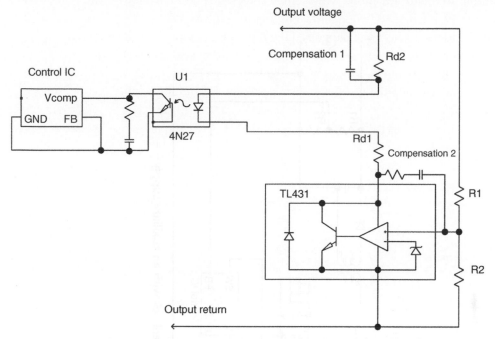

Figure 6.1: Representative optocoupler feedback with compensation circuits

and power amplifier in one convenient package. The feedback pin of the control IC is connected to the common of the input to force the IC to the largest duty cycle possible. The V_{COMP} pin is an open collector style output with a current source on most modern ICs. The resistor and capacitor to ground add more compensation and the optocoupler transistor reduces the error amplifier output to reduce the duty cycle.

Another method of feedback isolation is to use a small power line transformer to generate an isolated auxiliary supply for the IC. The IC then drives a pulse transformer to supply isolated drive for the switches. Even for fairly high output systems, the power required for switch drive and the control IC is only a few watts. The auxiliary transformer does not need to be especially large, but it must be able to switch from 110 VAC to 240 VAC. The main drawback of this method is that the transformer adds size to the supply. It is entirely possible that this auxiliary transformer could be larger than the switching transformer at the 100 W level! This method works for systems where 110 V or 240 V operation is selected manually. It is less desirable for universal input supplies because the transformer has to be able to handle 240 VAC/50 Hz nominal input but still generate enough power at 90 VAC. The only practical way to power the control IC in such a universal supply is to provide some form of linear regulation like a zener diode or three-terminal regulator. Figure 6.2 shows a representative transformer drive with an auxiliary supply. T1 is a small iron core power

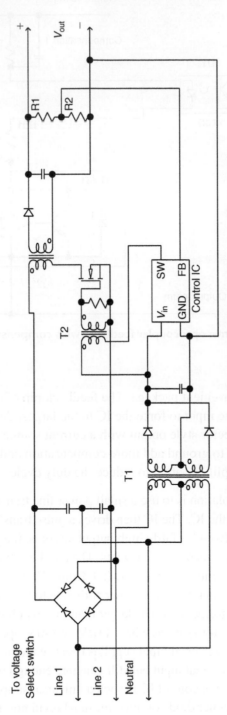

Figure 6.2: Representative isolation and feedback with an auxiliary supply

transformer and T2 is a pulse transformer for driving the MOSFET switch. Both T1 and T2 must meet safety agency isolation specifications.

TI produces ICs for use on the secondary side that use amplitude modulation of an AC signal to transfer the control signal across the isolation barrier. The UC1901 varies the amplitude of an RF carrier frequency, which is fed to a transformer and then rectified on the primary side to supply the feedback voltage. Figure 6.3 shows an application of this IC. The RF oscillator can operate up to 5 MHz. The high frequency allows the time constant of the rectifier filter (R4, C4) to be quite short so that there is minimal phase shift through the RF to DC part of the circuit. This IC also includes the error amplifier and other support circuitry. The error amplifier has a compensation pin that can be used to add poles or zeros to the loop response. TI describes this IC and applications in Application Note AN-94. The feedback transformer must meet safety agency isolation requirements similar to those imposed on the main power transformer.

An alternative to the UC1901 is to use a regular PWM control IC operating at a high frequency to drive a pulse transformer and pulse averaging circuit. Figure 6.4 shows an example. The high frequency allows the low pass filter (R1, R4, C4) to use a small capacitor so that the pulse averaging does not add significant time delay to the feedback circuit. A time delay corresponds to adding a pole to the loop response. We use C2 on the primary of the pulse transformer to avoid problems with current in the magnetizing inductance. Capacitors C3 and D3 form a DC restoration circuit. Without the DC restoration circuit, the DC level of the pulses will vary with duty cycle because the volt-microseconds of each portion of the pulse waveform will be equal.

Figure 6.5 shows the operation of a regular detector circuit and operation of a DC restoration circuit for three different duty cycles. In the regular detector, the output will be the height of the waveform above zero (the dark line). The area of the two shaded areas is equal, showing equal volt-seconds for the positive and negative portions of the AC waveform. The DC restored circuit below shows that the output is the peak amplitude minus the forward voltage of the diode for all three duty cycles.

Flyback converters maintain the voltage on the output circuits in proportion to the turns ratios of the inductor. The inductor winding charges each output capacitor to the voltage across the winding. This property allows use of a secondary circuit to provide both IC power and output voltage measurement. Figure 6.6 shows a representative flyback converter. D1 and C2 provide an auxiliary supply for the control IC. The feedback resistors (R3, R4) are chosen so that the control IC keeps the output voltage at 12.0 V. The filter capacitor (C2) on the IC supply adds a pole to the transfer function of the feedback loop, so the compensation becomes more complex. This control method is adequate for low power circuits where the regulation requirement is not too stringent. The voltage across D2 will vary with output current. As D2 drops more voltage, the output voltage

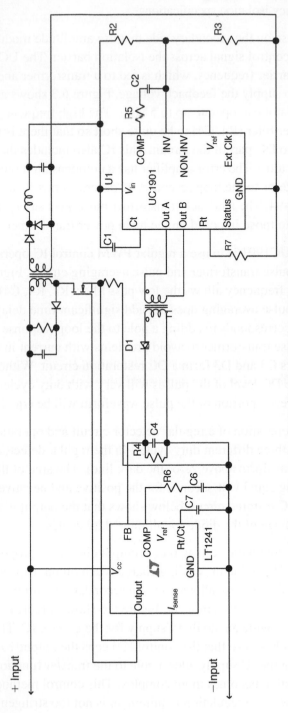

Figure 6.3: Isolated feedback using Texas Instrument's UC1901

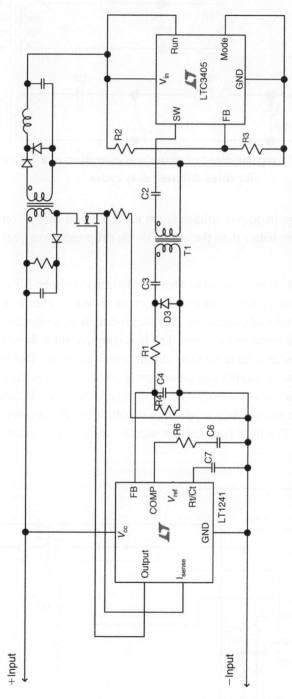

Figure 6.4: Using a standard PWM control IC for isolated feedback

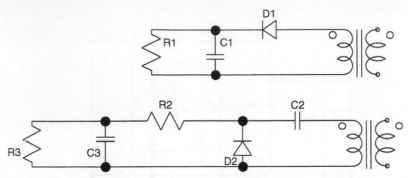

Figure 6.5: Operation of a regular detector circuit and operation of a DC restoration circuit for three different duty cycles

will go down. The change in output voltage is not reflected in a change on the voltage of C2, so the regulation is no better than the output diode drop variation over the range of the output current.

A bootstrap circuit (R2, C2) is required to provide the initial voltage for the IC when using an auxiliary winding on the main transformer, as shown in Figure 6.6. All five transformer circuits can take advantage of a bootstrap circuit in conjunction with powering the IC from the main transformer. The bootstrap circuit will work with any control IC that has an under-voltage lockout circuit with hysteresis. The bootstrap resistor will charge up the IC supply capacitor slowly until it reaches the under-voltage enable voltage. The capacitor must store enough energy to drive the IC and the switch for a few cycles until the main power supply can supply all of the current required by the IC and switch drive. The bootstrap resistor supplies the charge current as long as AC

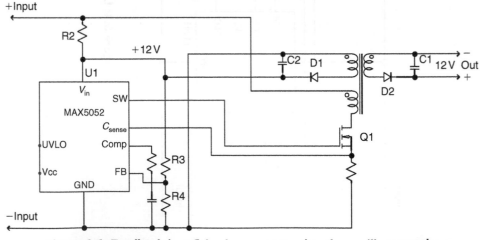

Figure 6.6: Feedback in a flyback converter using the auxiliary supply

power is supplied. This causes both heat and a reduction of efficiency. The advantage is that the resistor is an inexpensive part that is also very small compared to an iron core transformer as used in Figure 6.2. The bootstrap circuit is an excellent implementation for universal input supplies. ST produces a line of control ICs with the trade name VIPer that integrate the bootstrap circuit, as well as a high voltage MOSFET switch for very low part count operation in low power applications. National, Linear Technology, and other manufacturers also produce low power (under 20 W) fully integrated flyback circuits that require only a transformer and a few rectifiers and capacitors.

6.2 Flyback Circuits

A flyback converter works in a fashion similar to a boost converter, where energy is stored in the inductor while the switch is on and the energy is delivered to the load when the switch turns off.

Magnetic cores do not store magnetic energy very well. Efficient cores saturate at a low magnetizing force. A flyback circuit actually stores the energy of the inductor in an air gap. The core provides a low reluctance shielded path to couple the energy from the windings to the air gap. The energy storage is concentrated in the gap between the core faces.

Figure 6.7 shows a stylized ferrite core with the three windings for the circuit in Figure 6.6. The magnetic core concentrates almost all of the flux in the magnetic circuit inside the magnetic material. In a real core, there will be a very small amount of flux outside of the core in the vicinity of the windings, but all three windings will have essentially identical flux.

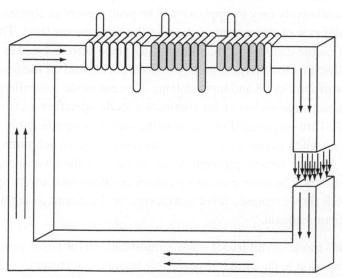

Figure 6.7: A stylized ferrite core with the three windings for the circuit in Figure 6.6

Recall the two equations for the voltage across an inductor:

$$V = L \, di/dt \quad \text{and} \quad V = N \, d\Phi/dt$$

When the switch in Figure 6.6 closes, the current and flux will begin changing in proportion to the applied voltage on the primary inductor. The change in flux creates a voltage on each secondary winding in proportion to the turns of each winding. Since the voltage induced is negative (notice the dots on the windings), the diodes will not allow current to flow. When the switch opens, $d\Phi/dt$ will change polarity instantaneously. As soon as $N \, d\Phi/dt$ is large enough to produce a voltage sufficient to forward bias one of the diodes, current will begin to flow in that secondary circuit. The consequence of this is that the secondary circuit with the lowest V/N ratio will hog all of the current from the collapsing magnetic field. Once the V/N ratio is equal for all secondary circuits, each will receive current from the collapsing field. This current hogging by the lowest V/N circuit is responsible for the close regulation of output voltage between all the secondary circuits. This is also why we can use the voltage on a secondary winding as a proxy for the voltage on the main output supply, as described above.

A flyback circuit can operate in either continuous or discontinuous mode. In continuous mode, current is always flowing in one of the windings of the inductor. In discontinuous mode, the current in all windings goes to zero for part of the cycle and the energy stored in the inductor goes to zero. Each mode has its advantages and disadvantages.

The primary advantage of continuous mode operation is that the relatively long current flow in the secondary requires a small filter capacitor (with larger allowable ESR). The primary inductance is relatively large with a small peak current requirement, so the inductance is relatively easy to implement. The peak current in continuous mode is roughly one-half that of discontinuous mode at the same power level. The principal disadvantage is that the control loop has a right-half plane zero that makes loop compensation difficult. However, the loop gain does not depend on the load current. It is only a factor of duty cycle and input voltage. Current mode controllers must also deal with slope compensation issues for continuous mode operation and duty cycle greater than 50%. Turn-on power dissipation in the switch is significant in continuous mode because the switch passes a large current as soon as the switch turns on with a large voltage applied. Another turn-on problem occurs because of the reverse recovery current in the output rectifiers. The reverse recovery causes an additional current spike during turn-on. Figure 6.8 shows representative waveforms for the circuit in Figure 6.6 when operated in continuous mode.

The discontinuous mode circuit trades many simplifications for larger peak currents. The turn-on dissipation in the switch is negligible because the current begins from zero and only the input voltage is applied to the switch. The output current goes to zero for

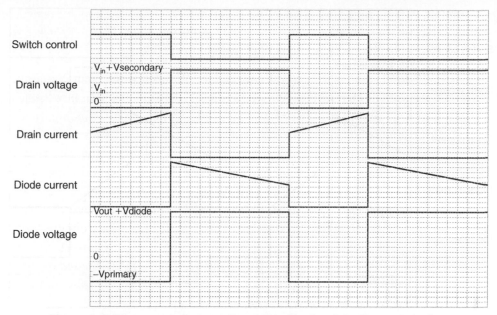

Figure 6.8: Representative waveforms for the circuit in Figure 6.6 when operated in continuous mode

part of the cycle, so there is no diode reverse recovery current to affect the switch during turn-on. The control loop is relatively straightforward in discontinuous mode. There is no right-half pole to deal with and slope compensation is never required. However, the load resistance is one of the factors in the loop equation. This makes open loop behavior less controlled than the continuous mode case. This is usually not a problem once proper compensation is accomplished and the loop is closed. The size of the gap in the inductor core becomes an issue for discontinuous mode because the higher peak current is likely to push the core closer to saturation. The AC flux in the core is quite large, so loss in the core is also an issue for discontinuous mode. The output ripple is typically larger in discontinuous mode because the AC current in the capacitor ESR is larger and the capacitor must supply the entire load current for a longer portion of the switching cycle. The simplicity in design, repeatability, and compensation makes discontinuous mode preferable, especially for low power circuits. Figure 6.9 shows representative waveforms for discontinuous operation. Discontinuous mode also has faster transient response and lack of load dump concerns as compared to continuous mode operation.

Switching circuits have parasitic inductances that are not associated with the energy storage inductor. These inductances are due to circuit traces and the leakage inductance of the main inductor. The parasitic inductances create a voltage that adds to the primary winding voltage, so the switch breakdown must be larger than the voltage implied by the

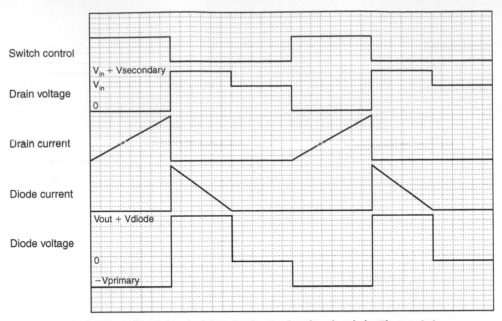

**Figure 6.9: Representative waveforms for the circuit in Figure 6.6
when operated in discontinuous mode**

reverse voltage plus input voltage. The turn-on time for the output diodes creates a short period of high secondary voltage, so there is a short time where *di/dt* becomes quite large. The extra *di/dt* from diode turn-on creates a spike on the primary.

Transformers and diodes have parasitic capacitances that can have undesirable consequences. The secondary capacitances, along with the secondary leakage inductance, can form a high frequency resonant circuit that is excited when the diode turns off. This effect is more pronounced with hard recovery diodes. The resonant circuit will ring and transfer the AC waveform back to the primary.

Clamp circuits are used to reduce the stress on the switch from parasitic inductance elements. Figure 6.10 shows clamp circuits that will limit the voltage at the switch. Circuit A shows a clamp winding that returns energy from the magnetizing inductance of a transformer to the input supply. The clamp winding has the same number of turns as the primary. This sets the maximum voltage on the switch to twice the input voltage. Notice that D1 is connected to the input supply rather than between the clamp winding and ground. This topology is important because of the capacitances between windings. Placing the diode between the winding and ground will cause the capacitance to interfere with turn-on of the switch.

Circuit B uses the voltage of the capacitor to clamp the voltage at the switch. The time constant of the RC circuit is set to several switching cycles. The capacitor charges up

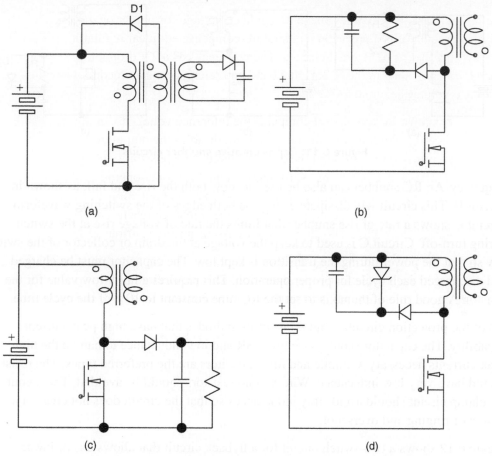

Figure 6.10: Clamp circuits that limit the voltage at the switch

to the reverse voltage created by the secondary windings plus any voltage from leakage inductances. This circuit is less efficient than the clamp winding because all of the energy stored in the leakage inductance and some of the energy of the primary inductance is dissipated in the resistor. Circuits C and D are variations of circuit B. The capacitor across the zener in circuit D may be necessary because zeners are not fast turn-on devices. The zener voltage must be set to a value larger than the normal voltage across the primary from secondary current flow.

Snubber circuits are similar to the clamp circuits. Figure 6.11 shows representative snubber circuits. The only interesting snubber circuits are those that dissipate energy in a resistor. Circuit A shows a simple RC snubber used on the output diode to dampen ringing when the diode turns off. The capacitor must be a small value so that the snubber provides a low impedance at the ringing frequency but a high impedance at the switching

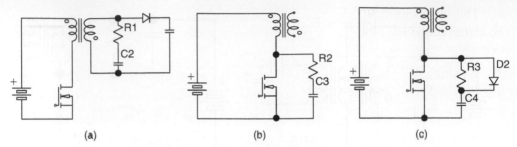

Figure 6.11: Representative snubber circuits

frequency. An RC snubber can also be used to slow both the rise and fall, as shown in Circuit B. This circuit will dissipate energy on both edges of the switching waveform. Circuit C shows a rate of rise snubber that limits the rate of voltage rise at the switch during turn-off. Circuit C is used to keep the voltage at the drain or collector of the switch low so that the power during the transition is kept low. The capacitor must be charged and discharged each cycle for proper operation. This requires a fairly low value for the resistor. A good rule of thumb is to set the RC time constant to 10% of the cycle time.

All of the protection circuits require fast turn-on diodes that have high peak current capability. The capacitor should have low ESR and low inductance to handle the high peak currents necessary. Ceramic and film capacitors are the preferred types. The resistor should have very low inductance. Wire-wound resistors should be avoided. The layout of the clamp circuit should avoid stray inductances so that the circuit does not create a new source of ringing and overshoot.

Figure 6.12 shows a two-switch circuit for a flyback circuit that allows use of lower voltage switches. The two diodes (D1, D2) clamp the primary winding to the input supply

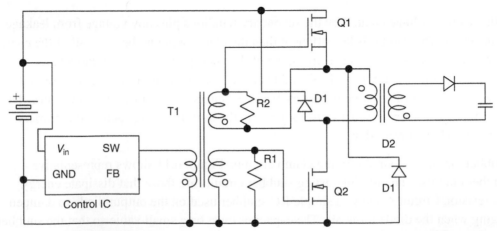

Figure 6.12: A two-switch circuit for a flyback circuit that allows use of lower voltage switches

rails. This allows us to use switches with breakdown voltage just above the input voltage. The clamp action is efficient since the energy is returned to the input supply. T1 is the price we pay for using lower voltage switches. The transformer and resistors provide the necessary floating drive to Q1. The transformer drives both transistors to ensure that the switching times are as equal as possible.

6.3 Practical Flyback Circuit Design

Flyback design is usually iterative. We make educated guesses when picking component values and refine these in later iterations. Steps for designing a flyback circuit are listed below:

1. Choose a controller IC based on power level and bill of material constraints.

2. Choose the switching frequency.

3. Choose continuous mode or discontinuous mode.

4. Use the input voltage range to select the maximum duty cycle goal.

5. Determine the maximum power and pick a switch.

6. Design the primary inductance.

7. Design the transformer winding ratios.

8. Verify that the switch is adequate, based on the worst-case voltage.

9. Choose the bootstrap capacitor based on the gate charge required if a bootstrap supply is used.

10. Choose the output capacitor, based on ripple requirements.

11. Design the ancillary IC components.

6.4 Off-Line Flyback Example

Our first example is a universal input flyback design that has a 12.0 V/1 A output. The output must have regulation of ±200 mV with 100 mV or less ripple. This is a circuit similar to the power supply for a number of consumer devices with an input specification of 100–240 VAC and output of 12 V/400 mA. Instead of a "wall wart" with an iron core transformer, these products integrate the entire switching power supply and power plug into a plastic housing about four times larger than a standard U.S. two-prong power plug. Figure 6.13 shows the circuit we are designing.

role. This allows us to use switches with breakdown voltage just above the input voltage. The clamp circuit is efficient since the energy is returned to the input supply. T1 is the pulse we pay for using lower voltage switches. The transistor drive resistors provide the necessary floating drive to Q1. The transformer drives are adjusted to ensure that the switching times are as equal as possible.

6.7 Practical Flyback Circuit Design

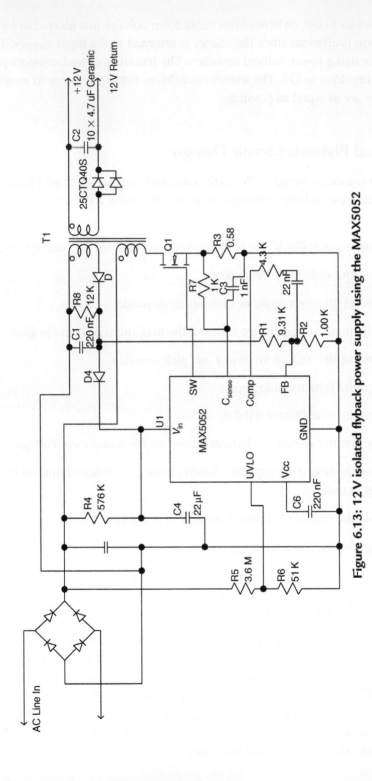

Figure 6.13: 12 V isolated flyback power supply using the MAX5052

My first step for the example was to check the websites of Maxim, TI, and Linear Technology to see which controller ICs came up on a search for the "flyback" keyword. The Maxim_NPP_PWM_Products.pdf Application Note came up as a match. I found the MAX5052 that is designed for precisely the type of circuit we will design. This IC is designed for low to moderate power operation in a universal input application. Its major advantage is a very large hysteresis in the under-voltage lockout circuit. The worst-case hysteresis is 9.25 V and typical is 11.86 V. The large difference between the wake-up level and the shutdown level means that we can use a smaller reservoir capacitor and lower wattage resistor for the IC bootstrap power supply. This IC has a fixed switching frequency of 262 kHz, which should be appropriate for our design. The cycle time is 3.82 μs.

Discontinuous operation seems to be a reasonable choice, since we are aiming for a simple design. We have two possible choices for maximum duty cycle. The MAX5052A has a maximum duty cycle of 50% and the MAX5052B has a maximum duty cycle of 75%. Our initial choice is to limit the duty cycle to 50%.

The voltage applied to the main output rectifier and bias supply rectifier is likely to be close to double the output voltage plus the forward voltage of the diode. The main output voltage is only 12.0 V, so the PRV rating of the main rectifier can be on the order of 40 V. This allows us to use a Schottky diode for minimum power dissipation. The peak-to-average ratio of current for discontinuous mode can be large, so it is likely that peak rectifier current may be on the order of 10 A. An IRF 30BQ040 Schottky diode has a 40 V PRV rating and 3.0 A average forward current. Looking at the forward voltage versus instantaneous forward current shows that the voltage drop changes from 0.8 V at 10 A to only 0.25 V at 100 mA, so we will have a problem meeting the regulation specification.

We have a couple of options that we can consider at this point to improve the voltage regulation over input voltage and output current. The first option is to look for a diode with better voltage drop at high current; the second option is to change course and go to a continuous mode converter to reduce the peak output current. In searching the IRF website we find that a 6CWQ03FN dual diode has less change in forward voltage versus current. Additionally, each diode will only carry one-half the total current, so we stay in the more vertical part of the curve. This diode seems to have more than adequate ratings with 3.5 A per diode and 30 V PRV. It is also a relatively small surface mount component. Worst-case forward voltage is 0.5 V at 5 A and 25°C. This should barely meet the regulation requirement.

It is reasonable to also use 12.0 V for the auxiliary supply. This allows the secondary windings to be identical. The IC draws 2.5 mA maximum, so the IC will consume 12.5 V × 2.5 mA = 30 mW. A guess for the power to drive the MOSFET switch is 70 mW (an educated guess of twice IC power plus a little more). The power for the main output will be approximately 12.5 V × 1.0 A = 12.5 W. This means the total power will be 12.6 W.

The highest switch current will occur at the lowest input voltage (85 VAC or 115 VDC with 10 V of ripple). The average current is 12.6 W/115 V = 110 mA. If we make a guess of a 10:1 peak-to-average ratio, the peak input current will be 1.1 A. A search of the IRF website indicates that the IRBF20S MOSFET has 900 V breakdown and 1.7 A average current. This should be a good choice for the switch. Now we can calculate the drive power for the switch. The switch is driven by a regulated 10.5 V supply from the control IC. The switch drive current is the total gate charge times frequency or 38 nC × 262 kHz = 10 mA. The drive power is 10 mA × 10.5 V = 105 mW. Our guess was close enough to approximate the actual power.

There is a trade-off between duty cycle, turns ratio, primary inductance, and switch voltage. Longer duty cycle will require a larger primary inductance, but it will allow a smaller turns ratio and lower switch voltage. We can choose the primary inductance so that the circuit is at the crossover between continuous and discontinuous operation at 50% duty cycle and the lowest input voltage. We know that we need 12.6 W of power delivered to the load during one-half of a cycle. The average current during the time the switch is on is one-half the peak current because of the triangular shape of the current waveform (refer to the waveforms for discontinuous operation in Figure 6.9).

The average current for the whole cycle is $I_{\text{Peak}} \times 0.5 \times$ Duty Cycle. From this, we calculate that I_{Peak} is 110/(0.5 × 0.5) = 440 mA.

We can use the inductor equation to calculate the primary inductance:

$$L = \frac{V\,\Delta t}{\Delta I} = \frac{110 \times (3.82\,\mu s \times 0.5)}{440\,\text{mA}} = 478\,\mu H$$

The energy stored in the core is transferred to the output when the switch opens. Flyback designs have an additional element of freedom compared to boost designs, since the inductance of the secondary and output voltage will set both peak current and di/dt. In discontinuous operation, we know that all of the energy will be transferred to the output circuit before the switch closes again. This puts an upper limit on dt. The largest secondary inductance possible will be a value that puts dt equal to (1 − Duty Cycle). We can make the secondary inductance smaller, if desired. The ratio of primary to secondary inductance sets the turns ratio of the inductor. Smaller secondary inductance gives a larger turns ratio that will also create a larger voltage requirement for the switch.

We will choose Δt equal to one-half of the cycle time. Now we can choose the secondary inductance. We use the highest current output to select the secondary inductance. We know that the main output will need 12.5 V at 1.0 A. The output waveform is a triangle as shown in Figure 6.9, so the average output current is $I_{\text{Peak}} \times 0.5 \times$ (1 − Duty Cycle).

From this, we can calculate that I_{Peak} is $1.0/(0.5 \times 0.5) = 4.0\,\text{A}$. Once again, we use the inductor equation to calculate the secondary inductance:

$$L = \frac{V\,\Delta t}{\Delta I} = \frac{12.5 \times (3.82\,\mu s \times 0.5)}{4.0\,\text{A}} = 5.96\,\mu H$$

We can calculate the turns ratio from the two inductances. The equation for inductance is

$$L = N^2 \times A_L,$$

so we can use this equation to develop the turns ratio in terms of inductance ratios:

$$\frac{L_P}{L_S} = \frac{N_P^2}{N_S^2}, \text{ and rearranging gives } \frac{N_P}{N_S} = \sqrt{\frac{L_P}{L_S}} = \sqrt{\frac{456}{5.96}} = 8.75:1$$

Going back to one of our inductor equations, $V = N\,dF/dt$, and recognizing that dF/dt is identical for all windings, we get the relationship of the voltages during switch on-time and switch off-time.

$$\frac{V}{N} = \frac{d\Phi}{dt}, \text{ so } \frac{V_P}{N_P} = \frac{V_S}{N_S} \text{ or } \frac{V_P}{V_S} = \frac{N_P}{N_S}$$

This appears to be the same as the transformer equation. It is similar, since the windings are coupled, but it is important to remember that it is the inductances of the windings that set the voltages. The transformer equations only apply when current is flowing in the primary and secondary at the same time.

During switch on-time the reverse secondary voltage is controlled by the input voltage, so the worst-case PRV on the secondary diode will be:

$$V_S = \frac{N_s \times V_P}{N_P} = \frac{1 \times 390}{8.75} = 45\,\text{V}$$

The worst-case switch voltage while the switch is off will equal the highest input voltage plus the reverse voltage on the primary:

$$390 + (12.5 \times 8.75) = 390 + 110 = 500\,\text{V}$$

We see that our choice for the secondary will overstress the output diodes and place minimal stress on the switch. We can shorten the output current time by reducing the secondary inductance and increasing the turns ratio. We can increase the turns ratio by 33% and see if the diode and switch characteristics are more reasonable. We set the turns ratio to 12.0. This yields a secondary inductance of $3.17\,\mu H$. The PRV for the diode becomes $390/12 = 32.5\,\text{V}$. The peak output current will be $7.5\,\text{A}$. The worst-case switch voltage will be $540\,\text{V}$. We need another iteration of choosing the output diode. A search of the IRF website yields the 25CTQ40S, which is in the same package as the 6CWQ03FN.

The 25CTQ40S dual diode has even better forward voltage characteristics and has enough margin with 40 V PRV.

The typical values for wake-up voltage (21.6) and shutdown voltage (9.74) give a change of 11.86 V. However, the worst-case operation of the IC bootstrap occurs when the IC wakes up at the lowest voltage and shuts down at the highest voltage. The lowest wake-up is 19.68 V; the highest shutdown voltage is 10.43 V. The current draw is relatively constant. The IC draws 2.5 mA and the gate charge draws an additional 10 mA. We allow 10 ms for the supply to charge the auxiliary bias supply above 10.43 V. 12.5 mA for 10 ms means we will use 125 μC of charge. We can use the capacitance equation Q = C × V and the change in charge to obtain an equation for capacitance:

$$Q_2 - Q_1 = 125\,\mu C - CV_1$$

$$C4 = \frac{Q_2 - Q_1}{V_2 - V_1} = \frac{125\,\mu C}{19.68 - 10.43} = 13.5\,\mu F$$

Rounding to the nearest value = 22 μF

This capacitance will be necessary because the bias supply will get no current from the switch until the main output voltage equals the bootstrap voltage (when both windings have equal V/N). The bootstrap resistance value (R4) is a compromise between fast startup and power dissipation. We can limit the power dissipation to 0.25 W to keep heat low and maintain high efficiency. The worst-case voltage is 390 V − 12.0 V = 378 V. The resistance required is 378^2/0.25 W = 571 k. The bootstrap charging current is 378 V/571 k = 660 μA. The nominal charge to reach the wakeup point is 22 μF × 20 V = 440 μC, so it will take 0.67 seconds to charge the bootstrap capacitor at high input (240 VAC) and 2.6 seconds at low input (100 VAC).

The next step is to choose the output capacitor. We are likely to encounter the same problem we saw with non-isolated circuits in Chapter 5, where the capacitance value is secondary to ESR in setting output ripple. Our goal is to assign 67% of ripple voltage to ESR and 33% to AC impedance, so we assign 67 mV of ripple to ESR.

$$ESR = \frac{67\,mV}{7.5\,A} = 8.9\,m\Omega$$

The target capacitance is :

$$X_C = \frac{33V}{7.5A} = 4.4\,m\Omega$$

$$C = \frac{1}{2\pi \times 262\,kHz \times 4.4\,m\Omega} = 140\,\mu F$$

A quick look in the Digi-Key catalog shows the Panasonic CD series polymer electrolytic would require seven 8.2 µF/16 WV capacitors to have low enough ESR and enough ripple capability. A search of the Panasonic website shows a 4.7 µF/16 WV MLC ceramic can handle 4 A of current and each capacitor has 9 mΩ for ESR. In the case of the ceramic capacitors, we will need multiple capacitors to have enough capacitance, and the ESR becomes quite small. Ten of these capacitors will probably make a better selection than the electrolytic capacitor. This would give only 0.9 mΩ ESR. This reduces the required capacitance to 45 µF. An aluminum electrolytic capacitor will be sufficient for the bias supply, since the total current is only 13 mA.

Notice that the bias supply has two stages of filtering isolated by diode D4. This allows the voltage at the feedback pin to follow the output voltage during startup so that the internal soft start circuit is not affected by the voltage from the bootstrap circuit. The time constant of the resistor in parallel with the feedback capacitor in the feedback portion is rather short (on the order of three cycle times). This allows the feedback to more closely follow a drop in the main output voltage.

The feedback voltage divider is calculated from the equation given in the data sheet:

$$V_{OUT} = \left(1 + \frac{R_1}{R_2}\right) \times 1.23\,V$$

The current sense resistor is calculated based on the worst-case peak current required. We calculated that the peak current in normal operation at 85 VAC input is 440 mA. We can set the current limit to a value slightly above this value to allow for additional current during startup. We choose 500 mA, so

$$R_{CS} = \frac{0.29\,V}{0.5\,A} = 0.58\,\Omega$$

We add a small amount of RC filtering (R7, C3) between the current sense resistor and current sense pin to allow for some transients when the switch turns on. This reduces false current limiting due to transients. The value of the capacitance can be adjusted in the lab. It is possible that this capacitor may not be needed.

The IC driver can sink and source more than 650 mA, so there is no need for a current limit between the gate of the switch and the IC.

The compensation components are taken from the data sheet. They serve only as a starting point. Actual compensation will need to be adjusted in the laboratory to ensure a stable loop.

Resistors R5 and R6 set the under-voltage lockout value. The voltage at this pin must be 1.28 V before the IC will operate. A reasonable input voltage is 95 V for this pin to be active. The value of R5 is very large, so the bias current of the UVLO pin will affect the value needed for R6. We can consider V_{IN} and R5 to be a constant current source, so we need to subtract the bias current from the current supplied by R5 when calculating R6. The data sheet also gives equations for calculating these resistors.

6.5 Non-isolated Flyback Example

Our next example shows the advantage of a non-isolated flyback design for automotive use. An automotive system can range from 11.5 V at low battery with the key off to 15.0 V when charging a drained battery. Some systems are designed to work at a nominal 13.6 V ± 0.5 V. This represents full voltage for a charged battery. Our example implements a system that produces 13.6 V at 10 A. The output ripple target is 300 mV. The regulation target is 400 mV. Figure 6.14 shows our circuit.

A reasonable choice for the control IC is the LT1680. This IC is designed for high power step-up DC–DC converters using an external MOSFET switch. It provides all of the necessary current mode PWM functions and will operate directly from the input supply.

A reasonable switching frequency is 167 kHz. The maximum frequency of the IC is 200 kHz, but we want to stay away from effects where we have no control. The cycle time is 6.0 μs. This frequency is low enough that parasitic effects at the high power level will be manageable. This frequency is also within the power range of reasonably priced inductor cores.

Continuous mode operation is a reasonable selection for this design. The output current will approximate the input current, since the input voltage range is +10/−20% of the output voltage range. Choosing continuous mode will allow the peak current to be only slightly larger than twice the output current. If we set the 50% duty cycle voltage to 10.5 V input, we will have enough margin when the voltage drops to 11.0 V to maintain control and avoid the need for slope compensation. This sets the target duty cycle for the lowest input voltage around 40%, as a first guess. We use the graph in the data sheet to choose the 3 K timing resistor based on our maximum duty cycle. Another graph in the data sheet gives us 2.2 nF based on the 167 kHz frequency and 3 K timing resistor.

A 60 V Schottky diode is a reasonable first try at the output rectifier. The turns ratio of the inductor is likely to be very close to 1:1. It is a reasonable guess that the turns ratio will be no larger than 1:2. The IRF 30CPQ060 has 60 V PRV and 30 A average current and is a dual diode package. The peak forward current is likely to be approximately 20 A, so this diode should fit our requirements. Each diode will pass one-half of the total current,

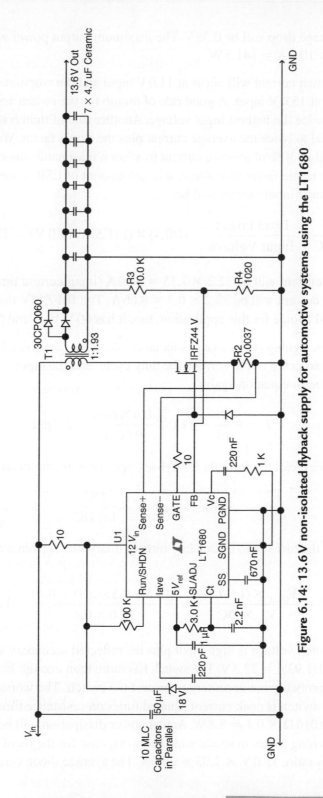

Figure 6.14: 13.6 V non-isolated flyback supply for automotive systems using the LT1680

so the forward voltage drop will be 0.55 V. The maximum output power will be $13.6\,V \times 10.0\,A + 0.55\,V \times 10.0\,A = 141.5\,W$.

The worst-case switch current will occur at 11.0 V input and the worst-case switch voltage will occur at 15.0 V input. A good rule of thumb for the switch voltage is to assume it will be twice the highest input voltage. Another rule of thumb is to pick the switch current equal to twice the average current plus the ripple factor. We will pick the ripple current equal to 30% of average current to allow a reasonable amount of dynamic response. This low ripple factor also allows a larger amount of ESR in the output capacitor. The average input current will be:

$$\frac{1}{DC} \times \frac{\text{Load Power}}{\text{Input Voltage}} = (1/0.4) \times (141.5\,W/11.0\,V) = 32.2\,A$$

The peak primary current will be $32.2 \times 1.15 = 37.0\,A$ (input current times ripple factor). The ripple current will be $32.2 \times 0.3 = 9.66\,A$. The IRFZ44V that we used in Chapter 5 is a good choice for this application, too. It has $60\,V\ V_{DSS}$ and $55A\ I_D$.

Now we can start designing the primary inductance. We have constrained the primary inductance by the expected ripple current, the duty cycle, and the input voltage. We use the rearranged inductor equation again:

$$L = V\frac{dt}{dI} = 11.0 \times \frac{0.4 \times 6\,\mu s}{9.66\,A} = 2.7\,\mu H$$

From Chapter 2, we recall the formula for flyback operation in continuous mode:

$$V_{OUT} = V_{IN} \times N \times \frac{DC}{1 - DC}$$

We can determine the turns ratio (secondary turns/primary turns) from our starting assumptions:

$$N = \frac{V_{OUT} \times (1 - DC)}{V_{IN} \times DC} = \frac{(13.6 + 0.55) \times (1 - 0.4)}{11.0 \times 0.4} = 1.93{:}1$$

The worst-case switch voltage is high input plus the reflected secondary voltage: $15.0 + (14.15 \times (1/1.93)) = 22.3\,V$. The switch has more than enough head room so that a clamp circuit is probably not necessary to protect the switch. The worst-case power dissipation for the switch is peak current squared times on-resistance times duty cycle: $(37A \times 37A) \times 0.016\,\Omega \times 0.4 = 8.8\,W$. Actual power dissipation will be slightly higher once we take switching losses in to account. The worst case for the rectifier is high input voltage times turns ratio: $15.0\,V \times 1.93 = 29.0\,V$. The average diode current when the switch is off is output current divided by $(1 - DC)$: $10\,A \times 0.6 = 16.7\,A$. The peak output current is peak input current times the turns ratio: $37.0\,A \times (1/1.93) = 19.2\,A$. These calculations show our choice of semiconductors is appropriate.

We again assign 67% of ripple voltage to the ESR of the output capacitor, so:

$$ESR = \frac{200\,mV}{19.2\,A} = 10.4\,m\Omega$$

$$X_C = \frac{100\,mV}{19.2\,A} = 5.2\,m\Omega$$

$$C = \frac{1}{2\pi \times 167\,kHz \times 5.2\,m\Omega} = 180\,\mu F$$

This value is similar to our previous example and will require multiple ceramic or aluminum capacitors to satisfy both the ESR requirement and the ripple current requirement. The capacitors for the previous example will be adequate for this design when enough are used in parallel. The most significant requirement for the capacitors is ripple current capability. Seven of the $4.7\,\mu F/16\,WV$ MLC capacitors will have only $1.3\,\Omega$ ESR, so the $33\,\mu F$ combined capacitance will be more than enough to meet our ripple voltage requirement. Using so many capacitors in parallel will invite problems with EMI and secondary problems that will increase ripple unless we pay strict attention to proper layout. The connections to the capacitors should be made with very wide but closely spaced conductors. This will reduce the inductance of the traces and minimize the loop area of the traces.

The current sense resistor is set by average current rather than peak current for this control IC. The equation is found in the data sheet:

$$R_{CS} = 120\,mV/I_{AVG} = 0.12\,V/32.2\,A = 3.7\,m\Omega$$

The average current limit is set by the combination of the current sense resistor and the current limit integration capacitor. The data sheet recommends setting this capacitor to 220 pF.

The output voltage is set by the equation:

$$V_{OUT} = \left(1 + \frac{R_1}{R_2}\right) \times 1.25\,V$$

Running the calculations requires a resistor ratio of 9.88:1.

We can set the soft start time to 100 ms, using the equation from the data sheet:

$$C_{SS} = 0.1\,s/150{,}000 = 670\,nF$$

Again, we start with the compensation values from the data sheet and will change them based on results in the lab. No slope compensation is necessary, since duty cycle is limited to 50%.

The large current pulses on the input will require very low ESR to maintain the voltage at the control IC. Selecting input capacitors equal to the output capacitors will provide the necessary low ripple. The very large input current pulses may make a forward converter a better choice for this application.

6.6 Forward Converter Circuits

A forward converter is a single switch converter that uses a transformer to transfer energy from the primary circuit to the secondary circuits. Energy flows from the primary to the secondary while the switch is conducting current. Figure 6.15 shows a representative circuit for a forward converter. A voltage clamp is necessary for a forward converter because all transformer current stops when the switch turns off. The clamp provides a path for the current in the magnetizing inductance of the transformer and the leakage inductance. In the flyback circuit, the current flow in the secondary provides a path for the flux of the core when the switch opens; the clamp is only necessary to reduce stress on the switch from leakage inductances.

Any of the clamp circuits in Figure 6.10 can be applied to the forward converter. The clamp circuits will have a voltage that is controlled by the secondary voltage when used in a flyback circuit because of the requirement that V/N is equal for all windings. This is not true for the forward converter. The clamp winding in Figure 6.10(a) guarantees that the switch voltage is twice the input voltage while the magnetizing current ramps down. Circuits B and C will have varying voltages depending on the amount of energy that is dissipated in the resistor. You must exercise care in designing the maximum duty cycle, transformer magnetizing inductance, and RC time constant when using Circuits B and C to ensure you do not exceed the voltage rating of the switch. Notice that circuit C is

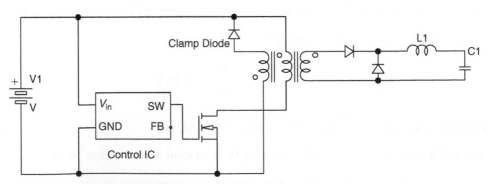

Figure 6.15: Representative circuit for a forward converter

identical to a boost regulator. International Rectifier Application Note AN-939 A gives a very good description of using dissipative clamp circuits in forward converters.

The clamp circuit design affects the maximum switch voltage required for a forward converter. The energy stored in the magnetizing inductance is proportional to the volt-seconds while the switch is on. The same number of volt-seconds is necessary to dissipate the energy stored in the magnetizing inductance of the transformer during the switch off-time. The voltage stress on the switch can be reduced by limiting the duty cycle. However, reducing the duty cycle will increase the primary peak current and the output peak current and voltage. The clamp winding generally has the same number of turns as the primary which sets the switch voltage to twice the input. However, the maximum duty cycle and clamp winding turns can be adjusted to set the switch voltage to any desired value. Our second example will show how to use a large switch voltage to reset the flux in the core when the duty cycle is greater than 50%. The clamp circuit only dissipates the energy in inductances inside the loop created by the clamp circuit. Any parasitic inductances outside the clamp circuit, such as switch lead inductances, will create voltages when the switch turns off and will add to the voltage stress on the switch.

The same two switch circuit we saw in Figure 6.12 can be used for a forward converter circuit by substituting a transformer for the flyback inductor. The maximum voltage on each switch will be slightly higher than the input voltage. The diodes again clamp the reverse voltage of the transformer inductance to the input voltage. Since the clamp voltage can be no larger than the input voltage, the duty cycle must be restricted to a value less than 50% to ensure that flux does not build up in the core and result in saturation.

6.7 Practical Forward Converter Design

Typical steps for designing a forward converter are listed below:

1. Choose a controller IC based on power level and bill of material constraints.

2. Choose the switching frequency.

3. Use the input voltage range and output ripple current goal to select the maximum duty cycle goal.

4. Pick the output diodes.

5. Design the transformer winding ratios.

6. Determine the maximum power and pick a switch.

7. Choose the bootstrap capacitor based on the gate charge required if a bootstrap supply is used.

8. Calculate the output inductor value.

9. Choose the output capacitor based on ripple requirements.

10. Design the auxiliary supply, if needed.

11. Design the ancillary IC components including the feedback circuit.

6.8 Off-Line Forward Converter Example

Our first example is a universal input off-line supply to provide 5.0 V at 20 A. (See Figure 6.16.) The ripple voltage is required to be below 100 mV and the regulation is required to be 200 mV. Even though the feature list of the MAX5052 says it is good for 50 W of output power, there is no reason it cannot be used above that power level as long as it is able to drive the switch. We will choose the MAX5052A for 50% maximum duty cycle. A 45% duty cycle is reasonable for the very lowest input voltage. This allows enough margin that the supply will start with the lowest input voltage for a 100 VAC power system. We will want to keep the output ripple current to a minimum to keep the ripple voltage low. We can choose an output ripple target of 10% or 2 A. The easiest clamp design is to use a winding on the power transformer and a diode (D3). D3 must be a fast-turn-on diode. The current through the diode will go to zero, so we are not concerned with turn-off characteristics.

There will be a constant diode drop in the output circuit because inductor current will flow for the whole switch cycle. Schottky diodes are the preferred components in low voltage supplies with modest power output. We can choose a dual diode that can handle the peak current. Our peak current is 20A + 1A ripple. The IRF 30CPQ060 is a dual diode in a TO-247AC package with a 30A average current rating and 60 V PRV rating. This diode has a forward voltage of 0.7 V at 20 A forward current.

We use a rearranged version of the buck converter equation from Chapter 2 to determine the required input voltage.

$$V_{IN} = (V_{OUT} + V_{Diode})/DC = 5.7\,V/0.45 = 12.7\,V$$

This voltage must be present at the secondary winding at the lowest input voltage. This gives the turns ratio of the transformer:

$$N = 100\,V/12.7\,V = 7.9$$

We can verify the required duty cycle at high input voltage. The input voltage will be:

$$390\,V/7.9 = 49.5\,V$$

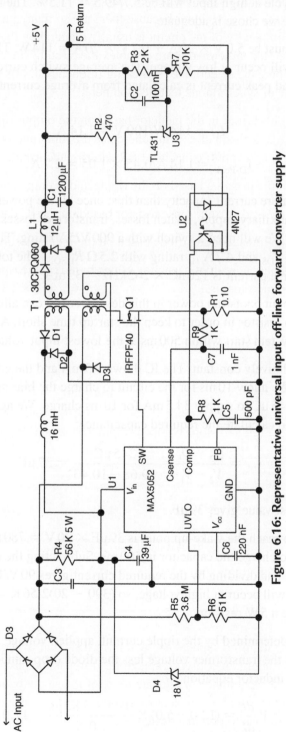

Figure 6.16: Representative universal input off-line forward converter supply

This means the duty cycle at high input will be 5.7/49.5 = 11.5%. The high input voltage confirms that the diode we chose is adequate.

The power delivered must be $5.0\,\text{V} \times 20\,\text{A} + 0.7\,\text{V} \times 20\,\text{A} = 114\,\text{W}$. The maximum current in the switch will occur at low voltage. The average switch current is calculated from average power and peak current is calculated from average current, duty cycle, and ripple factor:

$$I_D = 114\,\text{W}/100\,\text{V} = 1.14\,\text{A}$$

$$I_{D\text{-Peak}} = 1.14\,\text{A}/0.45 \times 1.05 = 2.7\,\text{A}$$

The switch will need more current capacity than this, once all the power consumption sources are factored (auxiliary supply, switch losses, transformer losses, inductor losses, capacitor losses, etc.). We will need a switch with a 900 V/5A rating. The IRF IRFPF40 MOSFET has a 900 V V_{DS} and 4.7 A I_D rating with 2.5 Ω R_{DSON}. The total gate charge is 120 nC, so the gate drive current is $120\,\text{nC} \times 262\,\text{kHz} = 32\,\text{mA}$.

We are less worried about bootstrap power in this design, so we can allow more dissipation in the bootstrap resistor in order to keep the startup time short. A good rule of thumb is to have the system start within 500 ms at the lowest input voltage.

The current draw is relatively constant. The IC draws 2.5 mA and the gate charge draws an additional 32 mA. We allow 10 ms for the circuit to charge the bias supply above 10.43 V. We will use 345 μC to supply 34.5 mA for 10 ms charge. We again use the capacitance equation to calculate the required capacitance:

$$C_4 = \frac{Q_2 - Q_1}{V_2 - V_1} = \frac{345\,\mu\text{C}}{19.68 - 10.43} = 37\,\mu\text{F}$$

Rounding to the nearest value gives 39 μF.

The nominal charge to reach the wake-up point is $39\,\mu\text{F} \times 20\,\text{V} = 780\,\mu\text{C}$. This means we will need 1.6 mA to charge the capacitor in 500 ms. Subtracting the capacitor voltage from the input voltage and dividing by the required current gives 90 V/1.6 mA = 56 K. The maximum power will occur at high voltage, so $(390 - 20)2/56\,\text{K} = 2.5\,\text{W}$. This resistor will need to be a 5 W resistor.

The inductor value is determined by the ripple current, applied voltage, and duty cycle. The applied voltage is the transformer voltage less the diode drop minus the output voltage. We apply the inductor equation:

$$L = V\frac{dt}{dI} = (12.0 - 5.0) \times \frac{0.45 \times 3.82\,\mu\text{s}}{1.0\,\text{A}} = 12.0\,\mu\text{H}$$

The output capacitor value is determined by the ripple voltage requirement. We have 100 mV of ripple and 1.0 A. We can choose ESR and the capacitor value using our one-third and two-thirds rule:

$$\text{ESR} = \frac{67\,\text{mV}}{1.0\,\text{A}} = 67\,\text{m}\Omega$$

The target capacitance is

$$X_C = \frac{33\,\text{mV}}{1.0\,\text{A}} = 33\,\text{m}\Omega$$

$$C = \frac{1}{2\pi \times 262\,\text{kHz} \times 33\,\text{m}\Omega} = 18\,\mu\text{F}$$

A good choice for the output capacitor is a Panasonic series FM Type A. There are no capacitors close in value to 18 µF in the 6.3 WV range. The closest value that has a low enough ESR and enough ripple capacity is the EEUFM0J122L 1200 µF capacitor that can handle 1.56 A of ripple and has 30 mΩ ESR.

The auxiliary supply needs to provide approximately 12 V for normal operation but must not go above 30 V. The diodes D1 and D2 can be small Schottky diodes with 60 PRV. The auxiliary supply is not regulated and there is no coupling between the main output and the auxiliary supply. It is very likely that the auxiliary supply will rise to a large voltage during startup and during large transients on the main output. The zener diode shunt (D4) is provided to ensure that extra current will keep the supply within the limits of the control IC. The zener voltage is set high enough that it normally will not draw current. We can choose a very low value for the inductor ripple because the current is essentially constant. There is no need for rapid transient response, and the low ripple will reduce fluctuations in output voltage during main output transients. We choose 5% ripple current for this supply, or 34.5 mA × 0.05 = 1.7 mA.

We calculate the inductor at the lowest input voltage. We also use lowest input voltage to calculate the turns ratio for this supply.

$$V_{\text{IN}} = (V_{\text{OUT}} + V_{\text{Diode}})/\text{DC} = 12.7\,\text{V}/0.45 = 28.2\,\text{V}$$
$$N = 100\,\text{V}/28.2\,\text{V} = 3.6$$
$$L = V\frac{dt}{dI} = (27.5 - 12.0) \times \frac{0.45 \times 3.82\,\mu\text{s}}{1.7\,\text{mA}} = 16\,\text{mH}$$

The last step is to design the feedback circuit. We will use a standard 4N27 optoisolator and a TL431 shunt regulator to provide feedback to the control IC. We choose a small amount of feed forward compensation to the TL431 and a small pole at the feedback pin of the control IC. The actual compensation values will need to be determined by taking the prototype supply into the laboratory and making measurements and adjustments.

The selection of current sense components and under-voltage components are the same as the MAX5052 example in the flyback converter section.

6.9 Non-isolated Forward Converter Example

The current levels in the flyback example for automotive use were quite high. The input current consists of very large, short pulses. The output also consists of very large, short pulses. A forward converter can reduce both the output ripple and the input ripple by allowing the duty cycle to be larger than 50%. Our next example, Figure 6.17, shows how to implement such a supply.

The duty cycle in an off-line forward converter is limited to 50% by the voltage required to reset the flux in the transformer and the switch breakdown voltage. At 50% duty cycle, the reverse voltage can be equal to the input voltage. In our automotive application, we can use a high voltage switch to advantage. The high reverse voltage will allow the flux in the transformer to reset in a very short period of time.

We start from the same set of requirements as the flyback example and use the same control IC. We choose the same 167-kHz operating frequency for a cycle time of 6 μs.

We can set the maximum duty cycle to 75% at 11.0 V input. The data sheet shows that the maximum duty cycle will vary from IC to IC, from about 70% to about 78%, when we set the nominal value to 75%. Our calculations will need to allow for 80% duty cycle as the worst case. The volt-seconds during switch on-time will need to equal the volt-seconds when the switch is off. The ratio of on-time to off-time is 80/20, so the reverse voltage on the transformer primary during off-time will be four times the input voltage. This sets the turns ratio for the clamp winding to 4:1. The switch withstand voltage will be five times the input voltage (4 × for the clamp plus 1 × for the input supply) at the highest input voltage. This gives a minimum value of 15.0 V × 5 = 75 V. A check of the International Rectifier website shows either 100 V or 150 V MOSFETs. It probably makes the most sense to choose a 150 V device to ensure margin in the presence of transients. The IRF3415 is a TO-220 package that has 150 V V_{DSS}, 42 mΩ on resistance, and 43 A I_{DSS}. The IRF3315 is a similar and less expensive part, but it only has 15 A I_{DSS} at 100°C.

A 150 V Schottky diode is a reasonable first try at the output rectifier. The turns ratio of the transformer is likely to be very close to 1.5:1 for primary to secondary, since our goal is to reduce the input ripple and output ripple. However, we are allowing the reverse voltage during transformer reset to be four times the input voltage. This means the reverse

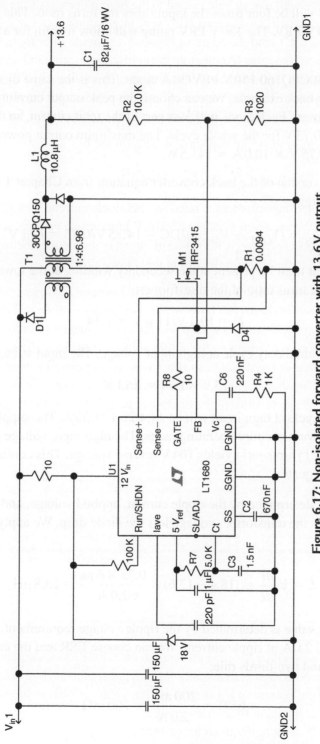

Figure 6.17: Non-isolated forward converter with 13.6 V output

voltage on the diodes will be four times the input times the turns ratio. This will require a diode with at least 90 V PRV. The 150 V PRV rating will allow margin for a transformer ratio up to 2.25:1.

We can use the IRF 30CPQ160 150 V PRV/30 A diode. This is the same diode family that we used in the flyback example. We can choose the peak output current as 11 A with 2 A of ripple current. Each diode will pass part of the total current, so the forward voltage drop will be 0.75 V for the whole cycle. The maximum output power will be 13.6 V × 10.0 A + 0.75 V × 10.0 A = 143.5 W.

We use a rearranged version of the buck converter equation from Chapter 1 to determine the required input voltage.

$$V_{\text{IN}} = (V_{\text{OUT}} + V_{\text{Diode}})/\text{DC} = 14.35 \text{ V}/0.75 = 19.1 \text{ V}$$

This is the voltage that must be present at the secondary winding at the lowest input voltage. This gives the turns ratio of the transformer:

$$N = 19.1 \text{ V}/11.0 \text{ V} = 1.74$$

We can verify the required duty cycle at high input voltage. The input voltage will be:

$$15 \text{ V} \times 1.74 = 26.1 \text{ V}$$

This means the duty cycle at high input will be 14.4/26.1 = 55%. The supply will require slope compensation over the entire operating range. The high input voltage of 15.0 V times the (4:1 × 1.74:1) turns ratio yields 104 V reverse voltage. This confirms that the diode we chose is adequate.

The inductor value is determined by the ripple current, applied voltage, and duty cycle. The applied voltage is the transformer voltage less the diode drop. We apply the inductor equation:

$$L = V \frac{dt}{dI} = (18.4 - 13.6) \times \frac{0.75 \times 6 \,\mu\text{s}}{2.0 \text{ A}} = 10.8 \,\mu\text{H}$$

The output capacitor value is determined by the ripple voltage requirement. We have 300 mV of ripple and 2.0 A of ripple current. We can choose ESR and the capacitor value using our one-third and two-thirds rule:

$$\text{ESR} = \frac{200 \,\text{mV}}{2.0 \text{ A}} = 100 \,\text{m}\Omega$$

The target capacitance is:

$$X_C = \frac{100\,\text{mV}}{2.0\,\text{A}} = 50\,\text{m}\Omega$$

$$C = \frac{1}{2\pi \times 167\,\text{kHz} \times 50\,\text{m}\Omega} = 19\,\mu\text{F}$$

The ripple current and ESR requirements are easily met with a single $82\,\mu\text{F}/16\,\text{WV}$ Panasonic WA series polymer electrolytic capacitor. This capacitor has $39\,\text{m}\Omega$ ESR and 2.5 A ripple current rating in a surface mount package. The RMS ripple current is approximately equal to one-half the P–P ripple for a triangular wave, so our output ripple current is approximately 1 A.

The average input current is $141\,\text{W}/11.0\,\text{V} = 12.8\,\text{A}$. The input current is essentially a rectangular pulse of $12.8\text{A}/0.75 = 17\text{A}$. The RMS current is

$$I_{\text{RMS}} = I_{\text{IN}}\,(\text{DC} - \text{DC}^2)^{1/2} = 12.8(0.75 - 0.56)^{1/2} = 5.6\,\text{A}$$

Two of the $150\,\mu\text{F}/20\,\text{WV}$ Panasonic WA series polymer electrolytic capacitors will do nicely for the input filter. This capacitor has $26\,\text{m}\Omega$ ESR and 3.7 A ripple current rating in a surface mount package. This is quite a contrast to the ripple requirements of the flyback design, where the input RMS ripple was 9 A RMS and the output ripple was 4.8 A RMS. We require fewer and less expensive filter capacitors by changing from a flyback circuit to a forward converter.

The current sense resistor is set by average current rather than peak current for this control IC. The equation is found in the data sheet:

$$R_{\text{CS}} = 120\,\text{mV}/I_{\text{AVG}} = 0.12\,\text{V}/12.8\,\text{A} = 9.4\,\text{m}\Omega$$

The average current limit is set by the combination of the current sense resistor and the current limit integration capacitor. The data sheet recommends setting this capacitor to 220 pF.

The output voltage and soft start calculations are the same as for the flyback example.

We want to restrict the maximum duty cycle to 75%, so we choose the 5 K timing resistor from the graph in the data sheet. Another graph on the data sheet indicates that 1.5 nF will yield 167 kHz operation for this timing resistor. A duty cycle above 50% requires slope compensation for all current mode controllers. The LT1680 provides internal slope compensation that should be adequate for our example supply.

6.10 Push-Pull Circuits

Push-pull circuits are not well suited to voltage mode IC controllers because any flux imbalance in one leg of the transformer primary will eventually saturate the transformer core. A current mode controller will control the imbalance and limit the current through both legs. One switch and one winding of the transformer may still carry more load than the other, but the total flux in the core is controlled by limiting the maximum current in each winding.

Figure 6.18 shows a representative push-pull converter. Notice that the secondary side uses a center-tapped full wave rectifier configuration. A push-pull circuit requires full wave rectification. Most practical circuits use a center-tapped transformer and a dual diode, since there is only one diode drop during each half-cycle and two diodes. It is possible to use a full wave bridge to simplify the transformer, but then the voltage drop for each half-cycle is two diode drops and it uses four diodes. Basically, copper is much cheaper than silicon.

Push-pull circuits fell out of favor when the only IC controllers were voltage mode because of the problems with transformer balance. They are more popular for moderate power circuits now that current mode controllers are readily available. Push-pull circuits are popular at all power levels for point of load applications where the voltage stress on the switch is not an issue.

The primary requires twice as many turns as a bridge circuit, so the transformer is more complicated than a half bridge transformer. The switches must withstand twice the input voltage where the switch voltage for a half bridge is equal to the input voltage. The biggest advantage of push-pull over half bridge is that neither switch requires isolated drive. A clamp circuit is not necessary in a push-pull circuit because one of the output diodes will continue to conduct when both switches are off. This allows magnetizing inductor current to flow while current in the output choke ramps down. The magnetizing inductance current will be forced to zero when the alternate switch closes.

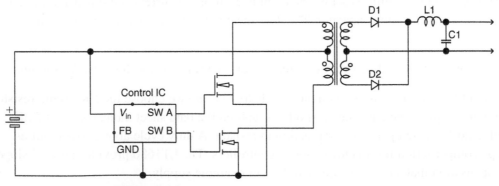

Figure 6.18: Representative push-pull converter

The effective switching frequency is double the oscillator frequency. Each switch provides the equivalent of a single switch forward converter. The bipolar drive doubles the effective duty cycle and the operating frequency of the output filter is double the switching frequency.

The control IC must provide two-phase output pulses to alternately drive the switches. Additionally, the circuit will behave badly if both switches conduct at the same time. The transformer will allow very large switch currents to flow if both switches conduct at the same time. A push-pull control IC must provide the ability to set a proper amount of dead time between the alternate phases. This will ensure that one switch is off before the other switch begins conducting.

6.11 Practical Push-Pull Circuit Design

Typical steps for designing a push-pull converter are listed below:

1. Choose a controller IC based on power level and bill of material constraints.

2. Choose the switching frequency.

3. Use the input voltage range goal to select the maximum duty cycle goal.

4. Pick the output diodes.

5. Calculate the output inductor value.

6. Design the transformer winding ratios.

7. Determine the maximum power and pick the switches.

8. Choose the output capacitor based on ripple requirements.

9. Design the auxiliary supply, if needed.

10. Design the ancillary IC components, including the feedback circuit.

Our push-pull example is a telecom supply that converts 48 V to an isolated 5 V/20 A supply with 100 mV of ripple. Figure 6.19 shows the circuit we are designing.

A search for control ICs designed specifically for push-pull or bridge operation yields very few parts. Most of the first- and second-generation current mode controllers (such as the 1846) provide the necessary functions, but they need a large number of external components for a working supply. There are not many modern control ICs designed for push-pull and bridge operation. Some manufacturers have only one or two parts for this application and many have no modern products at all for this market. This is

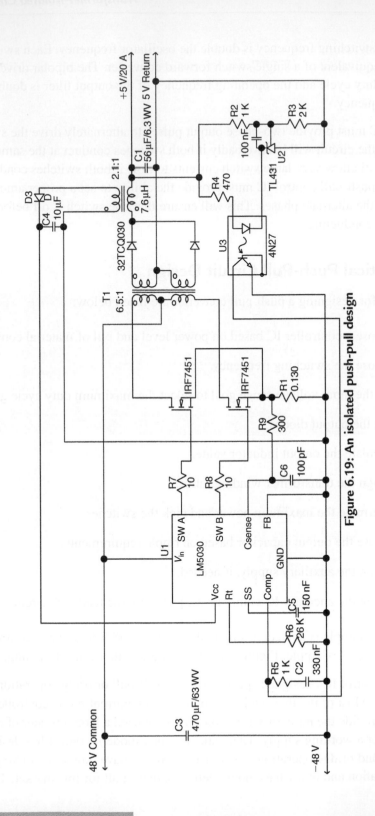

Figure 6.19: An isolated push-pull design

understandable, since only a very small portion of the power supply market includes designs above 200 W.

We will choose the National LM5030 for our example. This part is a 10-pin surface mount part designed for off-line or high voltage applications. Like most modern ICs, it integrates most of the functionality necessary for a low part count design. 200 kHz is a reasonable frequency for a high power system. Higher frequencies require more attention to switching times and second- and third-order effects. It is possible to produce high power transformer designs at higher frequencies, but you must put more design effort into controlling elements like layout, transformer design, and semiconductor selection. The LM5030 data sheet gives a graph of timing resistor versus frequency. The graph shows 26 K will give 200 kHz for the oscillator frequency. The switches will switch at 100 kHz and the output filter will operate at 200 kHz.

We saw in the non-isolated forward converter design that choosing a duty cycle above 50% significantly reduces the ripple on the input supply without any effect on the output ripple. The effective duty cycle will be twice the single switch duty cycle. We can choose 40% for the single switch duty cycle. This yields an 80% duty cycle for both input and output at 200 kHz. This large duty cycle is adequate because the 48 V input supply has reasonable regulation. We will choose 1.0 A of ripple current in the output inductor to minimize the ESR requirements of the output capacitor. The other advantage of using a large duty cycle is that it reduces the PRV rating required for the output diodes. We can expect the reverse voltage to be less than twice the output voltage, so 20 V PRV diodes should be adequate. International Rectifier Schottky diodes are available in either 15 V or 30 V ratings, so we will choose the 32TCQ030 dual diode. This device has 30 A rating and 30 V PRV. This diode has 0.5 V forward drop at 20 A forward current.

We use a rearranged version of the buck converter equation from Chapter 2 to determine the required input voltage.

$$V_{\text{IN}} = (V_{\text{OUT}} + V_{\text{Diode}})/\text{DC} = 5.5\,\text{V}/0.80 = 6.9\,\text{V}$$

This is the voltage that must be present at the secondary winding at the lowest input voltage. This gives the turns ratio of the transformer:

$$N = 48\text{V}/6.9\text{V} = 6.96$$

It will simplify the design significantly if we adjust this to 6.5 and use a slightly smaller duty cycle. This value will make the transformer design reasonable. We will probably design the transformer with either two or three turns for each leg of the secondary, which will require either 13 or 20 turns per primary winding.

The inductor value is determined by the ripple current, applied voltage, and duty cycle. We apply the inductor equation:

$$L = V \frac{dt}{dI} = (6.9 - 5.0) \times \frac{0.8 \times 5\,\mu s}{1.0\,A} = 7.6\,\mu H$$

The output capacitor value is determined by the ripple voltage requirement. We have 100 mV of ripple and 1.0 A of ripple current. We can choose ESR and the capacitor value using our one-third and two-thirds rule:

$$ESR = \frac{67\,mV}{1.0\,A} = 67\,m\Omega$$

The target capacitance is:

$$X_C = \frac{33\,mV}{1.0\,A} = 33\,m\Omega$$

$$C = \frac{1}{2\pi \times 200\,kHz \times 33\,m\Omega} = 24\,\mu F$$

A 56 μF/6.3 WV Panasonic S series surface mount polymer electrolytic capacitor has only 9 mΩ ESR, so the ripple will be significantly below the 100 mV target. This capacitor has 3A ripple current rating.

The input supply power is approximately 117 W (85% efficiency). The average input current is 2.4 A and the peak input current is 3.1 A with an 80% duty cycle. The RMS current is 1.24 A. A Panasonic FC series 470 μF/63 WV capacitor will provide low ESR (under 1 Ω) with adequate ripple current rating. The switches will need to handle at least twice the input voltage. The closest V_{DSS} rating with margin is 150 V. The IRF3415S is a D2PAK device that has more than enough current rating and adequate voltage rating. The IRF7451 is a device in an SO-8 package. It has a continuous drain current rating of 3.6 A.

Since the average current for each switch is one-half the total, this device might be adequate if small size is a design goal.

The feedback circuit uses a TL431 to drive a 4N27 optoisolator. There is loop compensation on both the TL431 and on the compensation pin of the control IC. The control IC implements internal slope compensation so no external slope compensation should be necessary.

The V_{CC} supply is different from any we have used so far. We put an auxiliary winding on the main filter inductor to use it the same way we used the inductor in the flyback supplies. Note that the auxiliary supply winding is polarized so that the supply charges while the current is discharging in the filter choke. The voltage across the inductor will vary depending on the 48 V input voltage while the filter choke is charging with current. However, when the filter choke is discharging, the diodes clamp the voltage across the choke to approximately the output voltage. Since the output voltage is highly regulated, we get a well regulated auxiliary supply for the control IC. This would seem to be the best possible source for IC power. The problem, especially for off-line supplies, is that the safety isolation between the windings of the filter inductor must be the same as the safety isolation of the main transformer. The supply voltage of our example will be (2.1 × 5 V − 0.7 V) or 9.8 V. The IC supply voltage will change only slightly with changes in output current level.

The current sense resistor is calculated using the information in the data sheet:

$$R = 0.5/I_{PK} = 0.5/3.8 = 0.130 \, \Omega$$

We implement a small RC filter at the current sense pin in order to eliminate false setting of the current sense because of transients.

The soft start pin provides a current source of $10 \, \mu A$. This current source charges the soft start capacitor to 0.5 V. If we use 30 ms for soft start, we need a $0.15 \, \mu F$ soft start capacitor.

6.12 Half Bridge Circuits

Half bridge circuits are the topology of choice for off-line converters between 200 W and 1000 W. Figure 6.20 shows a representative half bridge converter.

The capacitive voltage divider (C2, C3) is an integral part of the circuit. It provides a voltage equal to one-half of the input voltage. The switches alternately drive current in opposite directions through the transformer primary, as in the push-pull circuit. The advantage of the half bridge is that the switches must only withstand a voltage equal to the input voltage plus a little more for transients. The transformer primary is also simpler than the push-pull transformer, since only a single primary winding is necessary.

Notice that there is a small coupling capacitor (C4) between the switches and the transformer primary. This capacitor ensures that flux cannot build up in the primary winding and saturate the transformer. When the two reservoir capacitors are driven by a full wave voltage doubler for 115 V operation, the input supply diodes alternately charge up the capacitors to the full peak voltage of the input power. The voltage on each capacitor has a hard supply that ensures a hard center tap voltage regardless of capacitor symmetry. The coupling capacitor between the switches and the transformer is less likely to be necessary. However, if the capacitors are driven by a full wave bridge for a universal input or 240 V system, the voltage at the connection of the reservoir capacitors will be a factor

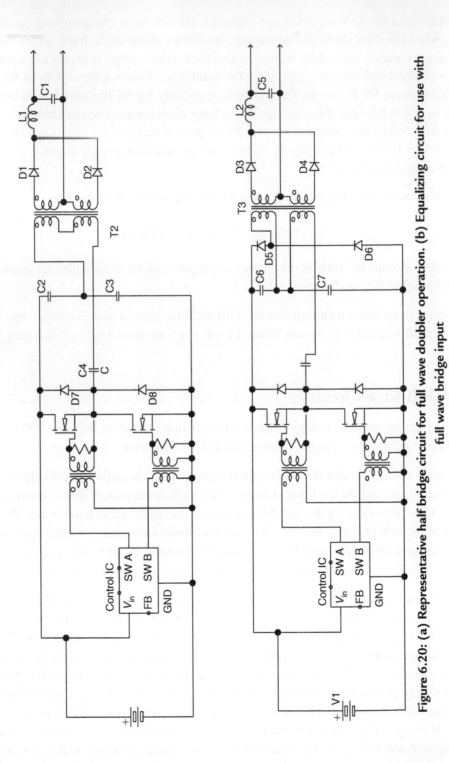

Figure 6.20: (a) Representative half bridge circuit for full wave doubler operation. (b) Equalizing circuit for use with full wave bridge input

of the relative values of the capacitors. The center voltage is now a "soft" value that will depend on the capacitor values and circuit operation. "Soft" operation requires that the coupling capacitor is used to ensure that the transformer does not saturate. The coupling capacitor has one-half of the input voltage applied and all of the primary current. This will require an AC capacitor that is rated for the full AC current of the power supply.

Figure 6.20 shows another method of ensuring that the center voltage of the capacitors stays symmetrical. A second primary winding (balancing winding) with the same number of turns is connected through diodes D5 and D6 to the input supply. Circuit operation places the two windings in series across the two capacitors. If the voltage across the windings is not identical, current flows from the balancing winding to equalize the voltages on the capacitors. The current in the balancing winding is typically on the order of 100 mA, so the winding can be a small gauge wire.

The half bridge circuit is more complicated than push-pull because the top switch requires isolated drive. Current mode control requires that a current transformer be placed in series with the primary winding. The current sense also requires full wave rectification to sense the current of each switch. Notice that there are clamp diodes across each of the switches. It is possible to use the body-drain diode of the MOSFETs, but these diodes have poor turn-on and turn-off characteristics. It is good practice to use high speed diodes to prevent the MOSFET diodes from conducting.

6.13 Practical Half Bridge Circuit Design

The design of half bridge and full bridge circuits contains the same steps. Typical steps for designing a bridge converter are listed below:

1. Choose a controller IC based on power level and bill of material constraints.

2. Choose the switching frequency.

3. Use the input voltage range goal to select the maximum duty cycle goal.

4. Pick the output diodes.

5. Calculate the output inductor value.

6. Design the transformer winding ratios.

7. Determine the maximum power and pick the switches.

8. Choose the output capacitor based on ripple requirements.

9. Design the auxiliary supply, if needed.

10. Design the ancillary IC components, including the feedback circuit.

Our half bridge example is a 12.0 V/40 A off-line universal power supply. The ripple goal is 100 mV. Figure 6.21 shows our example power supply. The National LM5030 is a good candidate for our supply. Once again, we choose 100-kHz operation to simplify design but still give good efficiency.

We must start our design at 100 VDC input to design the inductor for maximum duty cycle. We will choose a maximum duty cycle of 40%. We also set our ripple current goal at 4.0 A. The output diode reverse voltage at 100 V input is likely to be 18 V. The reverse voltage at 390 V input will then be approximately 70 V. The International Rectifier 80CNQ080 A has 80 A current rating and 80 V PRV. The forward voltage at 40 A is 0.8 V, so this diode will dissipate 32 W at full output.

We use a rearranged version of the buck converter equation from Chapter 1 to determine the required input voltage:

$$V_{IN} = (V_{OUT} + V_{Diode})/DC = 12.8\,V/0.80 = 16.0\,V$$

The maximum voltage at the rectifiers will be 62.4 V, so our diode selection is adequate. The duty cycle at high input voltage will be 20% in the output circuit at high input voltage.

The output inductor will be:

$$L = V\frac{dt}{dI} = (15.2 - 12.0) \times \frac{0.8 \times 5\,\mu s}{4.0\,A} = 3.2\,\mu H$$

Remember that the voltage across the transformer is only one-half of the input voltage. This gives the turns ratio of the transformer:

$$N = 50\,V/16.0\,V = 3.2$$

The losses in the circuit are quite large. The diode losses are the main contribution to power loss in the switching circuit. We should add at least 20 W more to account for other losses in the circuit. This gives a total switching power input of 532 W. The input current at low input voltage will be 5.32 A average or 6.65 A peak. The switches will need 450 V V_{DSS} and at least 7 A I_{DSS}. The IRFP344 has 450 V V_{DSS} and 9 A I_{DSS} with 0.63 Ω on resistance. The total gate charge is 60 nC for this switch, so gate current will be 12 mA.

The output capacitor value is determined by the ripple voltage requirement. We have 100 mV of ripple and 1.0 A of ripple current. We can choose ESR and the capacitor value using our one-third and two-thirds rule:

$$ESR = \frac{67\,mV}{4.0\,A} = 17\,m\Omega$$

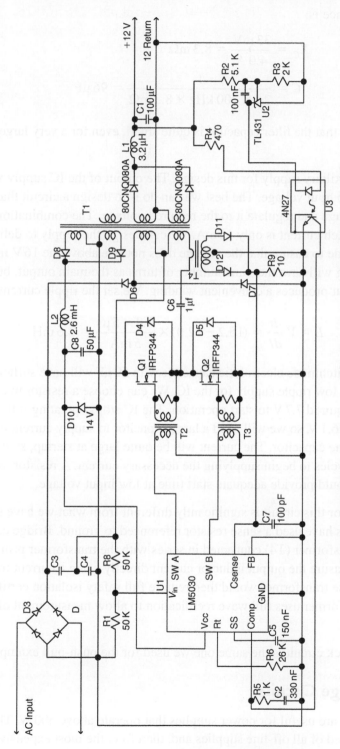

FigURE 6.21: Half bridge design 12.0 V/40 A off-line universal power supply

The target capacitance is:

$$X_C = \frac{33\,mV}{4.0\,A} = 8.3\,m\Omega$$

$$C = \frac{1}{2\pi \times 200\,kHz \times 8.3\,m\Omega} = 96\,\mu F$$

Once again, we see that the filter capacitor is quite small, even for a very large output current.

We will need an auxiliary supply for this design. The output of the IC supply will have no relationship to the output voltage. The best we can do is to design a circuit that is close to the required voltage and regulate it to the required voltage. The combination of the IC current and the switch current is only 15 mA. We can design the supply to deliver 12 V and use a zener diode to ensure that the voltage does not rise above the 16 V maximum of the IC. The winding will have the same number of turns as the main output, but we can use smaller wire that produces a convenient winding. We set the ripple current to 5 mA.

$$L = V\frac{dt}{dI} = (15.2 - 12.0) \times \frac{0.8 \times 5\,\mu s}{5\,mA} = 2.6\,mH$$

Any convenient switchmode electrolytic capacitor with 50 μF will have sufficiently low ESR and produce a low ripple supply for the IC. We can choose a resistor to charge the IC supply to the required 7.7 V to start operation. The IC stops operating if the IC supply voltage falls below 6.1 V, so we will need a large capacitor to supply current until the bootstrap charges the capacitor. The current will be quite large at startup, so it should only take two or three cycles to begin supplying the necessary current. A resistor to supply 1 mA of current should provide adequate start time at low input voltage.

The current sense for this circuit is significantly different from what we have seen so far. All of our examples have used a sense resistor referenced to ground. Bridge circuits need a current sense transformer (T4) connected in series with the transformer primary. It is also possible to measure the output inductor current directly using a current transformer, but the current sense transformer would then require full safety isolation certification. The current sense transformer uses full wave rectification to allow measurement of the current from both switches.

The voltage feedback circuit is the same one we used for the push-pull example.

6.14 Full Bridge Circuits

Full bridge circuits are useful for power supplies that operate above 500 W. They are the most complicated of all off-line supplies and, therefore, the most expensive. Full

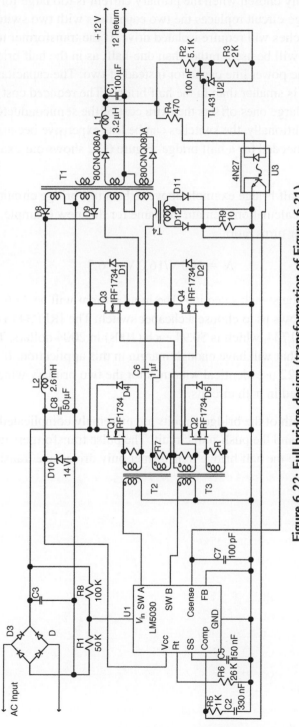

Figure 6.22: Full bridge design (transformation of Figure 6.21)

bridge operation is only chosen when the primary current is too large for two switches to handle. A full bridge circuit replaces the two capacitors with two switches and clamp diodes. Both top switches will require isolated drive. If the transformer turns calculations, the full input voltage will be used, rather than one-half, as in the half bridge. The full bridge design uses one power line capacitor instead of two. The capacitance of the single capacitor (C3) is smaller than in the half bridge. The reduced cost of one smaller capacitor versus two large ones offsets the extra cost of the semiconductors for full bridge operation. Additionally, the switches can be less expensive because the current level is one-half that needed for a half bridge. Figure 6.22 shows our example full bridge converter.

We can redesign the half bridge example above to be a full bridge circuit. The main design decisions and calculations remain the same for this new example. The first change will be the transformer turns ratio.

$$N = 100\,\text{V}/16.0\,\text{V} = 6.3$$

The next change will be switch selection. The input current will be 2.66 A or 3.33 A peak. This lower current allows us to choose a cheaper switch. The IRFP344 is $2.33 each (100s) versus the IRF1734, which is $0.94 each (100s) in 2004 dollars. The IRF1734 is a 450 V/3.4 A switch that will have enough margin in this application. If the transformers in Figures 6.21 and 6.22 use identical windings for the two primary windings, the same transformer can be used in both circuits.

The schematics for both of the bridge circuits are extremely complicated compared to the single switch circuits and the push-pull circuit. The pulse transformers must drive two transistors each, where the half bridge transformer only drives one transistor.

CHAPTER 7

Power Semiconductors

Nihal Kularatna

Throughout the early years in my career of designing switching power supplies, I always accumulated a small pile of power transistors on the bench while I was evaluating my power supply designs. I had developed a saying: "If you don't stick your fingers inside the circuit, you won't break it." Unfortunately, you always have to "stick your finger in." So order a lot more power transistors and other semiconductors than you think you need.

The semiconductors are the most fragile components within any switching power supply. Applying them correctly will make them operate reliably over all of the operating conditions of the system. They also have a direct effect on the overall supply.

Kularatna overviews all of the major power switches and rectifiers used in the switching power supply field today, some of which will not be used by the majority of power supply designers except in the very high power market (over a kilowatt). He summarizes each of the semiconductor's advantages and weaknesses.

—Marty Brown

7.1 Introduction

During the elapsed half century since the invention of the transistor, the power electronics world has been able to enjoy the benefit of many different types of power semiconductor devices. These devices are able to handle voltages from a few volts to several kilovolts and switching currents from a few milliamperes to kiloamperes. Within a decade from the invention of the transistor, the thyristor was commercialized.

Around 1968 power transistors began replacing the thyristors in switchmode power systems. The power MOSFET, as a practical commercial device, has been available since 1976. When "smart power" devices appeared in the market, designers were able to make use of the insulated gate bipolar transistor (IGBT) from the early 90s.

In 1992 MOS-controlled thyristors were commercially introduced, and around 1995 semiconductor materials such as GaAs and silicon carbide (SiC) opened new vistas for better performance power diodes for high frequency switching systems.

Presently, the spectrum of what are referred to as "power devices" spans a very wide range of devices and technology. Discrete power semiconductors will continue to be the leading edge for power electronics in the 1990s. Improvements on the fabrication processes for basic components such as diodes, thyristors, and bipolar power transistors have paved the way to high voltage, high current, and high speed devices. Some major players in the industry have invested in manufacturing capabilities to transfer the best and newest power semiconductor technologies from research areas to production.

Commercially available power semiconductor devices could be categorized into several basic groups, such as diodes, thyristors, bipolar junction power transistors (BJTs), power metal oxide silicon field effect transistors (power MOSFETs), insulated gate bipolar transistors (IGBTs), MOS-controlled thyristors (MCTs), and gate turn-off thyristors (GTOs). This chapter provides an overview of the characteristics, performance factors, and limitations of these device families.

7.2 Power Diodes and Thyristors

7.2.1 Power Diodes

The diode is the simplest semiconductor device, comprising a P-N junction. In attempts to improve its static and dynamic properties, numerous diode types have evolved. In power applications diodes are used principally to rectify, that is, to convert alternating current to direct current. However, a diode is used also to allow current freewheeling. That is, if the supply to an inductive load is interrupted, a diode across the load provides a path for the inductive current and prevents high voltages *Ldi/dt* damaging sensitive components of the circuit.

The basic parameters characterizing the diodes are the maximum forward average current $I_{F(Ave)}$ and the peak inverse voltage (PIV)1. This parameter is sometimes termed as *blocking voltage* (V_{rrm}). There are two main categories of diodes, namely general-purpose P-N junction rectifiers and fast recovery P-N junction rectifiers. General purpose types are used in circuits operating at the line frequencies such as 50 or 60 Hz. Fast recovery (or fast turn-off) types are used in conjunction with other power electronics systems with fast switching circuits.

Classic examples of the second type are switchmode power supplies (SMPS) or inverters, etc. Figure 7.1(a) indicates the capabilities of a power device manufacturer catering to very high power systems and Figure 7.1(b) indicates the capabilities of a manufacturer catering to a wide range of applications.

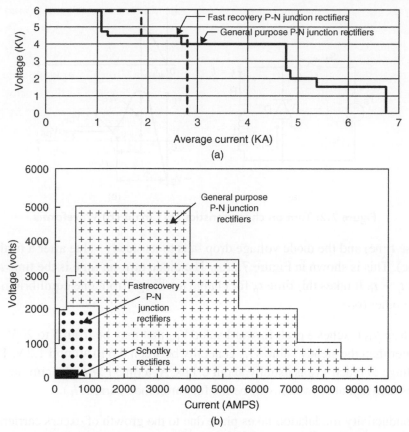

Figure 7.1: Rectifier capabilities (a) Rectifier capabilities of a major supplier of high power semiconductors (Reproduced by permission of GEC Plessey Semiconductors, UK.) (b) Rectifier capabilities of a manufacturer catering for wide range of applications (Reproduced by permission of International Rectifier Inc., USA.)

In high frequency situations such as inverters and SMPS, etc., two other important phenomena dominate the selection of rectifiers. Those are the "forward recovery" and the "reverse recovery."

7.2.1.1 Forward recovery

The turn-on transient can be explained with Figure 7.2. When the load time constant L/R is long compared to the time for turn on t_{fr} (**f**orward **r**ecovery time), load current will hardly change during this period. For the time $t < 0$, the switch S_w is closed. Steady conditions prevail and the diode D is reverse biased at $-V_S$. It is in the off-state, and $i_D = 0$.

At $t = 0$, the switch S_w is opened. The diode becomes forward biased, providing a path for the load current in R and L, so that the diode current i_D rises to I_F (RI_1) after a short

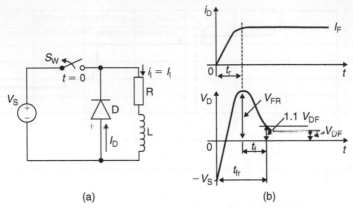

Figure 7.2: Turn on characteristics (a) Circuit (b) Waveforms

time t_r (**rise time**) and the diode voltage drop falls to its steady value after a further time t_f (**fall time**). This is shown in Figure 7.2(b). The diode turn-on time is the time t_{fr}, that comprises $t_r + t_f$. It takes this time t_{fr} for charge to change from one equilibrium state (off) to the other (on).

The total drop V_D reaches a peak forward value V_{FR} that may be from 5 to 20 V, a value much greater than the steady value V_{DF} generally between 0.6 to around 1.2 V. The time t_r for the voltage to reach V_{FR} is usually about 0.1 µs. At a time $t > t_r$, the current i_D becomes constant at I_1 (which will be the forward diode current I_F).

Further, conductivity modulation takes place due to the growth of excess carriers in the semiconductor accompanied by a reduction of resistance. Consequently, the $i_D R_D$ voltage drop reduces. In the equilibrium state, that may take a time of t_f, with a uniform distribution of excess carriers, the voltage drop v_D reaches its minimum steady-state value V_{DF}.

During the turn-on interval t_f, the current is not uniformly distributed so the current density can be high enough in some parts to cause hot spots and possible failure. Accordingly, the rate of rise of current di_D/dt should be limited until the conduction spreads uniformly and the current density decreases. Associated with the high voltage V_{FR} at turn-on, there is high current, so there is extra power dissipation that is not evident from the steady-state model. The turn-on time varies from a few nanoseconds to about 1 ms, depending on the device type.

7.2.1.2 Reverse recovery

Turn-off phenomenon can be explained using Figure 7.3. In Figure 7.3(a) except for the diode, the circuit elements of this simple chopper are considered to be ideal. Switching S_w at a regular frequency, the source of constant voltage, V_s maintains a constant

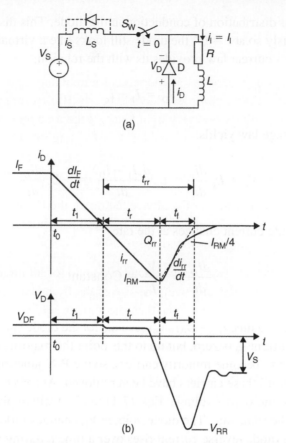

(a)

(b)

Figure 7.3: Diode turn-off (a) Chopper circuit (b) Waveforms

current I_1 in the RL load, because it is assumed that the load time constant (L/R) is long compared with the period of the switching.

While the switch S_w is closed, the load is being charged and the diode should be reverse biased. While the switch S_w is open, the diode D provides a freewheeling path for the load current I_1. The inductance L_S is included for practical reasons and may be the lumped source inductance and snubber inductance, that should have a freewheeling diode to suppress high voltages when the switch is opened.

Let us consider that steady conditions prevail. At the time $t = 0^-$ the switch S_w is open, the load current is $i_1 = I_1$, the diode current is $i_D = I_1 = I_F$, and the voltage drop V_D across the diode is small (about 1 V).

The important concern is what happens after the switch is closed at $t = 0$. Figure 7.3(b) depicts the waveforms of the diode current i_D and voltage v_D. At $t = 0^-$ there was the excess charge carrier distribution of conduction in the diode. This distribution cannot change instantaneously so at $t = 0^+$ the diode still looks like a virtual short circuit, with $v_D = 1$ V. Kirchhoff's current law provides us with the relation:

$$i_D = I_1 - i_s \qquad (7\text{-}1)$$

and Kirchhoff's voltage law yields:

$$V_s = L_s \frac{di_s}{di_t} = L_s \frac{d(I_1 - i_D)}{dt} = -L_s \frac{di_D}{dt} \qquad (7\text{-}2)$$

Accordingly, the diode current changes at the rate,

$$\frac{di_D}{dt} = -\frac{V_s}{L_s} = \text{Constant} \qquad (7\text{-}3)$$

This means that it takes a time $t_1 = L_s(I_1/V_s)$ seconds for the diode current to fall to zero. At the time $t = t_1$ the current i_D is zero, but up to this point the majority carriers have been crossing the junction to become minority carriers, so the P-N junction cannot assume a blocking condition until these carriers have been removed. At zero current, the diode is still a short circuit to the source voltage. Eqs. (7-1) to (7-3) still apply and the current i_D rises above I_1 at the same rate. The diode voltage V_D changes little while the excess carriers remain. The diode reverse current rises over a time t_r during which the excess charge carriers are swept out of the region.

At the end of the interval t_r the reverse current i_D can have risen to a substantial value I_{RR} (peak **r**everse **r**ecovery current), but, by this time, sufficient carriers have been swept out and recombined that current cannot be supported. Therefore, over a fall-time interval t_f the diode current i_D reduces to almost zero very rapidly while the remaining excess carriers are swept out or recombined.

It is during the interval t_f that the potential barrier begins to increase both to block the reverse bias voltage applied by the source voltage as i_D reduces, and to suppress the diffusion of majority carriers because the excess carrier density at the junction is zero. The reverse voltage creates the electric field that allows the depletion layer to acquire space charge and widen.

That is, the electric field causes electrons in the n region to be forced away from the junction towards the cathode and causes holes to be forced away from the junction towards the anode. The blocking voltage v_D can rise above the voltage V_s transiently because of the additional voltage $L_s(di_D/dt)$ as i_D falls to zero over the time t_f.

The sum of the intervals $t_r + t_f = t_{rr}$ is known as the reverse recovery time and it varies generally (between 10 ns to over 1 microsecond) for different diodes. This time is also known as the storage time because it is the time that is taken to sweep out the excess charge Q_{RR} from the silicon by the reverse current. Q_{RR} is a function of $I_D = I_1$, di_D/dt and the junction temperature. It has an effect on the reverse recovery dt current I_{RR} and the reverse recovery time t_{rr}, so it is usually quoted in the data sheets. The fall time t_f can be influenced by the design of the diode. It would seem reasonable to make it short to decrease the turn-off time, but the process is expensive. The bulk of the silicon can be doped with gold or platinum to reduce carrier lifetimes and hence to reduce t_f. The advantage is an increased frequency of switching.

There are two disadvantages associated with this gain in performance. One is an increased on-state voltage drop and the other is an increased voltage recovery overshoot V_{RR}, that is caused by the increased $L_s(di_s/dt)$ as i_D falls more quickly.

Of the two effects, reverse recovery usually results in the greater power loss, and can also generate significant EMI. However these phenomena were considered to be no big deal at 50 or 60 Hz. With the advent of semiconductor power switches, power conversion began to move into the multi-kilohertz range, and faster rectifiers were needed.

The relatively long minority carrier lifetime in silicon (tens of microseconds) causes a lot more charge to be stored than is necessary for effective conductivity modulation. In order to speed up reverse recovery, early "fast" rectifiers used various lifetime killing techniques to reduce the stored minority charge in the lightly doped region. The reverse recovery times of these rectifiers were dramatically reduced, down to about 200 ns, although forward recovery and forward voltage were moderately increased as a side effect of the lifetime killing process. As power conversion frequencies increased to 20 kHz and beyond, there eventually became a growing need for even faster rectifiers, which caused the "epitaxial" rectifier to be developed.

7.2.1.3 Fast and ultra fast rectifiers

The foregoing discussion reveals the importance of the switching parameters such as (i) forward recovery time (t_{fr}), (ii) forward recovery voltage (V_{FR}), (iii) reverse recovery time (t_{rr}), (iv) reverse recovery charge (Q_{rr}), and (v) reverse recovery current I_{RM}, etc., during the transition from forward to reverse and vice versa. With various process improvements fast and ultrafast rectifiers have been achieved within the voltage and current limitations shown in Figure 7.1.

The figure shows that technology is available for devices up to 2000 V ratings and over 1000 A current ratings which are mutually exclusive. In these diodes although cold t_{rr} values are good, at high junction temperature t_{rr} is three to four times higher, increasing switching losses and, in many cases, causing thermal runaway.

Several methods exist to control the switching characteristics of diodes and each leads to a different interdependency of forward voltage drop V_F, blocking voltage V_{RRM} and t_{rr} values. It is these interdependencies (or compromises) that differentiate the ultrafast diodes available on the market today. The important parameters for the turn-on and turn-off behavior of a diode are V_{FR}, V_F, t_{fr}, I_{RM} and t_{rr} and the values vary depending on the manufacturing processes.

Several manufacturers such as IXYS Semiconductors, International Rectifier, etc., manufacture a series of ultra fast diodes, termed *fast recovery epitaxial diodes* (FREDs), which gained wide acceptance during the 1990s. For an excellent description of these components see Burkel and Schneider (1994).

7.2.1.4 Schottky rectifiers

Schottky rectifiers occupy a small corner of the total spectrum of available rectifier voltage and current ratings illustrated in Figure 7.1(b). They are, nonetheless, the rectifier of choice for low voltage switching power supply applications, with output voltages up to a few tens of volts, particularly at high switching frequency. For this reason, Schottkys account for a major segment of today's total rectifier usage. The Schottkys' unique electrical characteristics set them apart from conventional P-N junction rectifiers, in the following important respects:

- Lower forward voltage drop

- Lower blocking voltage

- Higher leakage current

- Virtual absence of reverse recovery charge

The two fundamental characteristics of the Schottky that make it a winner over the P-N junction rectifier in low voltage switching power supplies are its lower forward voltage drop, and virtual absence of minority carrier reverse recovery.

The absence of minority carrier reverse recovery means virtual absence of switching losses within the Schottky itself. Perhaps more significantly, the problem of switching voltage transients and attendant oscillations is less severe for Schottkys than for P-N junction rectifiers. Snubbers are therefore smaller and less dissipative.

The lower forward voltage drop of the Schottky means lower rectification losses, better efficiency, and smaller heat sinks. Forward voltage drop is a function of the Schottky's reverse voltage rating. The maximum voltage rating of today's Schottky rectifiers is about 150 V. At this voltage, the Schottky's forward voltage drop is lower than that of a fast recovery epitaxial P-N junction rectifier by 150 to 200 mV.

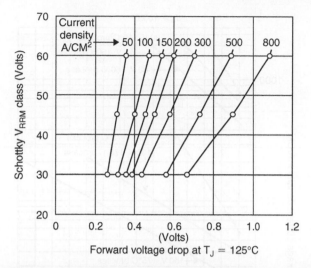

Figure 7.4: Relationships between Schottky V_{RRM} Class and forward voltage drop, for 150°C T_{jmax} class devices (Reproduced by permission of International Rectifier, USA)

At lower voltage ratings, the lower forward voltage drop of the Schottky becomes progressively more pronounced, and more of an advantage. A 45 V Schottky, for example, has a forward voltage drop of 0.4 to 0.6 V, versus 0.85 to 1.0 V for a fast epitaxial P-N junction rectifier. A 15 V Schottky has a mere 0.3 to 0.4 V forward voltage drop.

A conventional fast recovery epitaxial P-N junction rectifier, with a forward voltage drop of 0.9 V, would dissipate about 18 percent of the output power of a 5 V supply. A Schottky, by contrast, reduces rectification losses to the range of 8 to 12 percent. These are the simple reasons why Schottkys are virtually always preferred in low voltage high frequency switching power supplies. For any given current density, the Schottky's forward voltage drop increases as its reverse repetitive maximum voltage (V_{RRM}) increases. The basic hallmarks of any process are its maximum rated junction temperature—the T_{jmax} Class and the "prime" rated voltage, the V_{rrm} Class. These two basic hallmarks are set by the process; they in turn determine the forward voltage drop and reverse leakage current characteristics. Figure 7.4 indicates this condition for T_{jmax} of 150°C.

Leakage current and junction capacitance of Schottky diodes
Figure 7.5 shows the dependence of leakage current on the operating voltage and junction temperature within any given process. Reverse leakage current increases with applied reverse voltage, and with junction temperature. Figure 7.6 shows typical relationship between operating temperature and leakage current, at rated V_{RRM}, for the 150°C/45 V and 175°C/45 V Schottky processes.

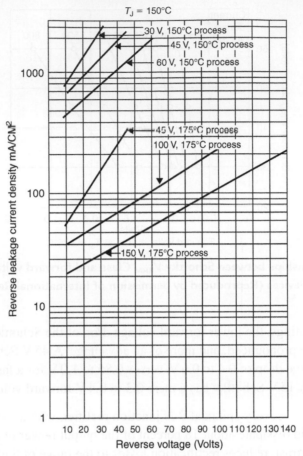

Figure 7.5: Relationships between reverse leakage current density, and applied reverse voltage (Reproduced by permission of International Rectifier, USA)

An important circuit characteristic of the Schottky is its junction capacitance. This is a function of the area and thickness of the Schottky die, and of the applied voltage. The higher the V_{RRM} class, the greater the die thickness and the lower the junction capacitance. This is illustrated in Figure 7.7. Junction capacitance is essentially independent of the Schottky's T_{jmax} Class, and of operating temperature.

7.2.1.5 GaAs power diodes

Efficient power conversion circuitry requires rectifiers that exhibit low forward voltage drop, low reverse recovery current, and fast recovery time. Silicon has been the material of choice for fast, efficient rectification in switched power applications. However, technology is nearing the theoretical limit for optimizing reverse recovery in silicon devices.

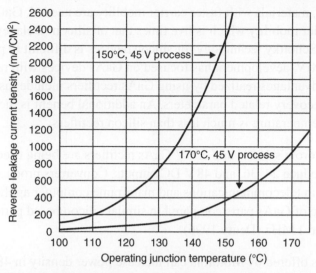

Figure 7.6: Typical relationships between reverse leakage current density, and operating junction temperature (Reproduced by permission of International Rectifier, USA)

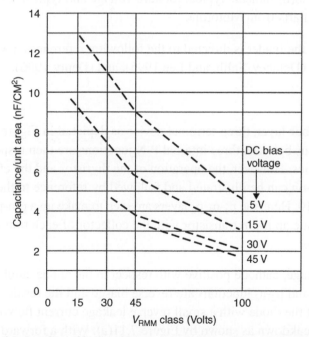

Figure 7.7: Typical Schottky self-capacitance versus V_{RRM} class, measured at various bias voltages (Reproduced by permission of International Rectifier, USA)

To increase speed, materials with faster carrier mobility are needed. Gallium arsenide (GaAs) has a carrier mobility which is five times that of silicon (Delaney, Salih, and Lee 1995). Since Schottky technology for silicon devices is difficult to produce at voltages above 200 V, development has focused on GaAs devices with ratings of 180 V and higher. The advantages realized by using GaAs rectifiers include fast switching and reduced reverse recovery related parameters. An additional benefit is the variation of parameters with temperature is much less than silicon rectifiers.

For example, Motorola's 180 V and 250 V GaAs rectifiers are being used in power converters that produce 24, 36, and 48 V DC outputs. Converters producing 48 V DC, specially popular in telecommunications and mainframe computer applications, could gain the advantage of GaAs parts compared to similar silicon based parts at switching frequencies around 1 MHz (Deuty 1996).

The 180 V devices offered by Motorola can increase power density in 48 V DC applications up to 90 W/in^3 (Ref. 21). These devices allow designers to switch converters at 1 MHz without generating large amounts of EMI.

Figure 7.8(a) and 7.8(b) indicate typical forward voltage and typical reverse current for 20 A, 180 V, GaAs parts from Motorola.

For further details, the reader is directed to the following references: (Ashkianazi, Lorch and Nathan 1995), (Delaney, Salih, and Lee 1995), and (Deuty 1996).

7.2.2 Thyristors

The thyristor is a four-layer, three-terminal device as depicted in Figure 7.9. The complex interactions between three internal P-N junctions are then responsible for the device characteristics. However, the operation of the thyristor and the effect of the gate in controlling turn-on can be illustrated and followed by reference to the two transistor model of Figure 7.10. Here, the p_1-n_1-p_2 layers are seen to make up a p-n-p transistor and the n_2-p_2-n_1 layers create an n-p-n transistor with the collector of each transistor connected to the base of the other.

With a reverse voltage, cathode positive with respect to the anode, applied to the thyristor the p_1-n_1 and p_2-n_2 junctions are reverse biased and the resulting characteristic is similar to that of the diode with a small reverse leakage current flowing up to the point of reverse breakdown as shown by Figure 7.11(a). With a forward voltage, applied and no gate current supported the thyristor is in the forward blocking mode. The emitters of the two transistors are now forward biased and no conduction occurs. As the applied voltage is increased, the leakage current through the transistors increases to the point at which the positive feedback resulting from the base/collector connections

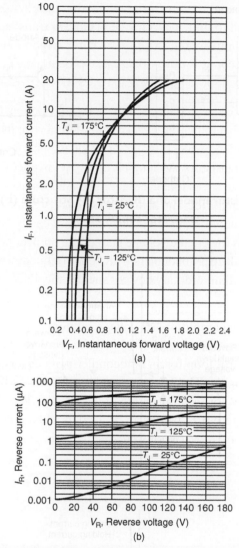

(a)

(b)

**Figure 7.8: Typical characteristics of GaAs power diodes with 20 A, 180 V ratings
(a) Forward voltage (b) Reverse current (Copyright of Motorola, used by permission)**

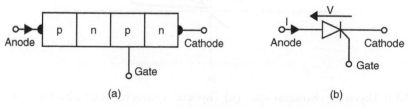

(a)   (b)

Figure 7.9: The thyristor (a) Construction (b) Circuit symbol

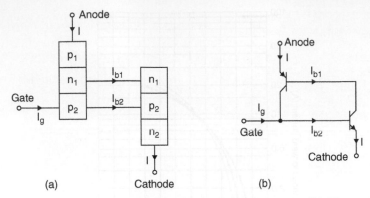

Figure 7.10: The two-transistor model of a thyristor (a) Structure (b) The p-n-p and n-p-n transistor combination

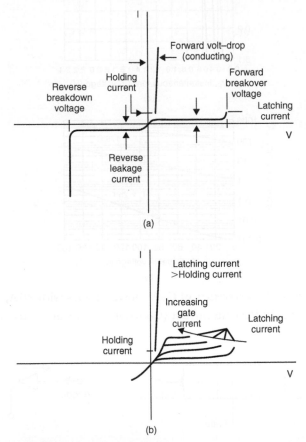

Figure 7.11: Thyristor characteristics (a) Thyristor characteristics with zero gate current (b) Switching characteristics

drives both transistors into saturation, turning them, and hence the thyristor, on. The thyristor is now conducting and the forward voltage drop across it falls to a value of the order of 1 to 2 V. This condition is also shown in the thyristor static characteristic of Figure 7.11(a).

If a current is injected into the gate at a voltage below the breakover voltage, this will cause the n-p-n transistor to turn on. The positive feedback loop will then initiate the turn on of the p-n-p transistor. Once both transistors are on, the gate current can be removed because the action of the positive feedback loop will be to hold both transistors, and hence the thyristor, in the on state.

The effect of the gate current is therefore to reduce the effective voltage at which forward breakover occurs, as illustrated by the Figure 7.11(b). After the thyristor has been turned on, it will continue to conduct as long as the forward current remains above the holding current level, irrespective of gate current or circuit conditions.

7.2.2.1 Ratings and different types of devices

The operation of all power semiconductors is limited by a series of ratings which define the operating boundaries of the device. These ratings include limits on the peak, average and RMS currents, the peak forward and reverse voltages for the devices, maximum rates of change of device current and voltage, device junction temperature and, in the case of the thyristor, the gate current limits.

The current ratings of a power semiconductor are related to the energy dissipation in the device and hence the device junction temperature. The maximum value of on-state current ($I_{av(max)}$) is the maximum continuous current the device can sustain under defined conditions of voltage and current waveform without exceeding the permitted temperature rise in the device. The rms current rating (IRMS) is similarly related to the permitted temperature rise when operating into a regular duty cycle load.

In the case of transient loads, as the internal losses and hence the temperature rise in a power semiconductor are related to the square of the device forward current, the relationship between the current and the permitted temperature rise can be defined in terms of an i^2dt rating for the device. On turn-on, current is initially concentrated into a very small area of the device cross-section and the device is therefore subject to a *di/dt* rating which sets a limit to the permitted rate of rise of forward current.

The voltage ratings of a power semiconductor device are primarily related to the maximum forward and reverse voltages that the device can sustain. Typically, values will be given for the maximum continuous reverse voltage ($V_{RC(max)}$), the maximum repetitive reverse voltage ($V_{RR(max)}$) and the maximum transient reverse voltage ($V_{RT(max)}$). Similar values exist for the forward voltage ratings.

The presence of a fast transient of forward voltage can cause a thyristor to turn on and a *dv/dt* rating is therefore specified for the device. The magnitude of the imposed *dv/dt* can be controlled by the use of a snubber circuit connected in parallel with the thyristor. Data sheets for thyristors always quote a figure for the maximum surge current I_{TSM} that the device can survive.

This figure assumes a half sine pulse with a width of either 8.3 or 10 ms, which are the conditions applicable for 60- or 50-Hz mains, respectively. This limit is not absolute; narrow pulses with much higher peaks can be handled without damage, but little information is available to enable the designer to determine a current rating for short pulses. Hammerton (1989) indicates guidelines in this area.

Ever since its introduction, circuit design engineers have been subjecting the thyristor to increasing levels of operating stress and demanding that these devices perform satisfactorily there. The different stress demands that the thyristor must be able to meet are:

(a) Higher blocking voltages

(b) More current carrying capability

(c) Higher *di/dt*'s

(d) Higher *dv/dt*'s

(e) Shorter turn-off times

(f) Lower gate drive

(g) Higher operating frequencies.

There are many different thyristors available today which can meet one or more of these requirements, but, as always, an improvement in one characteristic is usually only gained at the expense of another. As a result, different thyristors have been optimized for different applications. Modern thyristors can be classified into several general types, namely:

(a) Phase Control Thyristors

(b) Inverter Thyristors

(c) Asymmetrical Thyristors

(d) Reverse Conducting Thyristors (RCT)

(e) Light-Triggered Thyristors.

The voltage and current capabilities of phase control thyristors and inverter thyristors from a power device manufacturer are summarized in Figure 7.12.

Phase control thyristors

Phase control or *converter* thyristors generally operate at line frequency. They are turned off by natural commutation and do not have special fast-switching characteristics.

Current ratings of phase-control thyristors cover the range from a few amperes to about 3500 A, and the voltage ratings from 50 to over 6500 V. To simplify the gate-drive requirement and increase sensitivity, the use of an amplifying gate, which was originally developed for fast switching "inverter" thyristors, is widely adopted in phase control SCR.

Inverter thyristors

The most common feature of an inverter thyristor which distinguishes it from a standard phase control type is that it has fast turnoff time, generally in the range of 5 to 50 μs, depending upon voltage rating. Maximum average current ratings of over 2000 and 1300 A have been achieved with 2000 and 3000 V rated inverter thyristors, respectively.

Inverter thyristors are generally used in circuits that operate from DC supplies, where current in the thyristor is turned off either through the use of auxiliary comutating circuitry, by circuit resonance, or by "load" commutation. Whatever the circuit turn-off mechanism, fast turn-off is important because it minimizes sizes and weight of comutating and/or reactive circuit components.

Asymmetrical thyristors

One of the main salient characteristics of asymmetrical thyristors (ASCR) is that they do not block significant reverse voltage. They are typically designed to have a reverse blocking capability in the range of 400 to 2000 V.

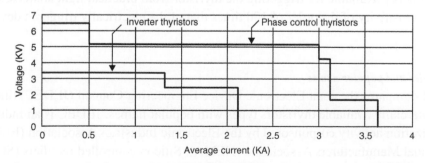

Figure 7.12: Thyristor rating capabilities (Reproduced with permission by GEC Plessey Semiconductors, UK)

The ASCR finds applications in many voltage-fed inverter circuits that require anti-parallel feedback rectifiers that keep the reverse voltage to less than 20 V. The fact that ASCR needs only to block voltage in the forward direction provides an extra degree of freedom in optimizing turn-off time, turn-on time, and forward voltage drop.

Reverse conducting thyristors

The reverse conducting thyristor (RCT) represents the monolithic integration of an asymmetrical thyristor with an anti-parallel rectifier. Beyond obvious advantages of the parts count reduction, the RCT eliminates the inductively induced voltage within the thyristor-diode loop (virtually unavoidable to some extent with separate discrete components). Also, it essentially limits the reverse voltage seen by the thyristor to only the conduction voltage of the diode.

Light-triggered thyristors

Many developments have taken place in the area of light-triggered thyristors. Direct irradiation of silicon with light created electron-hole pairs which, under the influence of an electric field, produce a current that triggers the thyristors.

The turn-on of a thyristor by optical means is an especially attractive approach for devices that are to be used in extremely high-voltage circuits. A typical application area is in switches for DC transmission lines operating in the hundreds of kilovolts range, which use series connections of many devices, each of which must be triggered on command. Optical firing in this application is ideal for providing the electrical isolation between trigger circuits and the thyristor which floats at a potential as high as hundreds of kilovolts above ground.

The main requirement for an optically-triggered thyristor is high sensitivity while maintaining high *dv/dt* and *di/dt* capabilities. Because of the small and limited quantity of photo energy available for triggering the thyristor from practical light sources, very high gate sensitivity of the order of 100 times that of the electrically triggered device is needed.

JEDEC titles and popular names

Table 7.1 compares the Joint Electronic Device Engineering Council (JEDEC) titles for commercially available thyristors types with popular names. JEDEC is an industry standardization activity cosponsored by the Electronic Industries Association (EIA) and the National Manufacturers Association (NEMA). Silicon controlled rectifiers (SCR) are the most widely used as power control elements. Triacs are quite popular in lower current (<40 A) AC power applications.

Table 7.1: Thyristor types and popular names

JEDEC Titles	Popular Names, Types
Reverse Blocking Diode Thyristor	* Four Layer Diode, Silicon Unilateral Switch (SUS)
Reverse Blocking Triode Thyristor	Silicon Controlled Rectifier (SCR)
Reverse Conducting Diode Thyristor	* Reverse Conducting Four Layer Diode
Reverse Conducting Triode Thyristor	Reverse Conducting SCR
Bidirectional Triode Thyristor	Triac
Turn-Off Thyristor	Gate Turn Off Switch (GTO)

* Not generally available

7.3 Gate Turn-Off Thyristors

A gate turn-off thyristor (GTO) is a thyristor-like latching device that can be turned off by application of a negative pulse of current to its gate. This gate turn-off capability is advantageous because it provides increased flexibility in circuit application. It now becomes possible to control power in DC circuits without the use of elaborate commutation circuitry.

Prime design objectives for GTO devices are to achieve fast turn-off time and high current turn-off capability and to enhance the safe operating area during turnoff. Significant progress has been made in both areas during the last few years, largely due to a better understanding of the turn-off mechanisms. The GTO's turn-off occurs by removal of excess holes in the cathode-base region by reversing the current through the gate terminal.

The GTO is gaining popularity in switching circuits, especially in equipment which operates directly from European mains. The GTO offers the following advantages over a bipolar transistor: high blocking voltage capabilities, in excess of 1500 V, and also high over-current capabilities. It also exhibits low gate currents, fast and efficient turn-off, as well as outstanding static and dynamic dv/dt capabilities.

Figure 7.13(a) depicts the symbol of GTO and Figure 7.13(b) shows its two transistor equivalent circuit. Figure 7.13(c) shows a basic drive circuit. The GTO is turned on by a positive gate current, and it is turned off by applying a negative gate cathode voltage.

A practical implementation of a GTO gate drive circuit is shown in Figure 7.14. In this circuit when transistor Q_2 is off, emitter follower transistor Q_1 acts as a current source pumping current into the gate of the GTO through a 12 V zener Z_1 and polarized capacitor C_1. When the control voltage at the base of Q_2 goes positive, transistor Q_2 turns on, while transistor Q_1 turns off since its base now is one diode drop more negative

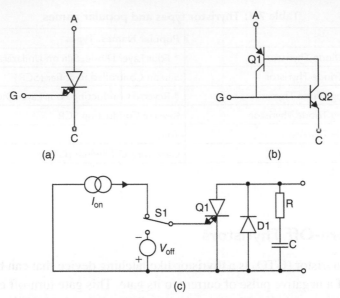

Figure 7.13: GTO symbol, equivalent circuit and basic drive circuit (a) Symbol of GTO
(b) Two-transistor equivalent of GTO (c) Basic drive circuit

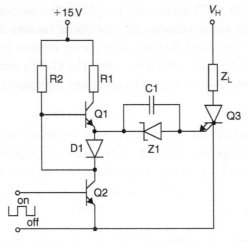

Figure 7.14: Practical realization of a GTO gate drive circuit

than its emitter. At this stage the positive side of capacitor C1 is essentially grounded, and C1 will act as a voltage source of approximately 10 V, turning the GTO off. Isolated gate drive circuits may also be easily implemented to drive the GTO.

With improved cathode emitter geometries and better optimized vertical structures, today's GTOs have made significant progress in turn-off performance (the prime

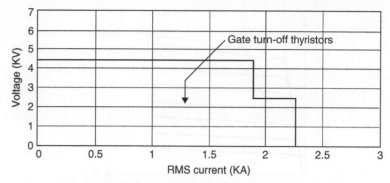

Figure 7.15: Ratings covered by available GTOs (Reproduced by permission of GEC Plessey Semiconductors, UK)

weakness of earlier day GTOs). Figure 7.15 shows the available GTO ratings and, as can be seen, they cover quite a wide spectrum. However, the main applications lie in the higher voltage end (>1200 V) where bipolar transistors and power MOSFETs are unable to compete effectively. In the present day market there are GTOs with current ratings over 3000 A and voltage ratings over 4500 V. For further details on GTOs see Coulbeck, Findlay, and Millington (1994) and Bassett and Smith (1989).

7.4 Bipolar Power Transistors

During the last two decades, attention has been focused on high-power transistors as switching devices in inverters, SMPS and similar switching applications. New devices with faster switching speeds and lower switching losses are being developed that offer performance beyond that of thyristors. With their faster speed, they can be used in an inverter circuit operating at frequencies over 200 KHz. In addition, these devices can be readily turned off with a low-cost reverse base drive without the costly commutation circuits required by thyristors.

7.4.1 Bipolar Transistor as a Switch

The bipolar transistor is essentially a current-driven device. That is, by injecting a current into the base terminal a flow of current is produced in the collector. There are essentially two modes of operation in a bipolar transistor: the linear and saturating modes. The linear mode is used when amplification is needed, while the saturating mode is used to switch the transistor either on or off.

Figure 7.16 shows the V-I characteristic of a typical bipolar transistor. Close examination of these curves shows that the saturation region of the V-I curve is of interest when the

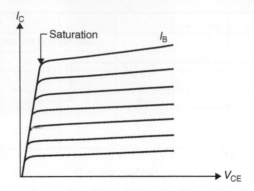

Figure 7.16: Typical output characteristics of BJT

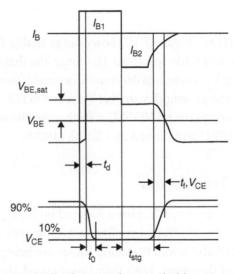

Figure 7.17: Bipolar transistor switching waveforms

transistor is used in a switching mode. At that region a certain base current can switch the transistor on, allowing a large amount of collector current to flow, while the collector-to-emitter voltage remains relatively small.

In actual switching applications a base drive current is needed to turn the transistor on, while a base current of reverse polarity is needed to switch the transistor back off. In practical switching operations certain delays and storage times are associated with transistors. In the following section are some parameter definitions for a discrete bipolar transistor driven by a step function into a resistive load.

Figure 7.17 illustrates the base-to-emitter and collector-to-emitter waveforms of a bipolar NPN transistor driven into a resistive load by a base current pulse I_B. The following are the definitions associated with these waveforms:

Delay Time, t_d

Delay time is defined as the interval of time from the application of the base drive current I_{BI} to the point at which the collector-emitter voltage V_{CE} has dropped to 90 percent of its initial off value.

Rise Time, t_r

Rise time is defined as the interval of time it takes the collector-emitter voltage V_{CE} to drop to 10 percent from its 90 percent off value.

Storage Time, t_{stg}

Storage time is the interval of time from the moment reverse base drive I_{B2} is applied to the point where the collector-emitter voltage V_{CE} has reached 10 percent of its final off value.

Fall Time, $t_{f,VCE}$

Fall time is the time interval required for the collector-emitter voltage to increase from 10 to 90 percent of its off value.

7.4.2 Inductive Load Switching

In the previous section, the definitions for the switching times of the bipolar transistor were made in terms of collector-emitter voltage. Since the load was defined to be a resistive one, the same definitions hold true for the collector current. However, when the transistor drives an inductive load, the collector voltage and current waveforms will differ. Since current through an inductor does not flow instantaneously with applied voltage, during turn-off, one expects to see the collector-emitter voltage of a transistor rise to the supply voltage before the current begins to fall. Thus, two fall time components may be defined, one for the collector-emitter voltage $t_{f,VCE}$, and the other for the collector current $t_{f,Ic}$ Figure 7.18 shows the actual waveforms.

Observing the waveforms we can define the collector-emitter fall time $t_{f,VCE}$ in the same manner as in the resistive case, while the collector fall time $t_{f,Ic}$ may be defined as the interval in which collector current drops from 90 to 10 percent of its initial value. Normally, the load inductance L behaves as a current source, and therefore it charges the base-collector transition capacitance faster than the resistive load. Thus, for the same base and collector currents the collector-emitter voltage fall time $t_{f,VCE}$ is shorter for the inductive circuit.

7.4.3 Safe Operating Area and V-I Characteristics

The output characteristics (I_C versus V_{CE}) of a typical npn power transistor are shown in Figure 7.19(a). The various curves are distinguished from each other by the value of the base current.

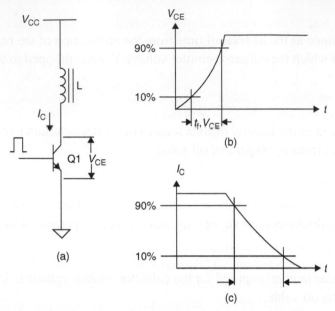

Figure 7.18: A bipolar switching transistor driving an inductive load with associated fall time waveforms (a) Circuit (b) Voltage waveform (c) Current waveform

Several features of the characteristics should be noted. First, there is a maximum collector-emitter voltage that can be sustained across the transistor when it is carrying substantial collector current. The voltage is usually labeled BV_{SUS}. In the limit of zero base current, the maximum voltage between collector and emitter that can be sustained increases somewhat to a value labeled BV_{CEO}, the collector-emitter breakdown voltage when the base is open circuited. This latter voltage is often used as the measure of the transistor's voltage standoff capability because usually the only time the transistor will see large voltages is when the base current is zero and the BJT is in cutoff.

The voltage BV_{CBO} is the collector-base breakdown voltage when the emitter is open circuited. The fact that this voltage is larger than BV_{CEO} is used to advantage in so-called open-emitter transistor turn-off circuits.

The region labeled *primary breakdown* is due to conventional avalanche breakdown of the collector-base junction and the attendant large flow of current. This region of the characteristics is to be avoided because of the large power dissipation that clearly accompanies such breakdown.

The region labeled *second breakdown* must also be avoided because large power dissipation also accompanies it, particularly at localized sites within the semiconductor. The origin of second breakdown is different from that of avalanche breakdown and will be considered in detail later in this chapter. BJT failure is often associated with second breakdown.

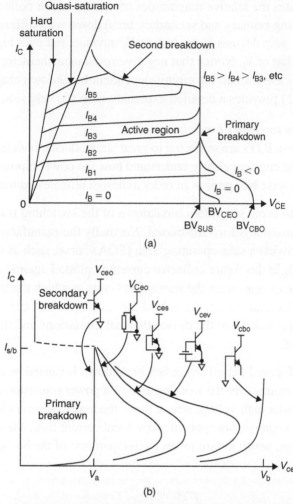

Figure 7.19: Current-voltage characteristics of a NPN power transistor showing breakdown phenomenon (a) indication of quasi saturation (b) relative primary and secondary breakdown conditions for different bias levels

The major observable difference between the I-V characteristics of a power transistor and those of a logic level transistor is the region labeled quasi-saturation on the power transistor characteristics of Figure 7.19(a). Quasi-saturation is a consequence of the lightly doped collector drift region found in the power transistor.

Logic level transistors do not have this drift region and so do not exhibit quasi-saturation. Otherwise all of the major features of the power transistor characteristic are also found on those of logic level devices.

Figure 7.19(b) indicates the relative magnitudes of npn transistor collector breakdown characteristics, showing primary and secondary breakdown with different base bias conditions. With low gain devices V_a is close to V_b in value, but with high gain devices V_b may be 2 to 3 times that of V_a. Notice that negative resistance characteristics occur after breakdown, as is the case with all the circuit-dependent breakdown characteristics. (B. W. Williams 1992) provides a detailed explanation on the behavior.

7.4.3.1 Forward-bias secondary breakdown

In the switching process **BJT**s are subjected to great stress, during both turn-on and turn-off. It is imperative that the engineer clearly understand how the power bipolar transistor behaves during forward and reverse bias periods in order to design reliable and trouble-free circuits.

The first problem is to avoid secondary breakdown of the switching transistor at turn-on, when the transistor is forward-biased. Normally the manufacturer's specifications will provide a safe-operating area (SOA) curve, such as the typical one shown in Figure 7.20. In this figure collector current is plotted against collector-emitter voltage. The curve locus represents the maximum limits at which the transistor may be operated. Load lines that fall within the pulsed forward-bias SOA curve during turn-on are considered safe, provided that the device thermal limitations and the SOA turn-on time are not exceeded.

The phenomenon of forward-biased secondary breakdown is caused by hot spots which are developed at random points over the working area of a power transistor, caused by unequal current conduction under high-voltage stress. Since the temperature coefficient of the base-to-emitter junction is negative, hot spots increase local current flow. More current means more power generation, which in turn raises the temperature of the hot spot even more.

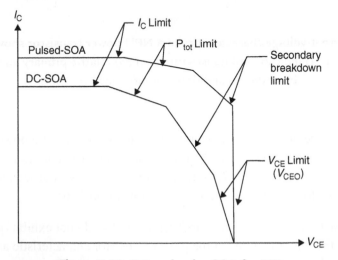

Figure 7.20: DC and pulse SOA for BJT

Since the temperature coefficient of the collector-to-emitter breakdown voltage is also negative, the same rules apply. Thus is the voltage stress is not removed, ending the current flow, the collector-emitter junction breaks down and the transistor fails because of thermal runaway.

7.4.3.2 Reverse-bias secondary breakdown

It was mentioned in previous paragraphs that when a power transistor is used in switching applications, the storage time and switching losses are the two most important parameters with which the designer has to deal extensively.

On the other hand the switching losses must also be controlled since they affect the overall efficiency of the system. Figure 7.21 shows turn-off characteristics of a high-voltage power transistor in resistive and inductive loads. Inspecting the curves we can see

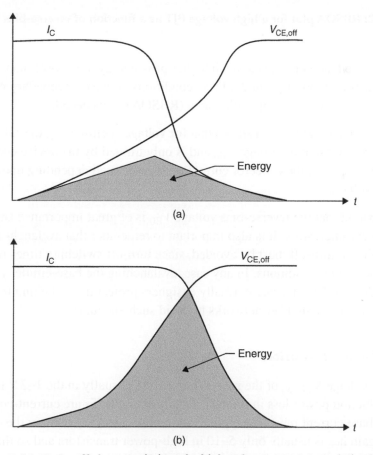

Figure 7.21: Turn-off characteristics of a high voltage BJT (a) Resistive load (b) Inductive load

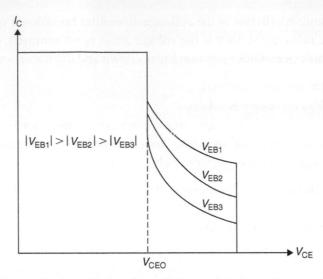

Figure 7.22: RBSOA plot for a high voltage BJT as a function of reverse-bias voltage V_{EB}

that the inductive load generates a much higher peak energy at turn-off than its resistive counterpart. It is then possible, under these conditions, to have a secondary breakdown failure if the reverse-bias safe operating area (RBSOA) is exceeded.

The RBSOA curve (Figure 7.22) shows that for voltages below V_{CEO} the safe area is independent of reverse-bias voltage V_{EB} and is only limited by the device collector current IC. Above V_{CEO} the collector current must be derated depending upon the applied reverse-bias voltage.

It is then apparent that the reverse-bias voltage V_{EB} is of great importance and its effect on RBSOA very interesting. It is also important to remember that avalanching the base-emitter junction at turn-off must be avoided, since turn-off switching times may be decreased under such conditions. In any case, avalanching the base-emitter junction may not be considered relevant, since normally designers protect the switching transistors with either clamp diodes or snubber networks to avoid such encounters.

7.4.4 Darlington Transistors

The on-state voltage $V_{CE(sat)}$ of the power transistors is usually in the 1–2 V range, so that the conduction power loss in the BJT is quite small. BJTs are current-controlled devices and base current must be supplied continuously to keep them in the on-state. The DC current gain h_{FE} is usually only 5–10 in high-power transistors and so these devices are sometimes connected in a Darlington or triple Darlington configuration as is shown in Figure 7.23 to achieve a larger current gain. However, some disadvantages accrue in

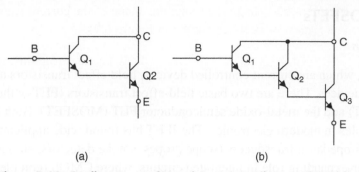

(a) (b)

Figure 7.23: Darlington configurations (a) Darlington (b) Triple Darlington

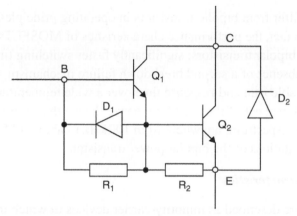

Figure 7.24: A practical monolithic Darlington pair

this configuration including slightly higher overall $V_{CE(sat)}$ values and slower switching speeds. The current gain of the pair h_{FE} is

$$h_{FE} = h_{FE1} \cdot h_{FE2} + h_{FE1} + h_{FE2} \qquad (7\text{-}4)$$

Darlington configurations using discrete BJTs or several transistors on a single chip (a Monolithic Darlington (MD)) have significant storage time during the turn-off transition. Typical switching times are in the range of a few hundred nanoseconds to a few microseconds.

BJTs including MDs are available in voltage ratings up to 1400 V and current ratings of a few hundred amperes. In spite of a negative temperature coefficient of on-state resistance, modern BJTs fabricated with good quality control can be paralleled provided that care is taken in the circuit layout and that some extra current margin is provided.

Figure 7.24 shows a practical monolithic Darlington pair with diode D_1 added to speed up the turn-off time of Q_1 and D_2 added for half and full bridge circuit applications. Resistors R_1 and R_2 are low value resistors and provide a leakage path for Q_1 and Q_7.

7.5 Power MOSFETs

7.5.1 Introduction

Compared to BJTs, which are current controlled devices, field effect transistors are voltage controlled devices. There are two basic field-effect transistors (FETs): the junction FET (JFET) and the metal-oxide semiconductor FET (MOSFET). Both have played important roles in modern electronics. The JFET has found wide application in such cases as high-impedance transducers (scope probes, smoke detectors, etc.) and the MOSFET in an ever-expanding role in integrated circuits, where CMOS (complementary MOS) is perhaps the most well-known.

Power MOSFETs differ from bipolar transistors in operating principles, specifications, and performance. In fact, the performance characteristics of MOSFETs are generally superior to those of bipolar transistors: significantly faster switching time, simpler drive circuitry, the absence of a second breakdown failure mechanism, the ability to be paralleled, and stable gain and response time over a wide temperature range. The MOSFET was developed out of the need for a power device that could work beyond the 20-kHz frequency spectrum, anywhere from 100 kHz to above 1 MHz, without experiencing the limitations of the bipolar power transistor.

7.5.2 General Characteristics

Bipolar transistors are described as minority-carrier devices in which injected minority carriers recombine with majority carriers. A drawback of recombination is that it limits the device's operating speed. Current-driven base-emitter input of a bipolar transistor presents a low-impedance load to its driving circuit. In most power circuits, this low-impedance input requires somewhat complex drive circuitry.

By contrast, a power MOSFET is a voltage-driven device whose gate terminal is electrically isolated from its silicon body by a thin layer of silicon dioxide (SiO_2). As a majority-carrier semiconductor, the MOSFET operates at much higher speed than its bipolar counterpart because there is no charge-storage mechanism. A positive voltage applied to the gate of an n-type MOSFET creates an electric field in the channel region beneath the gate; that is, the electric charge on the gate causes the p-region beneath the gate to convert to an n-type region, as shown in Figure 7.25(a).

This conversion, called the surface-inversion phenomenon, allows current to flow between the drain and source through an n-type material. In effect, the MOSFET ceases to be an n-p-n device when in this state. The region between the drain and source can be represented as a resistor, although it does not behave linearly, as a conventional resistor would. Because of this surface-inversion phenomenon, then, the operation of a MOSFET is entirely different from that of a bipolar transistor.

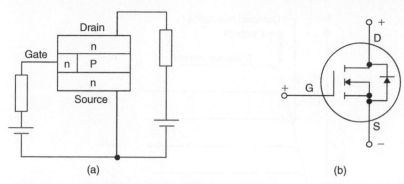

Figure 7.25: Structure of N-channel MOSFET and symbol (a) Structure (b) Symbol

By virtue of its electrically isolated gate, a MOSFET is described as a high-input impedance, voltage-controlled device, compared to a bipolar transistor. As a majority-carrier semiconductor, a MOSFET stores no charge, and so can switch faster than a bipolar device. Majority-carrier semiconductors also tend to slow down as temperature increases. This effect brought about by another phenomenon called carrier mobility makes a MOSFET more resistive at elevated temperatures, and much more immune to the thermal runaway problem experienced by bipolar devices. Mobility is a term that defines the average velocity of a carrier in terms of the electrical field imposed on it.

A useful byproduct of the MOSFET process is the internal parasitic diode formed between source and drain, Figure 7.25(b). (There is no equivalent for this diode in a bipolar transistor other than in a bipolar Darlington transistor.) Its characteristics make it useful as a clamp diode in inductive-load switching.

Different manufacturers use different techniques for constructing a power FET, and names like HEXFET, VMOS, TMOS, etc., have become trademarks of specific companies.

7.5.3 MOSFET Structures and On Resistance

Most power MOSFETs are manufactured using various proprietary processes by various manufacturers on a single silicon chip structured with a large number of closely packed identical cells. For example, Harris Power MOSFETs are manufactured using a vertical double-diffused process, called VDMOS or simply DMOS. In these cases, a 120-mil^2 chip contains about 5,000 cells and a 240-mil^2 chip has more than 25,000 cells.

One of the aims of multiple-cells construction is to minimize the MOSFET parameter $R_{DS(ON)}$ when the device is in the on-state. When $R_{DS(ON)}$ is minimized, the device provides superior power-switching performance because the voltage drop from drain to source is also minimized for a given value of drain-source current. Reference 6 provides more details.

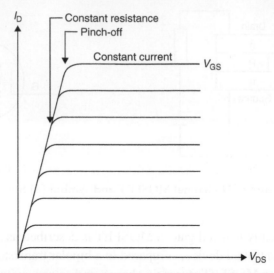

Figure 7.26: Typical output characteristics of a MOSFET

7.5.4 I-V Characteristics

Figure 7.26 shows the drain-to-source operating characteristics of the power MOSFET. Although the curve is similar to the case of the bipolar power transistor (Figure 7.15), there are some fundamental differences.

The MOSFET output characteristic curves reveal two distinct operating regions, namely, a "constant resistance" and a "constant current." Thus, as the drain-to-source voltage is increased, the drain current increases proportionally, until a certain drain-to-source voltage called *pinch off* is reached. After pinch off, an increase in drain-to-source voltage produces a constant drain current.

When the power MOSFET is used as a switch, the voltage drop between the drain and source terminals is proportional to the drain current; that is, the power MOSFET is working in the constant resistance region, and therefore it behaves essentially as a resistive element. Consequently the on-resistance $R_{DS(ON)}$ of the power MOSFET is an important figure of merit because it determines the power loss for a given drain current, just as $V_{CE,sat}$ is of importance for the bipolar power transistor.

By examining Figure 7.26, we note that the drain current does not increase appreciably when a gate-to-source voltage is applied; in fact, drain current starts to flow after a threshold gate voltage has been applied, in practice somewhere between 2 and 4 V. Beyond the threshold voltage, the relationship between drain current and gate voltage is approximately equal. Thus, the transconductance g_{fs}, which is defined as the rate of change of drain current to gate voltage, is practically constant at higher values of drain

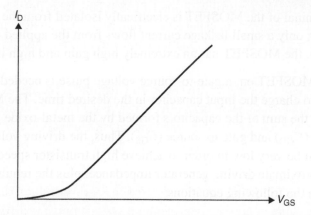

Figure 7.27: Transfer characteristics of a power MOSFET

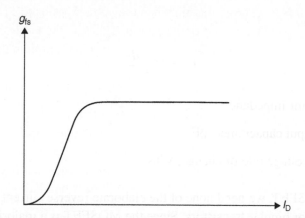

Figure 7.28: Relationship of transconductance (gfs) to ID of a power MOSFET

current. Figure 7.27 illustrates the transfer characteristics of I_D versus V_{DS}, while Figure 7.28 shows the relationship of transconductance g_{fs} to drain current.

It is now apparent that a rise in transconductance results in a proportional rise in the transistor gain, i.e., larger drain current flow, but unfortunately this condition swells the MOSFET input capacitance. Therefore, carefully designed gate drivers must be used to deliver the current required to charge the input capacitance in order to enhance the switching speed of the MOSFET.

7.5.5 Gate Drive Considerations

The MOSFET is a voltage controlled device; that is, a voltage of specified limits must be applied between gate and source in order to produce a current flow in the drain.

Since the gate terminal of the MOSFET is electrically isolated from the source by a silicon oxide layer, only a small leakage current flows from the applied voltage source into the gate. Thus, the MOSFET has an extremely high gain and high impedance.

In order to turn a MOSFET on, a gate-to-source voltage pulse is needed to deliver sufficient current to charge the input capacitor in the desired time. The MOSFET input capacitance C_{iss} is the sum of the capacitors formed by the metal-oxide gate structure, from gate to drain (C_{GD}) and gate to source (C_{GS}). Thus, the driving voltage source impedance R_g must be very low in order to achieve high transistor speeds. A way of estimating the approximate driving generator impedance, plus the required driving current, is given in the following equations:

$$R_g = \frac{t_r (\text{or } t_f)}{2.2 C_{iss}} \tag{7-5}$$

and

$$I_g = C_{iss} \cdot \frac{dv}{dt} \tag{7-6}$$

where R_g = generator impedance,

C_{iss} = MOSFET input capacitance, pF

dv/dt = generator voltage rate of change, V/ns

To turn off the MOSFET, we need none of the elaborate reverse current generating circuits described for bipolar transistors. Since the MOSFET is a majority carrier semiconductor, it begins to turn off immediately upon removal of the gate-to-source voltage. Upon removal of the gate voltage the transistor shuts down, presenting a very high impedance between drain and source, thus inhibiting any current flow, except leakage currents (in microamperes).

Figure 7.29 illustrates the relationship of drain current versus drain-to-source voltage. Note that drain current starts to flow only when the drain-to-source avalanche voltage is exceeded, while the gate-to-source voltage is kept at 0 V.

7.5.6 Temperature Characteristics

The high operating temperatures of bipolar transistors are a frequent cause of failure. The high temperatures are caused by hot spotting, the tendency of current in a bipolar device to concentrate in areas around the emitter. Unchecked, this hot spotting results in the mechanism of thermal runaway, and eventual destruction of the device. MOSFETs do not

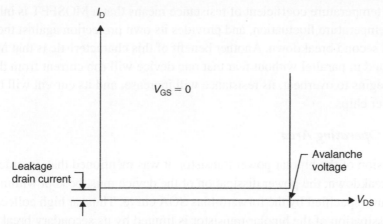

Figure 7.29: Drain-to-source blocking characteristics of the MOSFET

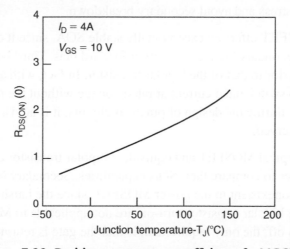

Figure 7.30: Positive temperature coefficient of a MOSFET

suffer this disadvantage because their current flow is in the form of majority carriers. The mobility of majority carriers in silicon decreases with increasing temperature.

This inverse relationship dictates that the carriers slow down as the chip gets hotter. In effect, the resistance of the silicon path is increased, which prevents the concentrations of current that lead to hot spots. In fact, if hot spots do attempt to form in a MOSFET, the local resistance increases and defocuses or spreads out the current, rerouting it to cooler portions of the chip.

Because of the character of its current flow, a MOSFET has a positive temperature coefficient of resistance, as shown by the curves of Figure 7.30.

The positive temperature coefficient of resistance means that a MOSFET is inherently stable with temperature fluctuation, and provides its own protection against thermal runaway and second breakdown. Another benefit of this characteristic is that MOSFETs can be operated in parallel without fear that one device will rob current from the others. If any device begins to overheat, its resistance will increase, and its current will be directed away to cooler chips.

7.5.7 Safe Operating Area

In the discussion of the bipolar power transistor, it was mentioned that, in order to avoid secondary breakdown, the power dissipation of the device must be kept within the operating limits specified by the forward-bias SOA curve. Thus, at high collector voltages the power dissipation of the bipolar transistor is limited by its secondary breakdown to a very small percentage of full rated power. Even at very short switching periods, the SOA capability is still restricted, and the use of snubber networks is incorporated to relieve transistor switching stress and avoid secondary breakdown.

In contrast, the MOSFET offers an exceptionally stable SOA, since it does not suffer from the effects of secondary breakdown during forward bias. Thus, both the DC and pulsed SOA are superior to that of the bipolar transistor. In fact, with a power MOSFET it is quite possible to switch rated current at rated voltage without the need of snubber networks. Of course, during the design of practical circuits, it is advisable that certain derating must be observed.

Figure 7.31 shows typical MOSFET and equivalent bipolar transistor curves superimposed in order to compare their SOA capabilities. Secondary breakdown during reverse bias is also nonexistent in the power MOSFET, since the harsh reverse-bias schemes used during bipolar transistor turn-off are not applicable to MOSFETs. Here, for the MOSFET to turn off, the only requirement is that the gate is returned to 0 V.

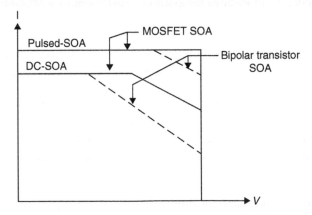

Figure 7.31: SOA curves for power MOSFET

7.5.8 Practical Components

7.5.8.1 High voltage and low on resistance devices

Denser geometries, processing innovations, and packaging improvements are resulting in power MOSFETs that have ever-higher voltage ratings and current-handling capabilities, as well as volumetric power-handling efficiency. See Travis (1989), Goodenough (1995), and Goodenough (1994) for more information. Bipolar transistors have always been available with very high voltage ratings, and those ratings do not carry onerous price penalties. Achieving good high-voltage performance in power MOSFETs, however, has been problematic, for several reasons.

First, the $R_{\text{DS(ON)}}$ of devices of equal silicon area increases exponentially with the voltage rating. To get the on-resistance down, manufacturers would usually pack more parallel cells onto a die. But this denser packing causes problems in high-voltage performance. Propagation delays across a chip as well as silicon defects, can lead to unequal voltage stresses and even to localized breakdown.

Manufacturers resort to a variety of techniques to produce high voltage ($>1000\,\text{V}$), low-$R_{\text{DS(ON)}}$ power MOSFETs that offer reasonable yields. Advanced Power Technology (APT), for example, deviates from the trend toward smaller and smaller feature sizes in its quest for low on-resistance. Instead, the company uses large dies to get $R_{\text{DS(ON)}}$ down. APT manufactures power MOSFETs using dies as large as 585×738 mil and reaching voltage ratings as high as $1000\,\text{V}$. APT10026JN, a device from their product range has a current rating of $1000\,\text{V}$ with $690\,\text{W}$ power rating. On resistance of the device is $0.26\,\Omega$.

When devices are aimed out to low power applications such as laptop and notebook computers, personal digital assistants (PDA), etc. extremely low $R_{\text{DS(ON)}}$ values from practical devices are necessary.

Specific on-resistance of double-diffused MOSFETs (DMOSFETs), more commonly known as power MOSFETs, has continually shrunk over the past two decades. In other words, the $R_{\text{DS(ON)}}$ per unit area has dropped. The reduced size with regard to low-voltage devices (those rated for a maximum drain-to-source voltage V_{DS} of under $100\,\text{V}$) was achieved by increasing the cell density.

Most power-MOSFET suppliers now offer low-voltage FETs from processes that pack 4,000,000 to 8,000,000 cells/in^2, in which each cell is an individual MOSFET. The drain, gate, and source terminals of all the cells are connected in parallel. Manufacturers such as International Rectifier (IOR) have developed many generations of MOSFETs based on DMOS technology. For example, the HEXFET family from IOR has gone through five generations, gradually increasing the number of cells per in^2 with almost tenfold decrease in the $R_{\text{DS(ON)}}$ parameter as per Figure 7.32. $R_{\text{DS(ON)}}$ times the device die area in this figure is a long-used figure of merit (FOM) for power semiconductors. This is called *specified on resistance*. For details Kinzer (1995) is suggested.

Because generation 5 die are smaller than the previous generation, there is room within the same package to accommodate additional devices such as a Schottky diode. The FETKY family from IOR (Davis 1997) which uses this concept of integrating a MOSFET with a Schottky diode is aimed at power converter applications such as synchronous regulators, etc.

In designing their DMOSFETs, Siliconix borrowed a DRAM process technique called the trench gate. They then developed a low voltage DMOSFET process that provides 12,000,000 cells/in^2, and offers lower specific on-resistance—$R_{DS(ON)}$ —than present planar processes.

The first device from the process, the n-channel Si4410DY, comes in Siliconix's data book "Little Foot" 8-pin DIP. The Si4410DY sports a maximum on-resistance of 13.5 mΩ, enhanced with 10 V of gate-to-source voltage (V_{GS}). At a V_{GS} of 4.5 V, $R_{DS(ON)}$ nearly doubles, reaching a maximum value of 20 mΩ.

During the year 1997 Temic Semiconductors (formerly Siliconix) has further improved their devices to carry 32 million cells/in^2 using their TrenchFET technology. These devices come in two basic families, namely (i) Low on-resistance devices and (ii) Low threshold devices. Maximum on resistance was reduced to 9 mΩ and 13 mΩ, compared to

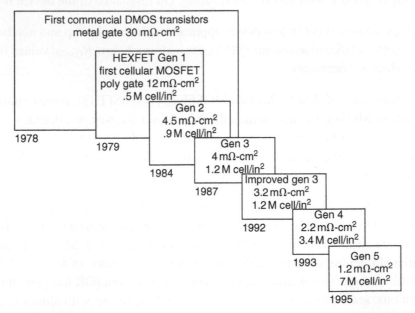

Figure 7.32: The power MOSFET generations (Courtesy: International Rectifier)

the case of Si 4410DY devices with corresponding values of $13.5\,m\Omega$ (for V_{GS} of $10\,V$) and $20\,m\Omega$ (for V_{GS} of $4.5\,V$), for these low on-resistance devices. In the case of the low threshold family, these values were $10\,m\Omega$ and $14\,m\Omega$ for gate source threshold values of $4.5\,V$ and $7.5\,V$, respectively. For details Goodenough (1997) is suggested.

7.5.8.2 P-Channel MOSFETs

Historically, p-channel FETs were not considered as useful as their n-channel counterparts. The higher resistivity of p-type silicon, resulting from its lower carrier mobility, put it at a disadvantage compared to n-type silicon.

Due to the approximately 2:1 superior mobility on n-type devices, n-channel power FETs dominate the available devices, because they need about half the area of silicon for a given current or voltage rating. However, as the technology matures, and with the demands of power-management applications, p-channel devices are starting to become available.

They make possible power CMOS designs and eliminate the need for special high-side drive circuits. When a typical n-channel FET is employed as a high-side switch running off a plus supply rail with its source driving the load, the gate must be pulled at least $10\,V$ above the drain. A p-channel FET has no such requirement. A high-side p-channel MOSFET and a low-side n-channel MOSFET tied with common drains make a superb high-current "CMOS equivalent" switch.

Because on-resistance rises rapidly with device voltage rating, it was only recently (1994/1995) high voltage p-channel power MOSFETs were introduced commercially. One such device is IXTH11P50 from IXYS Semiconductors with a voltage rating of $500\,V$ and current rating of $11\,A$ and an on-resistance of $900\,m\Omega$.

Such high-current devices eliminate the need to parallel many lower-current FETs. These devices make possible complementary high-voltage push-pull circuits and simplified half-bridge and H bridge motor drives.

Recently introduced low voltage p MOSFETs from the TrenchFET family of Temic Semiconductors (Goodenough, 1997) have typical $R_{DS(ON)}$ values between $14\,mN$ to $25\,mN$.

7.5.8.3 More advanced power MOSFETs

With the advancement of processing capabilities, industry benefits with more advanced power MOSFETs such as:

(a) Current-sensing MOSFETs

(b) Logic-level MOSFETs

(c) Current-limiting MOSFETs

(d) Voltage-clamping, current-limiting MOSFETs

The technique of current mirroring for source current-sensing purposes involves connecting a small fraction of the cells in a power MOSFET to a separate sense terminal. The current in this terminal (see Figure 7.33(a) is a fixed fraction of the source current feeding the load. Current sense lead provides an accurate fraction of the drain current that can be used as a feedback signal for control and/or protection.

It's also valuable if you must squeeze the maximum switching speed from a MOSFET. For example, you can use the sense terminal to eliminate the effects of source-lead inductance in high-speed switching applications. Several manufacturers such as Harris, IXYS, Phillips, etc. manufacture these components.

Another subdivision of the rapidly diversifying power-MOSFET market is a class of devices called logic-level FETs. Before the advent of these units, drive circuitry had to supply gate-source turn-on levels of 10 V or more. The logic-level MOSFETs accept drive signals from CMOS or TTL ICs that operate from a 5 V supply. Suppliers of these types include International Rectifier, Harris, IXYS, Phillips-Amperex, and Motorola. Similarly other types described under (b) and (d) above are also available in monolithic form and some of there devices are categorized under "Intelligent Discretes."

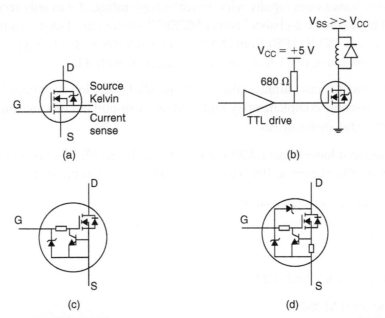

**Figure 7.33: Advanced Monolithic MOSFETs (a) Current-sensing MOSFET
(b) Application of a logic-level MOSFET (c) Current-limiting MOSFET
(d) Voltage-clamping, current-limiting MOSFET**

7.6 Insulated Gate Bipolar Transistor (IGBT)

MOSFETs have become increasingly important in discrete power device applications due primarily to their high input impedance, rapid switching times, and low on-resistance. However, the on-resistance of such devices increases with increasing drain-source voltage capability, thereby limiting the practical value of power MOSFETs to application below a few hundred volts.

To make use of the advantages of power MOSFETs and BJTs together a newer device, insulated gate bipolar transistor (IGBT), has been introduced recently. With the voltage-controlled gate and high-speed switching of a MOSFET and the low saturation voltage of a bipolar transistor, the IGBT is better than either device in many high power applications. It is a composite of a transistor with an n-Channel MOSFET connected to the base of the pnp transistor.

Figure 7.34(a) shows the symbol and Figure 7.34(b) shows the equivalent circuit. Typical IGBT characteristics are shown in Figure 7.34(c). Physical operation of the

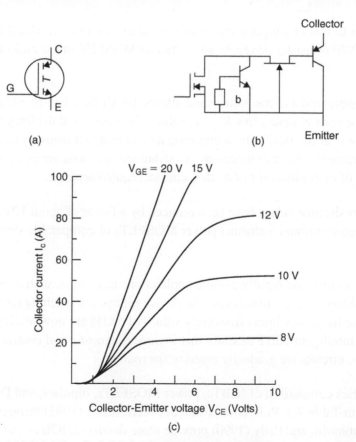

Figure 7.34: IGBT (a) Symbol (b) Equivalent circuit (c) Typical output characteristics

Table 7.2: Characteristics comparison of IGBTs, power MOSFETs, bipolars, and Darlingtons

	Power MOSFETs	IGBTs	Bipolars	Darlingtons
Type of Drive	Voltage	Voltage	Current	Current
Drive Power	Minimal	Minimal	Large	Medium
Drive Complexity	Simple	Simple	High (large positive and negative currents are required)	Medium
Current Density for Given Voltage Drop	High at Low Voltage—Low at High Voltages	Very High (small trade-off with switching speed)	Medium (severe trade-off with switching speed)	Low
Switching Losses	Very Low	Low to Medium (depending on trade-off with conduction losses)	Medium to High (depending on trade-off with conduction losses)	High

IGBT is closer to that of a bipolar transistor than to that of a power MOSFET. The IGBT consists of a PNP transistor driven by an n-channel MOSFET in a pseudo-Darlington configuration.

The JFET supports most of the voltage and allows the MOSFET to be a low voltage type, and consequently have a low $R_{DS(ON)}$ value. The absence of the integral reverse diode gives the user the flexibility of choosing an external fast recovery diode to match a specific requirement. This feature can be an advantage or a disadvantage, depending on the frequency of operation, cost of diodes, current requirement, etc.

In IGBTs on-resistance values have been reduced by a factor of about 10 compared with those of conventional n-channel power MOSFETs of comparable size and voltage rating.

IGBT power modules are rapidly gaining applications in systems such as inverters, UPS systems and automotive environments. The device ratings are reaching beyond 1800 V and 600 A. The frequency limits from early values of 5 kHz are now reaching beyond 20 kHz while intelligent IGBT modules that include diagnostic and control logic along with gate drive circuits are gradually entering the market.

A characteristics comparison of IGBTs, power MOSFETs, bipolars, and Darlingtons are indicated in Table 7.2. References listed between Russel (1992) through Clemente, Dubhashi, and Pelly (1990) provide more details on IGBTs and their applications.

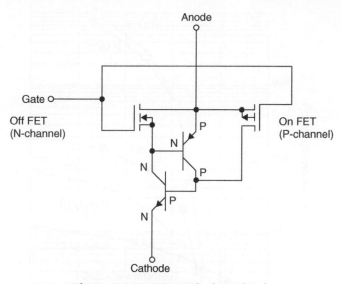

Figure 7.35: MCT equivalent circuit

7.7 MOS Controlled Thyristor (MCT)

MOS controlled thyristors are a new class of power semiconductor devices that combine thyristor current and voltage capability with MOS gated turn-on and turn-off. Various subclasses of MCTs can be made: P-type or N-type, symmetric or asymmetric blocking, one or two-sided Off-FET gate control, and various turn-on alternatives including direct turn-on with light.

All of these subclasses have one thing in common; turn-off is accomplished by turning on a highly interdigitated Off-FET to short out one or both of the thyristor's emitter-base junctions. The device, first announced a few years ago by General Electric's power semiconductor operation (now part of Harris Semiconductor, USA), was developed by Vic Temple. Harris is the only present supplier of MCTs; however, ABB has introduced a new device called insulated gate commutated thyristor (IGCT), which is in the same family of devices.

Figure 7.35 depicts the MCT equivalent circuit. Most of the characteristics of an MCT can be understood easily by reference to the equivalent circuit shown here. MCT closely approximates a bipolar thyristor (the two-transistor model is shown) with two opposite polarity MOSFET transistors connected between its anode and the proper layers to turn it on and off. Since MCT is a NPNP device rather than a PNPN device an output terminal or cathode must be negatively biased.

Driving the gate terminal negative with respect to the common terminal or anode turns the P channel FET on, firing the bipolar SCR. Driving the gate terminal positive with respect

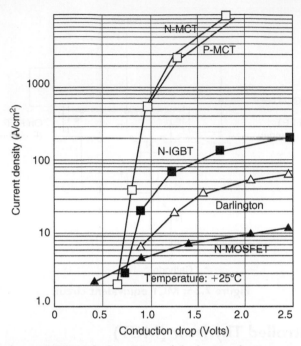

Figure 7.36: Comparison of 600 V devices (Copyright by Harris Corporation, reprinted with permission of Harris Semiconductor Sector)

to the anode turns on the N-channel FET shunting the base drive to PNP bipolar transistor making up part of the SCR, causing the SCR to turn off. It is obvious from the equivalent circuit that when no gate to anode voltage is applied to the gate terminal of the device, the input terminals of the bipolar SCR are unterminated. Operation without gate bias is not recommended.

In the P-MCT a P-channel On-FET is turned on with a negative voltage which charges up the base of the lower transistor to latch on the MCT. The MCT turns on simultaneously over the entire device area giving the MCT excellent *di/dt* capability. Figure 7.36 compares different 600 V power switching devices. Figure 7.37 compares the characteristics of 1000 V P-MCT device with a N-IGBT device of same voltage rating. Note that the MCT typically has 10 to 15 times the current capability at the same voltage drop.

The MCT will remain in the on-state until current is reversed (like a normal thyristor) or until the off-FET is activated by a positive gate voltage. Just as the IGBT looks like a MOSFET driving a BJT, the MCT looks like a MOSFET driving a thyristor (an SCR). SCRs and other thyristors turn on easily, but their turn-off requires stopping, or diverting virtually all of the current flowing through them for a short period of time. On the other hand, the MCT is turned off with voltage control on the high-impedance gate.

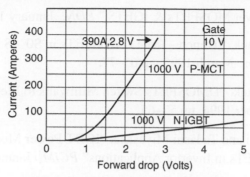

Figure 7.37: Comparison of forward voltage drop of 1000 V P-MCT and N-IGBT at 150°C (Copyright by Harris Corporation, reprinted with permission of Harris Semiconductor Sector)

The MCT offers a lower specific on-resistance at high voltage than any other gate-driven technology.

That is, just as the IGBT operates at a higher current density than the DMOSFET, the MCT (like all thyristors) operates at even higher current densities. In the future, the ultimate power switch may well be the MOS-controlled thyristor (MCT). References 16 to 20 provide details for designers.

References

[1]   Ashkianazi, G., J. Lorch, and M. Nathan. "Ultrafast GaAs Power Diodes Provide Dynamic Characteristics with Better Temperature Stability than Silicon Diodes." *PCIM*, April 1995, pp 10–16.

[2]   Hammerton, C. J. "Peak Current Capability of Thyristors." *PCIM*, November 1989, pp 52–55.

[3]   Coulbeck, L., W. J. Findlay, and A. D. Millington. "Electrical Trade-offs for GTO Thyristors." *Power Engineering Journal*, February 1994, pp 18–26.

[4]   Bassett, Roger J. and Colin Smith. "A GTO Tutorial: Part I." *PCIM*, July 1989, pp 35–39.

[5]   Bassett, Roger J. and Colin Smith. "A GTO Tutorial: Part II - Gate Drive." *PCIM*, August 1989, pp 21–28.

[6]   McNulty, Tom. "Understanding Power MOSFETs." Harris Semiconductor, Application note AN 7244.2, September 1993.

[7]   Travis, Bill. "Power MOSFETs & IGBTS." *EDN*, January 1989, pp 128–147.

[8]   Goodenough, Frank. "Trench – Gate DMOSFETs in S0 – 8 Switch 10A at 30 V." *Electronic Design*, March 1995, pp 65–77.

[9]   Goodenough, Frank. "DMOSFETs Switch Milliwatts to Megawatts." *Electronic Design*, September 1994, pp 57–65.

[10]   Furuhata, Sooichi and Tadashi Miyasaka. "IGBT Power Modules Challenge Bipolars, MOSFETs in Invertor Applications" *PCIM*, January 1990, pp 24–28.

[11]   Russel, J. P. et al. "The IGBTs—A new high conductance MOS-gated device." Harris Semiconductor, App. note AN 8607.1, May 1997.

[12]   Wojslawowicz, J. E. "Third Generation IGBTS Approach Ideal Switch Capability." *PCIM*, January 1995, pp 28–37.

[13]   Frank, Randy and John Wertz. "IGBTS Integrate Protection for Distributorless Ignition Systems." *PCIM*, February 1994, pp 42–49.

[14]   Dierberger, K. "IGBT Do's and Don'ts." *PCIM*, August 1992, pp 50–55.

[15]   Clemente, S., A. Dubhashi, and B. Pelly. "Improved IGBT Process Eliminates Latch–up, Yields Higher Switching Speed – Part I." *PCIM*, October 1990, pp 8–16.

[16]   Temple, V., D. Watrous, S. Arthur, and P. Kendle. "MOS-Controlled Thyristor (MCT) Power Switches – Part I – MCT Basics." *PCIM*, November 1992, pp 9–16.

[17]   Temple, V., D. Watrous, S. Arthur, and P. Kendle. "MOS-Controlled Thyristor (MCT) Power Switches – Part II: Gate Drive and Applications." *PCIM*, January 1993, pp 24–33.

[18]   Temple, V., D. Watrous, S. Arthur, and P. Kendle. "MOS-Controlled Thyristor (MCT) Power Switches – Part III: Switching, Applications and The Future." *PCIM*, February 1993, pp 24–33.

[19]   Temple, V.A.K., "MOS-Controlled Thyristors – A New Class of Power Devices." IEEE Trans., *Electron Devices*, vol ED–33, no 10, pp 1609–1618.

[20]   Temple, V. A. K. "Advances in MOS-Controlled Thyristor Technology." *PCIM*, November 1989, pp 12–15.

[21]   Burkel, R. and T. Schneider. "Fast Recovery Epitaxial Diodes Characteristics – Applications – Examples." IXYS Technical Information 33 (Publication No. D94004E, 1994).

[22] Williams, B.W. *Power Electronics: Devices, Drivers, Applications and Passive Components*. Macmillan, 1992.

[23] Kinzer, Dan. "Fifth-Generation MOSFETs Set New Benchmarks for Low On-Resistance." *PCIM*, August 1995, p 59.

[24] Delaney, S., A. Salih, and C. Lee. "GaAs Diodes Improve Efficiency of 500 kHz DC-DC converter." *PCIM*, August 1995, pp 10–11.

[25] Deuty, S. "GaAs Rectifiers Offer High Efficiency in a 1MHz. 400Vdc to 48 Vdc Converter." HFPC Conference Proceedings, September 1996, pp 24–35.

[26] Davis, C. "Integrated Power MOSFET and Schottky Diode Improves Power Supply Designs." *PCIM*, January 1997, pp 10–14.

[27] Goodenough, Frank. "Dense MOSFET enables portable power control." *Electronic Design*, April 14, 1997, pp 45–50.

Bibliography

Anderson, S., K. Gauen, and C. W. Roman. "Low Loss, Low Noise Diodes Improve High Frequency Power Supplies." *PCIM*, February 1991, pp 6–13.

Adler, Michael, et al. "The Evolution of Power Device Technology."*IEEE Transactions on Electronic Devices*, November 1984, Vol ED-31, No 11, pp 1570–1591.

Arthur, S. D. and V. A. K. Temple. "Special 1400 Volt N – MCT Designed for Surge Applications." Proceedings of EPE 93, (Vol 2) (1993), pp 266–271.

Barkhordarian, V. "Power MOSFET Basics." *PCIM*, June 1996, pp 28–39.

Bird, B. M., K. G. King, and D. A. G. Pedder. *An Introduction to Power Electronics* (2nd Edition), John Wiley, (1993).

Borras, R., P. Aloisi, and D. Shumate. "Avalanche Capability of Today's Power Semiconductors." Proceedings of EPE–93, (Vol 2), (1993), pp 167–171.

Bose, B. K. *Modern Power Electronics*. IEEE Press, 1997.

Bradley, D. A. *Power Electronics*. Chapman & Hall, 1995.

Consoli, A., et al. "On the selection of IGBT devices in soft switching applications." Proceedings of EPE–93, pp 337–343.

Deuty, Scott, Emory Carter, and Ali Salih. "GaAs Diodes Improve Power Factor Correction Boost Converter Performance." *PCIM*, January 1995, pp 8–19.

Driscoll, J. "Bipolar Transistors and High Side Switches in High Voltage, High Frequency Power Supplies." Proceedings of power conversion conference, October 1990.

Driscoll, J. C. "High current fast turn-on pulse generation using Power Tech PG–5xxx series of "Pulser" gate assisted turn-off thyristors (GATO's)." Power Tech App Note 1990.

Eckel, H. G. and L. Sack. "Optimization of the turn-off performance of IGBT at overcurrent and short-circuit current." Proceedings of EPE-93, 1993, pp 317–321.

Frank, Randy and Richard Valentine. "Power FETS Cope with the Automotive Environment." *PCIM*, February 1990, pp 33–39.

Gauen, K. and W. Chavez. "High Cell Density MOSFETs: Low on Resistance Affords New Design Options." Proceedings of PCI, October 1993, pp 254–264.

Goodenough, Frank "DMOSFETs, IGBTS Switch High Voltage." *Electronic Design*, 7 November 1994, pp 95–105.

Heumann, K. and M. Quenum. "Second Breakdown and Latch-up Behavior of IGBTs." EPE-93, 1993, pp 301–305.

International Rectifier. "Schottky Diode Designer's Manual." 1997.

Lynch, Fernando. "Two Terminal Power Semiconductor Technology Breaks Current/Voltage/Power Barrier." *PCIM*, October 1994, pp 10–14.

Locher, R. E. "1600V BIMOSFET™ Transistors Expand High Voltage Applications." *PCIM*, August 1996, pp 8–21.

Mitlehner, H. and H. J. Schulze. "Current Developments in High Power Thyristors." *EPE Journal*, March 1994 (Vol 4, No.1), pp 36–47.

Mitter, C. S. "Introduction to IGBTs." *PCIM*, December 1995, pp 32–39.

Mohan, N., T. M. Undeland, and W. P. Robbins. *Power Electronics: Converter, Applications and Design.* John Wiley, 1989.

Nilsson, T. "The insulated gate bipolar transistor response in different short circuit situations." Proceedings of EPE-93, 1993, pp 328–331.

Peter, Jean Marie. "State of The Art and Development in the Field of Medium Power Devices." *PCIM*, May 1986, pp 14–27.

Polner, Alex. "Characteristics of Ultra High Power Transistors." Proceeding of First National Solid State Power Conversion Conference, March 1995.

Ramshaw, R. S. *Power Electronics Semiconductor Switches.* Chapman & Hall, 1993.

Rippel, Wally E. "MCT/FET Composite Switch. Big Performance with Small Silicon." *PCIM*, November 1989, pp 16–27.

Roehr, Bill. "Power Semiconductor Mounting Considerations." *PCIM*, September 1989, pp 8–18.

Sasada, Yorimichi, Shigeki Morita, and Makato Hideshima. "High Voltage, High Speed IGBT Transistor Modules." *Toshiba Review*, No 157, Autumn 1986, pp 34–38.

Schultz, Warren. "Ultrafast – Recovery Diodes Extend the SOA of Bipolar Transistors." *Electronic Design*, 14 March, 1985, pp 167–174.

Serverns, Rudy and Jack Armijos. *"MOSPOWER Applications Handbook.".* Siliconix Inc., 1984.

Smith, Colin and Roger Bassett. "GTO Tutorial Part III - Power Loss in Switching Applications." *PCIM*, September 1989, pp 99–105.

Travis, B. "MOSFETs and IGBTs Differ in Drive Methods and Protections Needs." *EDN*, March 1, 1996, pp 123–137.

CHAPTER 8

Conduction and Switching Losses

Sanjaya Maniktala

Identifying and minimizing the losses within a switching power supply is one of the biggest challenges facing the power supply designer. The losses are primarily centered within the power semiconductors, with smaller losses occurring within the magnetics. Appreciating their nature can help one to select the optimum MOSFETs and rectifiers as well as the optimum frequency of operation. One can even predict on paper the amount of loss within each component and determine the best heatsinking methods.

There are three major types of losses within the power supply. Conduction losses are those voltage and current products that occur when a MOSFET (or power switch) or a rectifier is conducting current. This is duty-cycle dependent. Switching losses occur when the power switch or rectifier is transitioning between the ON state to the OFF state and vice versa. These are frequency dependent since there are more transitions per second at higher frequencies. Finally, we have gate drive losses, where there is some conduction loss in the MOSFET driver's output stage with that current exiting back through the input stage and not going to the output as output power.

To observe and calculate the losses within the actual switching power supply, one needs some fairly specialized tools: A voltage oscilloscope probe, an oscilloscope current probe and a contact thermal sensor. The waveforms can then be multiplied either graphically or by an intelligent oscilloscope with graphical math functions. Plus, it is always a good idea to isolate the earth ground connections on the AC power plug of all test equipment used to probe the power supply.

Be very suspicious of the plots displayed on the oscilloscope. Radiated noise is frequently picked up by the loop created between the ground clip and the probe tip. It is better to use coaxial cable with less than 3 cm of lead length at one end, which is soldered directly to the terminals of the component being examined.

Sanjaya Maniktala presents a very thorough examination of the losses within a typical switching power supply and how to calculate the resulting losses.

—**Marty Brown**

As switching frequencies increase, it becomes of paramount importance to reduce the *switching losses* in the converter. These are the losses associated with the transition of the switch from its on-state to off-state, and back. The higher the switching frequency, the greater the number of times the switch changes state per second. Therefore, these losses are proportional to the switching frequency. Further, of these frequency-dependent loss terms, the most significant are usually those that take place within the switch itself. Therefore, understanding the underlying sequence of events in the switch during each transition, and thereby quantifying the losses associated with each of these events, has become a key expectation of any power supply designer.

In this chapter, we are going to focus mainly on the MOSFET, since that is the most widely accepted "switch" in most high-frequency designs today. We will split its turn-on and turn-off transitions into small, well-defined subintervals, and explain what happens in each of these. The associated design equations will also be presented. Note, however, that as in most related literature, we too will be resorting to certain simplifications, since modeling the MOSFET (and its interplay with the board that it is mounted on) is certainly not a trivial task, to say the least. As a result, it is possible that theoretical estimates can end up underestimating the actual switching losses by a large margin (typically 20 to 50%). The designer should keep that in mind, and may need to eventually incorporate some sort of a "fudge factor" to correspond with reality. However, in our analysis, we have included a scaling factor to try to minimize this error.

We will also show how to estimate driver requirements and will demonstrate the importance of correctly matching driver capability to the MOSFET in a given application. That should ultimately help not only applications engineers to pick better MOSFETs for their applications, but also IC designers involved in the process of designing driver stages for target applications.

A cautionary note with regard to the terminology: in most of our switching analysis, what we are calling the "load" is the load as seen by the transistor. It is not the load of the *DC-DC converter* stage. Similarly the "input voltage" is only the voltage across the MOSFET when it is OFF; it is not the input to the DC-DC converter stage. We will eventually make the required connections into the area of power conversion, but it should be clear that, initially at least, the discussion is more from the standpoint of the MOSFET, not the topology that it may be a part of.

8.1 Switching a Resistive Load

Before we take up inductors, it is instructive to first understand what happens when we switch a resistive load.

For simplicity, we are considering an ideal situation. So we start with a "perfect" n-channel MOSFET in Figure 8.1. It behaves in the following manner:

- It has zero on-resistance.

- With zero gate-to-source voltage V_{gs} applied at its gate, it is completely nonconducting.

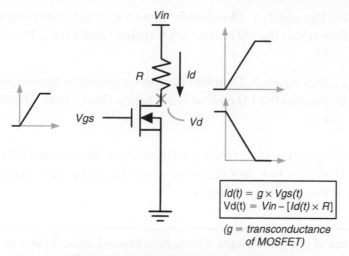

Figure 8.1: Switching a resistive load

- As we raise the gate-to-source voltage *Vgs* slightly above ground, it starts conducting, and so a drain current *Id* flows from the drain to the source terminal.

- The ratio of the drain current to the gate voltage is defined as the transconductance *g* of the MOSFET. It is expressed in mhos, that is, ohms spelled backward. Nowadays, however, mhos is being increasingly called *siemens*, abbreviated "S."

- We are assuming that *g* is a *constant*, equal to 1 s for this particular MOSFET. So, for example, if we apply 1 V at the gate, the MOSFET will pass 1 A. If we apply 2 V, it will pass 2 A, and so on.

The application circuit shown in Figure 8.1 works as follows:

- The applied input voltage is 10 V.

- The external resistance (in series with the drain) is 1 Ω.

- The gate voltage is ramped up *linearly* with respect to time. So at *t* = 1 s it is 1 V, at *t* = 2 s it is at 2 V, at *t* = 3 s it is at 3 V, and so on.

The analysis proceeds as follows (*Vds* is the drain-to-source voltage at any given moment, *Vgs* is the gate-to-source voltage, and *Id* is the drain-to-source current):

- At *t* = 0, *Vgs* equals 0 V. Therefore, from the transconductance equation, *Id* is 0 A. So the drop across the 1 Ω resistor is 0 V (using Ohm's law). Therefore the voltage at the drain of the MOSFET, *Vds*, equals 10 V.

- At $t = 1\,\text{s}$, *Vgs* equals 1. Therefore from the transconductance equation, *Id* is 1 A. So the drop across the 1 Ω resistor is 1 V (using Ohm's law). Therefore *Vds* equals $10 - 1 = 9\,\text{V}$.

- At $t = 2\,\text{s}$, *Vgs* equals 2. Therefore from the transconductance equation, *Id* is 2 A. So the drop across the 1 Ω resistor is 2 V (using Ohm's law). Therefore *Vds* equals $10 - 2 = 8\,\text{V}$.

We proceed ramping up the gate voltage progressively in this manner. When 10 s have elapsed, *Vgs* is 10 V, *Id* is 10 A, and *Vds* is 0 V. *After* 10 s, no further change in *Vds* or *Id* can occur, even if *Vgs* is increased further.

> **Note:** In general, if the gate voltage is increased beyond what it takes to deliver a specified maximum load current, we say that, in effect, we are applying "overdrive." This is usually considered wasteful in that sense, but in practice, overdrive helps reduce the on-resistance of the *MOSFET*, and thereby decrease its conduction losses.

The maximum load current in our example is therefore 10 A, and is *Idmax* in Figure 8.2. If we plot the drain current and drain voltage with respect to time, we see that the crossover time, *tcross*, is 10 s here. Note that this time is by definition the time for *both* the voltage and the current to complete their transitions.

The energy lost in the MOSFET during the transition is

$$E = \int_0^{\text{tcross}} Vd(t)Id(t)\,dt \text{ joules} \tag{8-1}$$

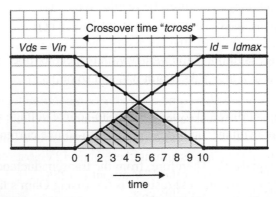

Figure 8.2: **The voltage and current waveforms when switching a resistive load**

A conceptual point to keep in mind here is that in related literature, it is often stated (rather inaccurately as we will see) that the "area (jointly) enclosed by the voltage, current, and the time axis is the energy lost in the switch" (during the transition). This is the gray isosceles triangle in Figure 8.2. Half of this gray area has been hatched. We thus see that within the "crossover interval rectangle," there are eight triangles (in all) with the same area as the hatched triangle. Therefore the total gray area is one-fourth the area of the crossover interval rectangle. So if the statement about energy being equal to the enclosed area is true, we would have gotten

$$E = \frac{1}{4} \cdot Vin \cdot Idmax \cdot tcross \text{ joules}$$

This is *not* correct. In fact, we would have reached the same unfortunate conclusion had we argued on the grounds that during the crossover duration, the *average* voltage is *Vin*/2 and the *average* current is *Idmax*/2, and therefore the average cross-product is equal to (*Vin* × *Idmax*)/4. This is fallacious too. In general,

$$A_{\text{AVG}} \times B_{\text{AVG}} \neq (A \times B)_{\text{AVG}}$$

So yes, this *could* in fact have turned out to be true, if, while the voltage was falling, the current had remained fixed, and vice versa. That is what happens with an inductive load, as we will soon see. However, in the case of a resistive load, both the voltage and the current change *simultaneously* during the crossover interval. We clearly need another (better) way to calculate the switching loss for the resistive case.

Let us compute the instantaneous cross-product $Vds(t) \times Id(t)$ at $t = 1, 2, 3, 4 \dots$ seconds. If we plot these points out, we get the bell-shaped curve shown in Figure 8.3. So, to get the energy lost during the crossover, we need to find the net area under this curve. But we can see that is not going to be easy, because this curve is rather oddly shaped. In fact, there is

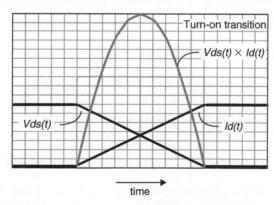

Figure 8.3: The instantaneous energy dissipation curve for resistive switching

no other way than to carry out a formal integration/summation procedure. And for that we have to revert to the basic equations for voltage and the current (as presented in Figure 8.1).

We then integrate their product over time, and we get

$$E = \frac{1}{6} \cdot Vin \cdot Idmax \cdot tcross \text{ joules} \qquad (8\text{-}2)$$

This is the correct result for the energy lost in the switch, during a resistive turn-on transition.

If we now turn the MOSFET OFF in the same way (with the crossover time kept fixed), we will get exactly the same energy loss term again, though this time with the voltage rising and the current falling.

We can thus also conclude that if we switch repetitively at the rate of *fsw* Hz, the net dissipation, that is, total energy lost per unit time as heat, is equal to

$$Psw = \frac{1}{3} \cdot Vin \cdot Idmax \cdot tcross \cdot fsw \text{ watts} \qquad (8\text{-}3)$$

This is therefore the *switching loss* (in the switch) for the case of a resistive load.

> **Note:** Note that to be precise, this particular term more correctly should be called the "crossover loss," as was first pointed out in *Chapter 1*. The crossover loss (i.e., specifically attributable to the *V-I overlap*) is not necessarily the entire switching loss taking place in the switch, as we will see.

Now, suppose we had ramped up the gate voltage at a rate of 1 V per second as before, but ramped down faster, say, at the rate of 2 V per second. Then the turn-on time and the turn-off transition times would be different. So in that case we need to split up the crossover loss *Psw* as follows:

$$\begin{aligned} Psw &= Pturnon + Pturnoff \\ &= \frac{1}{6} \cdot Vin \cdot Idmax \cdot tcross_{\text{on}} \cdot fsw \\ &+ \frac{1}{6} \cdot Vin \cdot Idmax \cdot tcross_{\text{off}} \cdot fsw \end{aligned} \qquad (8\text{-}4)$$

where $tcross_{\text{on}}$ and $tcross_{\text{off}}$ are the crossover times during turn-on and turn-off, respectively.

Now suppose the value of the external resistor was made larger, say 2 ohms instead of 1 ohm. Then the voltage at the drain would have swung from 10 V to 0 V in only 5 s.

And by that time, the drain current would have reached only 5 A. The gate voltage would at that moment be only at 5 V. However, no further change in *Id* is possible (even if we increase *Vgs* further). Therefore, though the crossover interval has become half of what it was, the rise time of the current is still equal to the fall time of the voltage (i.e., 5 s). This is a characteristic only of resistive loads (since *V = IR* applies to them).

The rules of the game change considerably, when we have an inductive load. In fact the calculation becomes simpler—ironically because the simplicity (and predictability) of Ohm's law is lost.

8.2 Switching an Inductive Load

When we switch an inductive load (with a freewheeling path present of course!), we will get the waveforms shown in Figure 8.4 (idealized). At first sight they may seem similar to the resistive load waveforms of Figure 8.2. But on closer examination, they are very different. In particular, we see that when the current is swinging, the voltage remains fixed, and when the voltage is swinging, the current remains fixed.

Let us calculate the crossover loss under these conditions. We can do a formal integration as before. But this time, we realize there is in fact an easy way out! Since one of the parameters (*V* or *I*) is fixed when the other is varying, we can now justifiably take the average value of the current, *Idmax*/2, and the *average* value of the voltage, *Vin*/2, to find the average cross-product. In this manner, we arrive at the energy lost (in joules) during the turn-on transition

$$E = \left[\frac{Vin}{2} \cdot Idmax \cdot \frac{tcross}{2} \right] + \left[Vin \cdot \frac{Idmax}{2} \cdot \frac{tcross}{2} \right]$$
$$= \frac{1}{2} \cdot Vin \cdot Idmax \cdot tcross$$

(8-5)

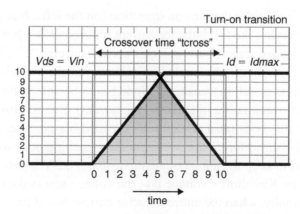

Figure 8.4: The voltage and current waveforms when switching an inductive load

Note that for the same reason as indicated above, we can now justifiably think in terms of the area enclosed. By simple geometry, the gray area in Figure 8.4 is half the rectangular area, and so we get the same result as above.

We realize that our ability to avoid integration (and use simpler arguments to calculate the crossover loss) is just a piece of "good luck" here, specific to the case of an inductive load.

Finally, when we switch repetitively, the inductive switching loss is

$$Psw = Vin \cdot Idmax \cdot tcross \cdot fsw \text{ watts} \tag{8-6}$$

> **Note:** We may superficially conclude that switching an inductive load leads to a dissipation three times greater than a resistive load. That is indeed true, but only *under the exact same conditions*. In reality, the value of Idmax is *fixed* for the case of a resistive load (depending on the value of the resistance used). But for an inductive load, the current can be virtually anything; there is no set "Idmax" as such anymore. It is whatever current that happens to be flowing through the inductor at the instant of switching (either just before or after).

A basic question still remains—*why* are the inductive waveforms so different from the resistive case? To answer that, we have go back to our previous analysis of the resistive load case. There we will see that we had invoked Ohm's law to find the voltage across the switch. But with an inductor, Ohm's law clearly does not apply. So to get the waveforms shown in Figure 8.4, we have to recollect something we learned earlier—when we turn the switch OFF, the inductor will create whatever voltage is necessary to maintain the continuity of current through it. Let us now show this principle at work in an actual buck converter, for example (see Figure 8.5).

In Figure 8.5, we first consider the turn-on transition (on the left). Just prior to this, the diode is obviously carrying the full inductor current (circled "1"). Then the switch starts to turn ON, trying to share some of this inductor current (circled "2"). The diode current therefore must fall correspondingly (circled "3"). However, the important point is that while the switch current is still in transit, the diode has to be able to pass some current (the remainder, or leftover amount of the inductor current). But, to provide even some of the inductor current, the diode must remain fully forward-biased. Therefore, nature (i.e., induced voltage in this case) forces the voltage at the switching node to remain slightly below ground—so as to keep the anode of the diode about 0.5 V higher than the cathode (circled "4"). Then, by Kirchhoff's voltage law, the voltage across the switch stays high (circled "5"). Only finally, when the entire inductor current has shifted to the switch, does the diode "let go." With that, the switching node is released, and it flies up close to the

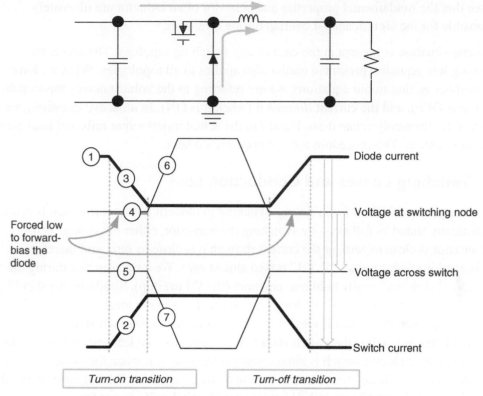

Figure 8.5: Analyzing the transitions in a buck converter

input voltage (circled "6")—and so now, the voltage across the switch is allowed to fall (circled "7").

- We therefore see that at turn-on, the voltage across the switch does not change until the current waveform has ***completed*** its transition. We thus get a significant V-I overlap.

If we do a similar analysis for the turn-off transition (right side of Figure 8.5), we will see that for the switch current to start decreasing by even a small amount, the diode must first be "positioned" to take up *any* current coming its way. So the voltage at the switching node must *first* fall close to zero, so as to forward-bias the diode. That also means the voltage across the switch must first transit fully, *before* the switch current is even allowed to decrease slightly (see Figure 8.5).

- We therefore see that at turn-off, the current through the switch does not change until the voltage waveform has ***completed*** its transition. We thus get a significant V-I overlap.

We see that the fundamental properties and behavior of an inductor are ultimately responsible for the significant V-I overlap during crossover.

The same situation is present in the case of any switching topology. Therefore, the switching loss equation presented earlier also applies to all topologies. What we have to remember is, that in our equations, we are referring to the voltage *across* the switch (when it is OFF), and the current *through* it (when it is ON). In an actual converter, we will need to ultimately relate these V and I to the actual input/output rails and load current of the application. The procedure for that is described later.

8.3 Switching Losses and Conduction Loss

The underlying motivation for initiating switching in modern power conversion is often simplistically stated as follows—by switching the transistor, either the voltage across the transistor is close to zero, or the current through it is close to zero, and therefore the dissipation cross-product "V × I" is also almost zero. We have seen that during the transition, that doesn't really hold true anymore (the V-I overlap). Similarly, we should keep in mind that though the V × I losses are much closer to the ideal or "expected" value of zero when the switch is OFF, there are considerable losses when the switch is ON. That is because when the switch is OFF, it is really so—the leakage current through a modern semiconductor switch is almost negligible. However, when the switch is ON, the voltage across it is not even close to zero in many cases. One of the highest reported forward drops is in the "Topswitch®" (an integrated switcher IC meant for medium off-line flyback applications)—over 15 V (over rated current and temperature)! In general, there will remain a significant V × I loss term even after the inductor current has shifted entirely from the diode to the switch. This particular loss term is clearly the conduction loss, P_{COND} (of the switch). It can in fact be comparable to, or even greater than, the crossover loss.

However, unlike the crossover loss, the conduction loss is *not* frequency-dependent. It does depend on *duty-cycle*, but not on frequency. For example, suppose the duty cycle is 0.6; then in a measurement interval of say, one second, the *net* time spent by the switch in the ON-state is equal to 0.6 seconds. But we know that conduction loss is incurred only when the switch is ON. So in this case, it is equal to a × 0.6, where "a" is an arbitrary proportionality constant. Now suppose the frequency is doubled. Then the net time spent in the on-state (in 1 second) is still 0.6 seconds. So the conduction loss remains a × 0.6. But now, suppose the duty cycle changes from 0.6 to 0.4 (the frequency can be even doubled in the process), the conduction loss is reduced to a × 0.4. So we realize that conduction loss can't possibly depend on frequency, only on duty cycle.

We can pose a rather philosophical question—why is it that the switching loss is frequency dependent, but *not* the conduction loss? That is simply because the conduction loss coincides with the interval in which power is being processed in the converter.

Therefore, as long as the application conditions do not change (duty cycle fixed, input and output power fixed), neither can the conduction loss.

The equation to calculate the conduction loss of a MOSFET is simply

$$P_{\text{COND}} = I_{\text{RMS}}^{2} \times R_{\text{ds}} \text{ watts} \tag{8-7}$$

where R_{ds} is the on-resistance of the MOSFET. I_{RMS} is the RMS of the switch current waveform. It is equal to

$$I_{\text{RMS}} = I_{\text{O}} \times \sqrt{D \times \left(1 + \frac{r^2}{12}\right)} \text{ (buck)} \tag{8-8}$$

$$I_{\text{RMS}} = \frac{I_{\text{O}}}{1 - D} \times \sqrt{D \times \left(1 + \frac{r^2}{12}\right)} \text{ (boost and buck-boost)} \tag{8-9}$$

where I_{O} is now the load current of the *DC-DC converter* stage, and D is its duty cycle. Note that to a first approximation (current ripple ratio assumed very small), this is equal to

$$I_{\text{RMS}} \approx I_{\text{DC}} \times \sqrt{D} \text{ (buck, boost, and buck-boost)} \tag{8-10}$$

where I_{DC} is the average inductor current and I_{RMS} is the RMS of the switch current waveform.

The *diode conduction loss* is the other major conduction loss term in a power supply. It is equal to $V_{\text{D}} \times I_{\text{DAVG}}$, where V_{D} is the diode forward-drop. I_{DAVG} is the average current through the diode—equal to I_{O} for the boost and the buck-boost, and $I_{\text{O}} \times (1 - D)$ for the buck. It too is frequency independent.

We realize that the way to reduce conduction losses is by lowering the forward-drops across the diode and switch. So we look for diodes with a low drop—like the Schottky diode. Similarly, we look for MOSFETs with a low on-resistance R_{ds}. However, there are compromises involved here. The leakage current in a Schottky diode can become significant as we try to choose diodes with very low drops. We can also run into significant body capacitance, which will end up being more dissipative. Similarly, the speed at which the MOSFET switches can be adversely affected as we try to reduce its R_{ds}.

8.4 A Simplified Model of the MOSFET for Studying Inductive Switching Losses

In Figure 8.6, on the left, we have the basic (simplified) model of the MOSFET. In particular, we observe that it has three parasitic capacitances, between its drain, source,

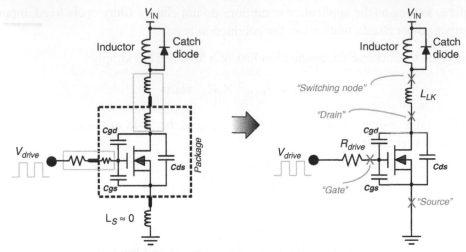

Figure 8.6: Simplified model of MOSFET

and gate. These "small" *interelectrode capacitances* are the key to maximizing switcher efficiency, especially at higher switching frequencies. Their role in the switching transition needs to be understood clearly.

We have seen that the basic reason why we get *any* crossover loss in the first place is because there is an unavoidable V-I overlap during every switching transition. That overlap occurs because the inductor keeps trying to force current, and tries to create suitable conditions for that to happen seamlessly, as we switch. But the reason why this overlap lasts as long as it does is mainly because these three interelectrode capacitors are demanding to be charged or discharged (as the case may be) at every switching event—so that they can reach their new DC levels, commensurate with the altered state of the switch. So crudely stated, if these capacitances are "big," they take a longer time to charge or discharge, thus increasing the crossover (overlap) time. And that in turn increases the crossover loss. Further, since the charging and discharging paths of these capacitors often include the *gate resistor*, the value of the gate resistance also considerably impacts the transition time, and thereby the switching loss.

On the right side of Figure 8.6, we have further simplified our simple model. So we have lumped the internal and external inductances present at the drain into a single leakage inductance *Llk*. Note that we are ignoring any gate-to-source inductance, thus implicitly assuming the PCB layout is very good in this regard. We also lump the small resistor present internally inside the MOSFET, along with the *external* gate resistor (if present), and the *driver resistance* (its internal pull-up or pull-down)—to give a single effective *Rdrive*, or *drive resistance*.

Note that in Figure 8.6, the main inductor is "coupled," because it has a freewheeling path available. But the leakage (or parasitic) inductance is "uncoupled," because it has no path to send forth its energy. It therefore expectedly "complains"—in the form of a voltage spike (whenever we try to change the current through it). However, in our analysis, we will be assuming this leakage inductance is very small (though not necessarily negligible either). We will find that this results in certain artifacts in the switching waveforms, which makes them appear slightly different, as compared to the idealized inductive switching waveforms shown in Figure 8.4 and Figure 8.5. However, it turns out that these artifacts are mainly of academic interest (provided of course that Rdrive is "small"). In addition, the artifacts in question typically help decrease the crossover losses slightly. Therefore, the idealized waveforms are more "conservative" in that sense, and we would do well just sticking to them.

Turning our attention to the "circuit" shown in Figure 8.6, we should be clear that this circuit doesn't really *work!* We know from previous discussions that we can never hope to achieve a steady state without at least an output capacitor present, to charge up and thereby help stabilize the voltseconds across the inductor. So this circuit is clearly an idealization— it only helps us to perform a paper analysis of a particular switching transition.

Note that, ultimately, the switch cares only about the voltage that appears across it when it turns OFF, and the current passing through it when it is ON. That is why this simple circuit can be safely accepted as representative of what happens in *any* topology, at the moment of transition. For instance, we could take both the leakage and the main inductor in Figure 8.6, and place them on the *source*-side of the MOSFET instead. As long as the gate drive is still well coupled to the source (i.e., no inductance between gate and source), nothing really changes. That is no surprise, because we know that if a certain component (or circuit block) "A" is in series with "B," we can always interchange their positions and make B in series with A, without changing a thing.

Finally, we should keep in mind that what we are calling the "drain" in our analysis is not necessarily the *pin* of the *package* (of the same name). Nor the switching node! The inductance *Llk* separates these points as indicated in Figure 8.6. Therefore, for example, though the switching node is necessarily clamped close to the "*Vin*" rail when the diode is freewheeling, the drain of the device may momentarily show a slightly different voltage (clearly equal to the voltage appearing across *Llk*).

8.5 The Parasitic Capacitances Expressed in an Alternate System

We will now progress to a detailed study of the inductive switching transitions of a MOSFET. For that, we will be splitting up the turn-on and turn-off into several subintervals of interest. We will learn that for most of these subintervals, the gate behaves as a simple input capacitance, that is being charged (or discharged) through the resistor

Rdrive. The situation is identical to the simple RC circuit we discussed previously. In effect, the gate is "blind" to what all may be happening between drain and source (on account of the transconductance of the MOSFET).

If we look *into* the gate, from the viewpoint of the AC drive signal, the effective input charging capacitance is the parallel combination (arithmetic sum) of *Cgs* and *Cgd*. We are going to call this simply the gate or input capacitance *Cg* in our discussion. So

$$Cg = Cgs + Cgd \tag{8-11}$$

The *time constant* of the charging/discharge cycles of the gate is therefore

$$Tg = Rdrive \times Cg \tag{8-12}$$

> **Note:** Here we seem to be indirectly suggesting that the drive resistance is the same for turn-on and turn-off. That need not be so. All the equations we will present can easily take any existing difference in the turn-on and turn-off drive resistances into account. So in general, we will have *different* crossover times for the turn-on and turn-off transitions. Also note that, in general, *within* a certain crossover interval (turn-on or turn-off), the actual time it takes for the voltage to transit need not be the same as the time the current takes (unlike the case of a resistive load).

An alternative system of writing the capacitances is in terms of the effective input, output, and reverse transfer capacitances—that is, *Ciss*, *Crss*, and *Coss,* respectively. These are related to the interelectrode capacitances as follows:

$$
\begin{aligned}
Ciss &= Cgs + Cgd \equiv Cg \\
Coss &= Cds + Cgd \\
Crss &= Cgd
\end{aligned}
\tag{8-13}
$$

So we can also write

$$
\begin{aligned}
Cgd &= Crss \\
Cgs &= Ciss - Crss \\
Cds &= Coss - Crss
\end{aligned}
\tag{8-14}
$$

In most vendors' datasheets, we can usually find *Ciss*, *Coss*, and *Crss* under the section "typical performance curves." We will then see that these parasitic capacitances are a function of voltage. Clearly, that can significantly complicate any analysis. So as an approximation, we are going to assume that the interelectrode capacitances are all constants. We will consult the typical performance curves of the MOSFET, and then pick the value of the capacitance corresponding to the voltage that appears across the

MOSFET when it is OFF (in our given application). Later, we will show how to minimize this error, by the use of a certain "scaling factor."

8.6 Gate Threshold Voltage

The "perfect MOSFET" we talked about earlier (Figure 8.1) started conducting the moment we raised the gate voltage above ground (i.e., source). But an actual MOSFET has a certain gate threshold voltage Vt. This is typically 1 to 3V for logic-level MOSFETs, and about 3 to 5V for high-voltage MOSFETs. So basically, we have to exceed the stated threshold voltage to get the MOSFET to conduct at all ("conduction" defined typically as a current in excess of 1 mA).

Because Vt is not zero, the definition of transconductance also needs to be modified slightly from

$$g = \frac{Id}{Vgs} \Rightarrow g = \frac{Id}{Vgs - Vt} \qquad (8\text{-}15)$$

Note that, in our analysis, we are making another simplifying assumption—that the transconductance too is a constant.

Finally, with all this background information, we can start looking closely at what actually happens during the turn-on and turn-off transitions.

8.7 The Turn-on Transition

We have divided this interval into *four* subintervals as detailed individually in Figures 8.7 through 8.10. For quick reference and ease of understanding, the relevant explanations and comments for each subinterval are also provided within their respective figures.

Briefly, the interval $t1$ is just the time to get to the threshold Vt. During this time, we just have a simple RC charging circuit. In $t2$ also, the exponential rise continues, but this time, the drain current starts ramping up. But for all practical purposes, the gate doesn't "know" anything has changed, because the transconductance is fully responsible for the drain current (and further, there is no change in the drain voltage). But in $t3$, the diode is allowed to stop conducting (since all the inductor current has by now shifted over into the switch). So now the drain voltage swings. But in doing so, it injects a current through Cgd. Note that this capacitance, despite being usually rather small, has probably the greatest effect on the crossover time—because of the fact that it directly injects current from a high switching voltage node (drain) on to the gate. Just prior to the interval $t3$, Cgd has a relatively high voltage across it. But when the switch is fully ON, the voltage across Cgd must decrease to its new final low value. Therefore, during $t3$, Cgd

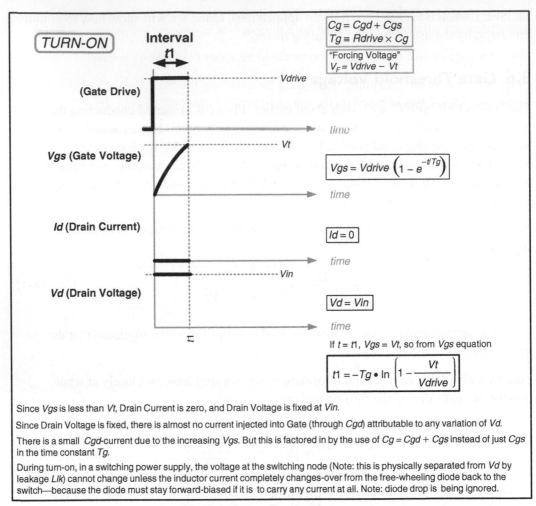

Figure 8.7: First interval of turn-on

is essentially discharging. So the question is: what is the path the Cgd discharge current takes? We can analyze that as follows: having reached the gate, this discharge current has two choices—either to go through Cgs and/or through $Rdrive$. But the gate is already at the constant level of $Vt + Io/g$, that being the gate voltage level required by the MOSFET to support the full inductor current Io. So, to a first approximation, the voltage across Cgs (gate voltage) need not and does not change. And, further, since the general equation for the current through any capacitor is $I = CdV/dt$, the current through Cgs must be zero, because there is no change in the voltage across it during this subinterval. Therefore we conclude that all the current coming through Cgd into the gate node gets diverted through $Rdrive$! But the voltage across $Rdrive$ is fixed—one end of it is at $Vdrive$, the other at $Vt + Io/g$. Therefore, the current through it is predetermined by Ohm's law. Which means

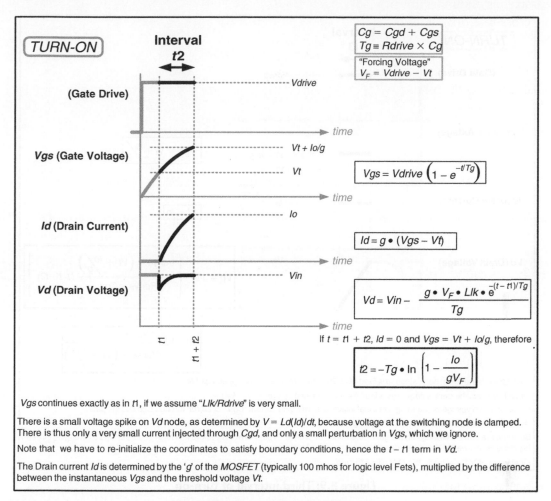

Figure 8.8: Second interval of turn-on

that *Rdrive* is actually in full control of the current through *Cgd* during the interval *t3*. However, the current through *Cgd* also obeys the equation $I = C \times dV/dt$. So if *I* is fixed at a certain value (by *Rdrive*), we can calculate the corresponding dV/dt across *Cgd*, and thereby calculate *Vd*. In effect, this means that *Cgd* and *Rdrive* are together determining the rate of fall of drain voltage during *t3* (and thus the transition time of the voltage). The plateau in the gate voltage waveform during *t3* is called the *Miller plateau*, referring to the effect of the reverse transfer capacitance *Cgd*. Finally, after the voltage too has completed its swing, the current through *Cgd* stops completely, and so once again, the gate behaves as a simple RC charging circuit. Note that during *t4*, the gate is in effect being overdriven—there is no change in the drain current anymore (which is already at its maximum possible value). However, driver dissipation continues during *t4*.

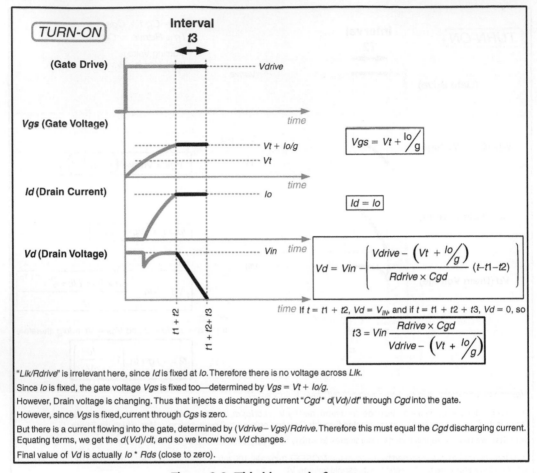

Figure 8.9: Third interval of turn-on

The *crossover time*, being the time during which both the current and voltage are transiting, is $t2 + t3$. As indicated, to know the driver dissipation, we need to consider the entire duration $t1 + t2 + t3 + t4$. Note that, by definition, at the end of $t4$, the gate voltage is at 90% of its asymptotic level (*Vdrive*). So we can safely assume that, for all practical purposes, the driver does very little after this point. Therefore, at the end of $t4$, the transition is considered complete—from the viewpoint of the switch, and also the driver.

8.8 The Turn-off Transition

In a similar manner as for turn-on, we have divided the turn-off interval into *four* subintervals, as shown in Figures 8.11 through 8.14.

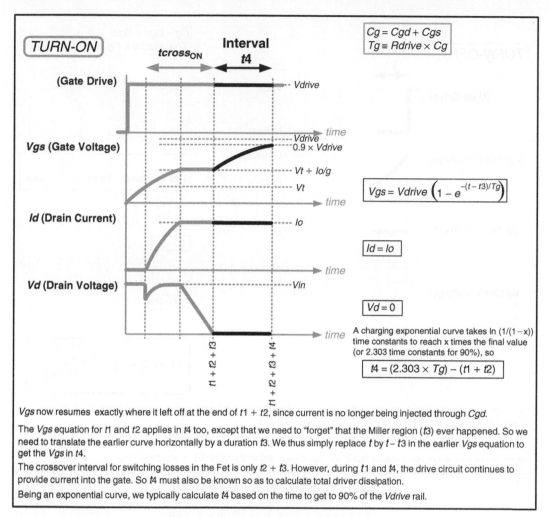

Figure 8.10: Fourth interval of turn-on

Briefly, the interval T1 is the time for the "overdrive" to cease; that is, the gate returns to the sustaining level $Vt + Io/g$ (the minimum gate voltage required to support the full drain current Io). During this time, there is no change in the drain current, nor in the drain voltage, and so in effect we once again have a simple RC discharging circuit. In T2, the gate voltage again plateaus. The reason for that is that the drain voltage must first swing close to Vin, and thereby "position" the diode to get forward-biased and be ready to start taking up the current that the switch will progressively shed (see Figure 8.5). So T2 is the time for the voltage transition to complete. During T1 and T2 therefore, no change in the drain current occurs. And with logic similar to what we presented for the turn-on subinterval $t3$, during T2 the rate of rise of the voltage Vds is once again determined

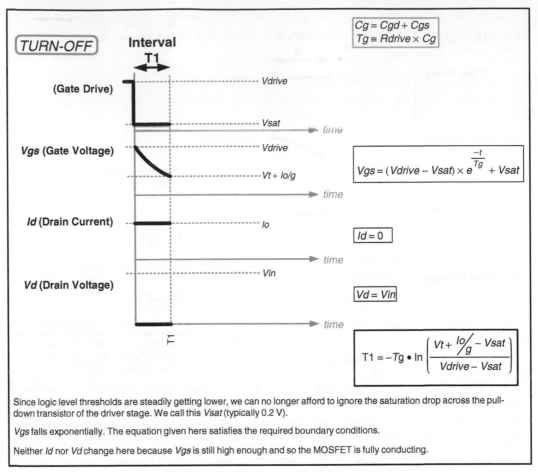

Figure 8.11: First interval of turn-off

(only) by *Rdrive* and *Cgd*. Finally, in T3, the current starts falling toward zero. The gate voltage falls exponentially (as an RC circuit)—down to *Vt*, at which moment, the end of subinterval T3 is declared. The transition is now complete as far as the switch is concerned. But after that, during T4, the RC exponential discharge continues down to 10% of the initial gate drive amplitude. As before, driver dissipation occurs over T1 + T2 + T3 + T4, whereas crossover occurs during T2 + T3.

8.9 Gate Charge Factors

A more recent way of describing the parasitic capacitor-based effects in a MOSFET is in terms of *gate charge factors*. In Figure 8.15, we show how these charge factors, *Qgs*, *Qgd*, and *Qg*, are defined. On the right column of the table in the figure, we have given

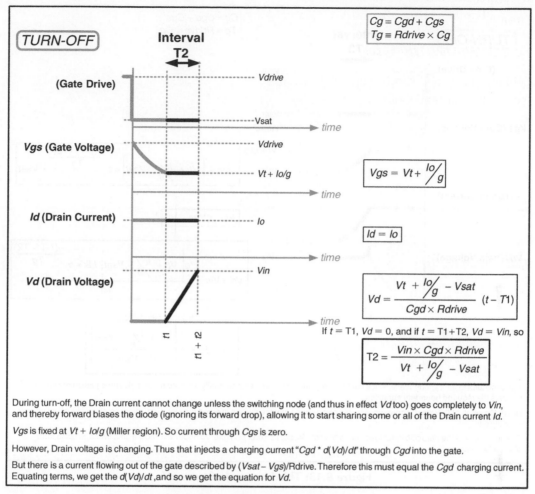

$$Cg = Cgd + Cgs$$
$$Tg \equiv Rdrive \times Cg$$

TURN-OFF

Interval T2

(Gate Drive)

Vdrive

Vsat → time

Vgs (Gate Voltage)

Vdrive

$Vt + Io/g$ → time

$$Vgs = Vt + \frac{Io}{g}$$

Id (Drain Current)

Io → time

$$Id = Io$$

Vd (Drain Voltage)

Vin → time

$$Vd = \frac{Vt + \frac{Io}{g} - Vsat}{Cgd \times Rdrive}(t - T1)$$

If $t = T1$, $Vd = 0$, and if $t = T1+T2$, $Vd = Vin$, so

$$T2 = \frac{Vin \times Cgd \times Rdrive}{Vt + \frac{Io}{g} - Vsat}$$

During turn-off, the Drain current cannot change unless the switching node (and thus in effect *Vd* too) goes completely to *Vin*, and thereby forward biases the diode (ignoring its forward drop), allowing it to start sharing some or all of the Drain current *Id*.

Vgs is fixed at *Vt* + *Io/g* (Miller region). So current through *Cgs* is zero.

However, Drain voltage is changing. Thus that injects a charging current "*Cgd* * d(*Vd*)/d*t*" through *Cgd* into the gate.

But there is a current flowing out of the gate described by (*Vsat* – *Vgs*)/Rdrive. Therefore this must equal the *Cgd* charging current. Equating terms, we get the d(*Vd*)/d*t*, and so we get the equation for *Vd*.

Figure 8.12: Second interval of turn-off

the relationships between the gate charge factors and the capacitances, assuming the latter are constants. Gate charge factors represent a more accurate way of proceeding, since the interelectrode capacitances are such strong functions of the applied voltage. However, our entire analysis of the turn-on and turn-off intervals so far has been implicitly based on the assumption that the interelectrode capacitances are constants. A possible way out of this, one that also helps reduce the error in our switching loss estimates, is detailed in Figure 8.16, using the Si4442DY (from Vishay) as an example.

Basically, we are using the gate charge factors to tell us what the effective capacitances are (and the voltage swings from 0 to *Vin*). We see that the effective input capacitance (*Ciss*), for example, is about 50% greater than the single-point *Ciss* value that we would

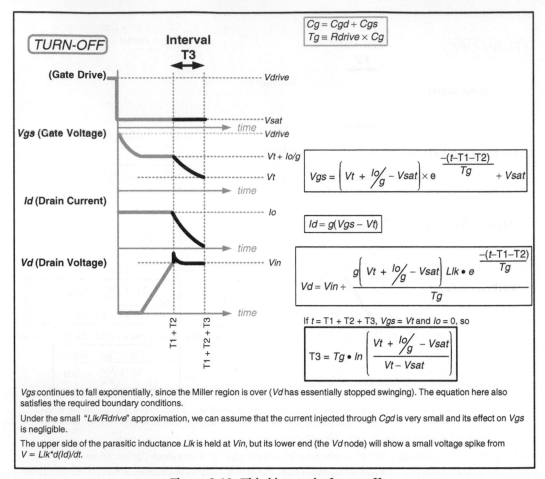

Figure 8.13: Third interval of turn-off

have read off from the typical performance curves (i.e., 6300 pF instead of 4200 pF). That factor accounts for the fact that as the voltage falls, the capacitance increases. Note that we could have calculated a scaling factor individually, for each capacitance. But it is simpler to use, say *Ciss*, to first find a "universal" scaling factor—and then apply it across the board to *all* the capacitances. In this manner, we arrive at the effective interelectrode capacitances quoted in Figure 8.16. These are the values we should use for our switching loss calculations (in preference to those provided by directly reading off *Ciss*, *Coss*, and *Crss* from their curves). Note that for finding the scaling factor, if we had looked at *Crss* (*Cgd*) instead of *Ciss*, then we would find that the calculated effective capacitance is only 40% higher (than what we would read directly from the curves). So the scaling factor can, in general, be fixed at around 1.4 to 1.5 typically.

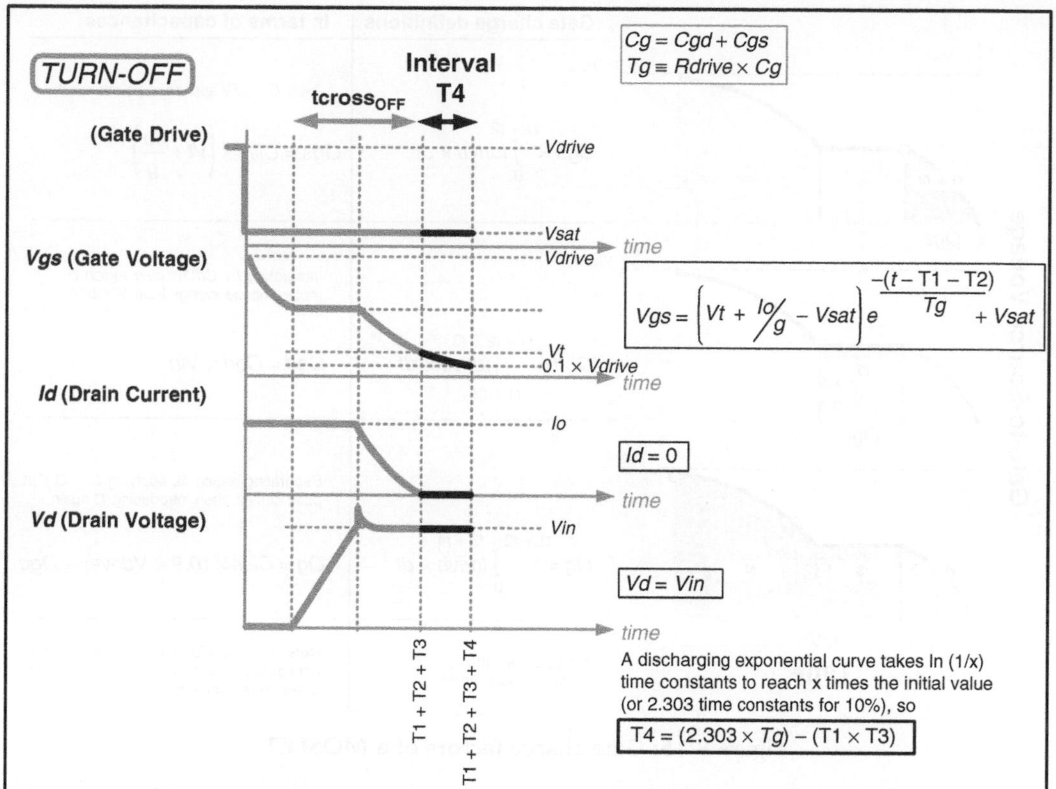

TURN-OFF
Interval T4
tcross$_{OFF}$
Cg = Cgd + Cgs
Tg ≡ Rdrive × Cg
(Gate Drive)
Vdrive
Vsat
time
Vgs (Gate Voltage)
Vdrive
$Vgs = \left(Vt + \frac{Io}{g} - Vsat\right) e^{\frac{-(t - T1 - T2)}{Tg}} + Vsat$
Vt
0.1 × Vdrive
time
Id (Drain Current)
Io
Id = 0
time
Vd (Drain Voltage)
Vin
Vd = Vin
time
T1 + T2 + T3
T1 + T2 + T3 + T4
A discharging exponential curve takes ln (1/x) time constants to reach x times the initial value (or 2.303 time constants for 10%), so
T4 = (2.303 × Tg) − (T1 × T3)

Vgs continues to fall exponentially (same equation as in T3). The equation here also satisfies the required boundary conditions.

Being an exponential curve, we typically calculate T4 based on the time to get to10% of the starting value (*Vdrive* rail). Note we have ignored *Vsat* for the T4 equation. Note also that if we take *Vsat* to be greater than 10% of *Vdrive*, there is no valid definition for T4 (the way we have defined it).

The crossover interval for switching losses in the FET is only T2 + T3. However, during T1 and T4, the drive circuit continues to provide current into the gate. So T4 must also be known so as to calculate total driver dissipation.

Figure 8.14: Fourth interval of turn-off

8.10 Worked Example

We are switching 22 A at 15 V through a Si4442DY MOSFET, at 500 kHz. The total pull-up drive resistance, by which the gate is driven by a pulse of amplitude 4.5 V, is 2 ohms. At turn-off, it is pulled-down (to source) by a total drive resistance of 1 ohm. Estimate the switching losses and the dissipation in the drive.

From Figure 8.16, we have $Cg = Cgs + Cgd = 6300$ pF.

8.10.1 Turn-on

The time constant is

$$Tg = Rdrive \times Cg = 2 \times 6300\,\text{pF} = 12.6\,\text{ns}$$

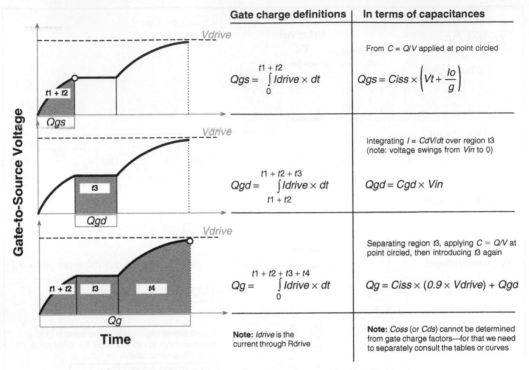

Gate charge definitions	In terms of capacitances
$Qgs = \int_{0}^{t1+t2} Idrive \times dt$	From $C = Q/V$ applied at point circled $Qgs = Ciss \times \left(Vt + \dfrac{Io}{g}\right)$
$Qgd = \int_{t1+t2}^{t1+t2+t3} Idrive \times dt$	Integrating $I = CdV/dt$ over region t3 (note: voltage swings from Vin to 0) $Qgd = Cgd \times Vin$
$Qg = \int_{0}^{t1+t2+t3+t4} Idrive \times dt$	Separating region t3, applying $C = Q/V$ at point circled, then introducing t3 again $Qg = Ciss \times (0.9 \times Vdrive) + Qgd$
Note: *Idrive* is the current through Rdrive	**Note:** *Coss* (or *Cds*) cannot be determined from gate charge factors—for that we need to separately consult the tables or curves

Figure 8.15: Gate charge factors of a MOSFET

The time for the current to transit is

$$t2 = -Tg \times \ln\left(1 - \frac{Io}{g \times (Vdrive - Vt)}\right) = -12.6 \times \ln\left(1 - \frac{22}{100 \times (4.5 - 1.05)}\right)$$

$$t2 = 0.83\,\text{ns}$$

The time for the voltage to transit is

$$t3 = Vin \times \frac{Rdrive \times Cgd}{Vdrive - \left(Vt + \dfrac{Io}{g}\right)} = 15 \times \frac{2 \times 0.75}{4.5 - \left(1.05 + \dfrac{22}{100}\right)}$$

$$t3 = 6.966\,\text{ns}$$

So the crossover time during turn-on is

$$tcross\_turnon = t2 + t3 = 0.83 + 6.966 = 7.8\,\text{ns}$$

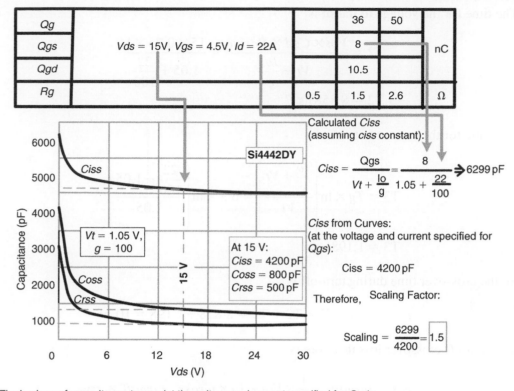

Final values of capacitance to use (at the voltage and current specified for *Qgs*):

$Ciss = 4200\,pF \times Scaling = 6300\,pF$
$Coss = 800\,pF \times Scaling = 1200\,pF$
$Crss = 500\,pF \times Scaling = 750\,pF$

$Cgd = Crss = 750pF$
$Cgs = Ciss - Cgd = 6300 - 750 = 5550\,pF$
$Cds = Coss - Cgd = 1200 - 750 = 450\,pF$

Figure 8.16: Estimating the *effective* interelectrode capacitances from the gate charge factors (Si4442DY as an example)

The turn-on crossover loss therefore is

$$Pcross\_turnon = \frac{1}{2} \times Vin \times Io \times tcross\_turnon \times fsw$$

$$= \frac{1}{2} \times 15 \times 22 \times 7.8 \times 10^{-9} \times 5 \times 10^{5}$$

$$Pcross\_turnon = 0.64\ watts$$

8.10.2 Turn-off

The time constant is now

$$Tg = Rdrive \times Cg = 1 \times 6300\,pF = 6.3\,ns$$

The time for the voltage to transit is

$$T2 = \frac{Vin \times Cgd \times Rdrive}{Vt + \dfrac{Io}{g}} = \frac{15 \times 0.75 \times 1}{1.05 + \dfrac{22}{100}}$$

$$T2 = 8.858\,ns$$

The time for the current to transit is

$$T3 = Tg \times \ln\left(\frac{\dfrac{Io}{g} + Vt}{Vt}\right) = 6.3 \times \ln\left(\frac{\dfrac{22}{100} + 1.05}{1.05}\right)$$

$$T3 = 1.198\,ns$$

So the crossover time during turn-off is

$$tcross\_turnoff = T2 + T3 = 8.858 + 1.198 = 10\ ns$$

The turn-off crossover loss therefore is

$$pcross\_turnoff = \frac{1}{2} \times Vin \times Io \times tcross\_turnoff \times fsw$$

$$= \frac{1}{2} \times 15 \times 22 \times 10 \times 10^{-9} \times 5 \times 10^{5}$$

$$pcross\_turnoff = 0.83\ watts$$

So finally, the **total crossover loss** is

$$Pcross = Pcross\_turnon + Pcross\_turnoff = 0.64 + 0.83 = 1.47\ watts$$

Notice that we have not even used *Cds* so far! This particular capacitance does not affect the V-I overlap (since it is not connected to the gate). But it still needs to be considered! Every cycle, it charges up during turn-off, and then during turn-on it dumps its stored energy inside the MOSFET. This is, in fact, the additional loss term that needs to be added to the crossover loss term, so as to get the total switching loss in a MOSFET. Note that in low-voltage applications, this additional term may seem insignificant, but in high-voltage/off-line applications, it does affect the efficiency noticeably. Let us calculate what it is in our case:

$$P\_Cds = \frac{1}{2} \times Cds \times Vin^2 \times fsw = \frac{1}{2} \times 450 \times 10^{-12} \times 15^2 \times 5 \times 10^5 = 0.025\ watts$$

So the **total switching loss** (in the switch) is

$$Psw = Pcross + P\_Cds = 1.47 + 0.025 = 1.5 \text{ watts}$$

The **driver dissipation** is

$$Pdrive = Vdrive \times Qg \times fsw = 4.5 \times 36 \times 10^{-9} \times 5 \times 10^5 = 0.081 \text{ watts}$$

Note that, typically, the above driver dissipation equation underestimates the actual driver dissipation by almost 20%, as can be confirmed by integrating the product of the drive current and the voltage across it, over each subinterval. The reason for the error is simply the Miller plateau—because during this interval, some additional current (other than from the stored charge Qg), gets injected into the drive resistor. So our corrected driver dissipation estimate is $1.2 \times 0.081 = 0.097 \text{ W}$. The driver supply rail current is $0.081/4.5 = 18 \text{ mA}$.

8.11 Applying the Switching Loss Analysis to Switching Topologies

Now we try to understand how our preceding analysis pertains to an actual switching regulator application—in particular, what Vin and I_O are, with respect to the topology.

For a **buck**, we know that at turn-on, the instantaneous switch (and inductor) current is $I_O \times (1 - r/2)$, where r is the current ripple ratio, and I_O is the load current of the DC-DC converter. At turn-off, the current is $I_O \times (1 + r/2)$. Usually, we can ignore the current ripple ratio and take the current as I_O for both the turn-on and the turn-off analysis. So the load current of the DC-DC converter, I_O, becomes the same as the I_O used so far in the switching loss analysis. Similarly, in a **boost and buck-boost**, the current I_O in our switching loss analysis, is actually the average inductor current $I_O/(1 - D)$.

Coming to the voltage across the MOSFET when it turns OFF (i.e., Vin in the switching loss analysis)—for the **buck**, this is almost equal to the input rail of the DC-DC converter V_{IN} (a diode drop more in reality). Similarly, for a **buck-boost**, the voltage Vin is almost exactly equal to $V_{IN} + V_O$, where V_O is the output rail of the DC-DC converter. For a **boost**, the voltage "Vin" is equal to V_O, that is, the output rail of the converter. Note that if we are dealing with an isolated **flyback**, the voltage at turn-off really is $V_{IN} + V_Z$, where V_Z is the voltage of the zener clamp (placed across the primary winding). However, at turn-on, the voltage across the MOSFET is only $V_{IN} + V_{OR}$ (V_{OR} being the reflected output voltage, i.e. $V_O \times n_P/n_S$). In a single-ended **forward converter**, we have $2 \times V_{IN}$ at turn-off, and only V_{IN} at turn-on. Note that in all cases discussed above, we are assuming CCM.

We have tabulated these results in Table 8.1 for convenience.

Note that if we were in DCM, there is in principle *no* switching loss at turn-on—because there is no current flowing in the inductor by that time. At turn-off, the current at transition is $I_{PK} = \Delta I$, which can be found using $V = L \times \Delta I/\Delta t$.

8.12 Worst-case Input Voltage for Switching Losses

We must return now to the all-important question—when we have a wide-input voltage range, what specific input voltage point represents the worst case for calculating switching losses?

The switching loss equation is generically

$$Psw = Vin \cdot Io \cdot tcross \cdot fsw \text{ watts}$$

We note that, in all cases, this loss depends on the product of *Vin* and *Io*. But by now, we know what *Vin* and *Io* are—from Table 8.2. So we can analyze the situation for each topology as follows

- For a **buck**, "$Vin \times Io$" $= V_{IN} \times I_O$. So the maximum loss will obviously occur at V_{INMAX}.

- For a **boost**, "$Vin \times Io$" $= V_O \times I_O/(1 - D)$. So the maximum loss will occur at D_{MAX}, that is, at V_{INMIN}.

- For a **buck-boost**, "$Vin \times Io$" $= (V_{IN} + V_O) \times I_O/(1 - D)$. We also know that $D = V_O/(V_{IN} + V_O)$. So plotting "$Vin \times Io$," we get Figure 8.17 (a typical case). Note that the curve is *symmetrical* around $D = 0.5$—and that is the point of *minimum* switching losses. Below that point, the *voltage* increases significantly,

Table 8.1: Connecting the switching loss analysis with actual topologies

	"Vin"		"Io"	
	Turn-on	**Turn-off**	**Turn-on**	**Turn-off**
Buck	V_{IN}		I_O	
Boost	V_O		$I_O/(1-D)$	
Buck-Boost	$V_{IN} + V_O$		$I_O/(1-D)$	
Flyback	$V_{IN} + V_{OR}$	$V_{IN} + V_Z$	$I_{OR}/(1-D)$	
Forward	V_{IN}	$2 \times V_{IN}$	I_{OR}	
$V_{OR} = V_O \times n$, $I_{OR} = I_O/n$ where $n = n_P/n_S$				

and above that, the *current* increases significantly. Either way, the switching losses *increase* as we move away from $D = 0.5$. Therefore, in general, we must first examine the input range of our application, and see which of its ends is *furthest* from $D = 0.5$. For example, if in our application, the input range corresponds to a duty cycle range of 0.6 to 0.8, we need to do the switching loss calculation at $D = 0.8$, that is, at V_{INMIN}. However, if the duty cycle range is say, 0.2 to 0.7, we need to do the calculation at $D = 0.2$, that is, at V_{INMAX}.

8.13 How Switching Losses Vary with the Parasitic Capacitances

In Figure 8.18, we have taken the Si4442DY, and "varied" its *Ciss* just to see what can happen as a result of that. On the right vertical axis, we have the corresponding

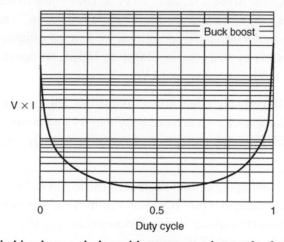

Figure 8.17: Switching loss variation with respect to duty cycle, for the buck-boost

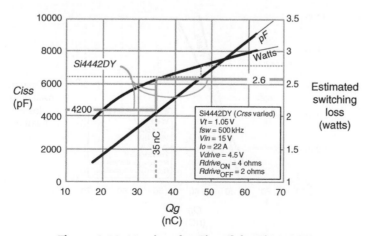

Figure 8.18: Varying the *Ciss* of the Si4442DY

(estimated) switching loss. Note that in computing the loss curve, a "scaling factor" of 1.5 has been applied to the *Ciss* values given on the left vertical axis (though this is not obvious).

The gray vertical dashed line (annotated "35 nC") represents the Si4442DY *as it is*. So under the stated conditions, we have an estimated switching loss of 2.6 W. If we increase *Ciss* by 50%, that is, 4200 pF to 6300 pF, we see that *Qg* will go up to about 47 nC, and the loss to 2.8 W only.

Note: In the actual calculations, using the scaling factor of 1.5, "4200 pF" is actually 6300 pF, and "6300 pF" is actually 9450 pF.

In Figure 8.19, we take the Si4442DY, and "vary" its *Crss*—just to see what can happen as a result of that. The gray vertical dashed line (annotated "35 nC") represents the Si4442DY *as it is*. So under the stated conditions, we have an estimated switching loss of 2.6 W. If we increase *Crss* by 50%, that is, 500 pF to 750 pF, we see that *Qg* will go up to about 39 nC only, but the loss goes up to 3.1 W.

In other words, *Qg* will certainly affect driver dissipation, but it is not necessarily a good indicator of the *switching losses*—it is more *helpful to try and minimize Qgd (or Crss) when selecting* MOSFETs, rather than just looking for a "low-Qg" MOSFET.

Note: In the worked example, we had estimated the losses to be 1.5W. There we had a pull-up of 2 ohms, and a pull-down of 1 ohm. Whereas in Figure 8.18, we have basically doubled the pull-up and pull-down resistors. However, the switching loss has not doubled—it is only 73% more.

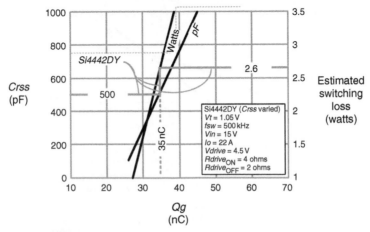

Figure 8.19: Varying the *Crss* of the Si4442DY

8.14 Optimizing Driver Capability vis-à-vis MOSFET Characteristics

In Figure 8.20, we have two separate graphs. The one on the left has a fixed pull-up of 4 ohms. On the *x*-axis, we are therefore, in effect, varying only the pull-down. So if for example, the *x*-axis is at 2, the pull-down resistor is 4 ohms/2 = 2 ohms. If the *x*-axis is at 4, the pull-down resistor is at 4 ohms/4 = 1 ohm. We see that as expected, the losses decrease as the pull-down is improved. We also see the effect of "varying" the threshold voltage. So, lower threshold voltages also help lower the switching losses—provided the pull-down is not too "weak." On the right graph, similarly, we have the results for a fixed pull-up of 10 ohms. We can thereby estimate the effect of varying the pull-up too, on the overall losses.

Finally, in Figure 8.21, we are keeping the pull-up + pull-down constant, as we vary the ratio of the pull-up and pull-down resistors. This is from the IC designer's viewpoint—suppose he or she has roughly allocated a certain die area for the driver stage, say simplistically fixed the pull-up + pull-down. Then the question is: how should the available drive capability be distributed between the pull-up and the pull-down sections. For example, if pull-up + pull-down = 6 ohms, is it better to split this as pull-up = 4 ohms and pull-down = 2 ohms, or say, pull-up = 3 ohms and pull-down = 3 ohms, or pull-up = 2 ohms and pull-down = 4 ohms, and so on? We see that the answer to that depends on the threshold voltage. So we need to have an idea of the MOSFETs we are planning to use, *before* we decide on the optimum ratio. From Figure 8.21 we see that if the threshold is greater than 2 V, improving the pull-up (at the expense of the pull-down) will help, and so, for example, pull-up = 4 ohms and pull-down = 2 ohms will be preferable to pull-up = 5 ohms and pull-down = 1 ohm. However, if the threshold voltage is below 2 V, we see that the reverse is true—so now, improving the pull-down (at the expense of the pull-up) will help.

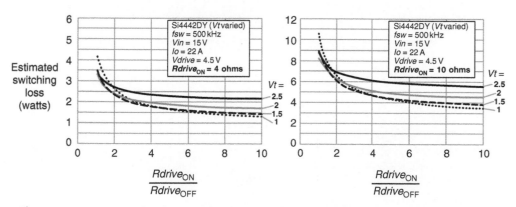

Figure 8.20: Varying the threshold voltage of the Si4442DY and the drive resistances (keeping pull-up resistance fixed)

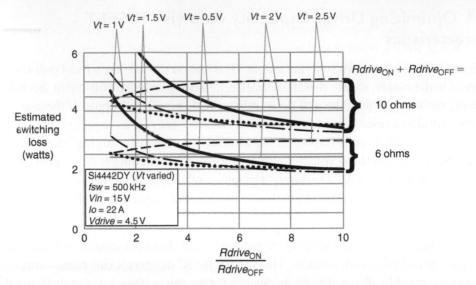

Figure 8.21: Varying the threshold voltage of the Si4442DY and the drive resistances (keeping total drive resistance, i.e., pull-up + pull-down, fixed)

Note: Some vendors provide a rather wide range ("MIN" to "MAX") for threshold voltage. Often, they do not even provide a "TYP" value. But, surprisingly, some do not even provide the threshold voltage at all! They simply state that their *MOSFET* is "capable of 4.5 V drive" (as, for example, most of the *MOSFET*s from www.renesas.com).

CHAPTER 9

Power Factor Correction

When I was writing my first book about designing switching power supplies in 1989, I purposely left this topic till last. Sure, I had taken the control theory classes in college many years before, but they did not relate well to the real-world applications facing me now. So I read all of the papers and books of the day on feedback loop compensation and I was still puzzled. So I took a moment and stepped back and asked myself, "What are they really trying to accomplish?" I formulated the four basic goals of a well-compensated switching power supply. Then I read a paper from Dean Vererable that presented a step-by-step method of determining the location of the compensating poles and zeroes and the resistor and capacitor values fell out at the end. I liked the procedure, but his procedure did not yield the bandwidth I needed for the power supplies. So I borrowed his step-by-step methodology and modified it to fit my purposes.

The following chapter is the result of a lot of work, of which I am very proud. This chapter makes a scary process into a very easy, step-by-step procedure which should only require 15 minutes. It has always worked for me in actual power supply designs.

I start out with a short review of the behavior of elemental circuits in frequency domain using Bode plots. Then we select the compensation method appropriate for your switching power supply topology and control method. We work through the equations and the resistor and capacitor values fall out at the end. The only other area that could cause instability within current-mode supplies is from the current feedback loop where slope compensation is needed.

I hope this chapter demystifies this topic for you.

—**Marty Brown**

Power factor correction is becoming a very important area in the power world. Adding more generating capacity to the world's electrical pool is very costly and would consume additional resources. One method of creating about 30 percent excess generating capacity is to use the AC power more efficiently through the broad use of power factor correction. Motors, electronic power supplies, and fluorescent lighting consume about 40 percent of the power in the world and each of these would benefit from power factor correction. From the mid-1990s to the present, many of the countries of the world are adopting requirements for power factor correction for the new products marketed within their borders. The added circuitry will add about 20 to 30 percent to the cost of power supplies, but the near-term energy savings will greatly outweigh the initial costs.

The term "power factor" in the field of power supplies is a slight departure from the traditional usage of the term, which applied to reactive ac loads, such as motors, powered from the AC power line. Here, the current drawn by the motor would be displaced in phase with respect to the voltage. The resulting power being drawn would have a very large reactive component and little power is actually used for producing work. Since power meters do not measure phase, the power measured is the scaler voltage times current. Capacitor banks are typically used to bring the phase more towards zero degrees if motors are the primary load.

In switching power supplies, the problem lies in the input rectification and filter network. The typical input circuit and its associated waveforms are shown in Figure 9.1. As one can see, the input rectifiers can only conduct current when the AC line voltage exceeds the voltage on the input filter capacitor. This typically occurs within 15 degrees of the crest of the AC voltage waveform. The result is that current pulses are 5 to 10 times higher than the expected average current draw. This also can lead to distortion of the AC voltage waveform and an imbalance of the three-phase power lines feeding the circuits. This produces a neutral line current where no current flow is expected. Another drawback is that no current is drawn when the rectifiers are not conducting, thus throwing away a significant portion of the power system's energy capability.

Power factor correction circuits are intended to increase the conduction angle of the rectifiers and to make the AC input current waveform sinusoidal and in phase with the voltage waveform. The input waveforms can be seen in Figure 9.2. This means that all the power drawn from the power line is real power and not reactive. The net result is that

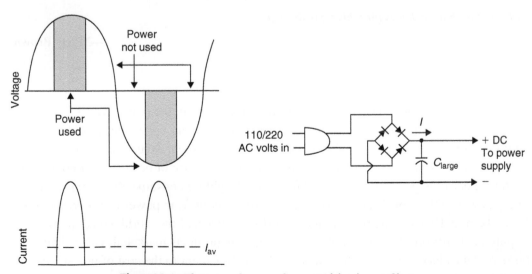

Figure 9.1: The waveforms of a capacitive input filter

the peak and RMS current drawn from the line is much lower than that drawn by the capacitive input filter circuit traditionally used.

Active power factor correction circuits can take the form of nontransformer isolated switching power supply topologies, such as buck, boost, and buck/boost. The buck topology in Figure 9.3 produces an output DC voltage lower than found at its input, whenever the PFC stage is operating ($V_{in} > V_{out}$). In other words, the output voltage is typically in the 30 to 50 VDC range. This can present a problem for higher powered loads which would then draw a large amount of current from the PFC circuit. The boost and the

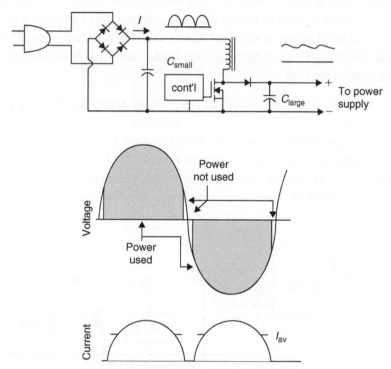

Figure 9.2: Power factor corrected input

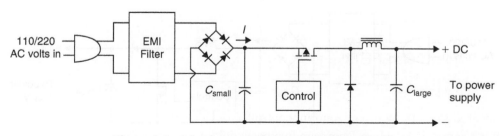

Figure 9.3: A buck power factor correction circuit

buck/boost topologies are popular within the field, since they produce a higher DC output voltage than the peak input voltage, which means lower average output currents. These are seen in Figures 9.4 and 9.5.

The buck/boost develops an output voltage that is negative with respect to the input ground following the rectifiers. The cascaded power supply and the PFC voltage sense networks must work with a negative voltage, but the DC output voltage can be independent from the values of the rectified input AC waveform. The major disadvantage is the need for a high-side power switch and high breakdown voltage requirements for the semiconductors. The boost topology has become the most popular topology. It has a low-side power switch that is easy to drive. Its only restriction is that the DC output voltage must be higher than the highest expected AC crest voltage. This means that for a PFC circuit to be useful in all power grids in the world, the output voltage must be greater than 390 VDC, and will pass voltage surges onto the load. Otherwise, it requires the fewest parts and hence costs the least.

Control of the power factor correction stage is a point of debate and battling patents. There are three general methods of control: fixed on-time, critical conduction-mode (just discontinuous), and continuous-mode. Fixed on-time has the minimum amount of circuitry, but limits the instantaneous current that can be drawn from the input line. Critical conduction-mode has no output rectifier reverse recovery loss, but is limited in output to about 300 to 600 W. Continuous-mode boost PFC circuits can go much higher

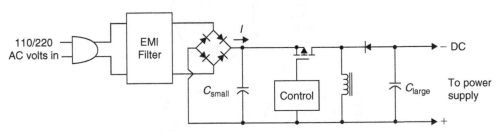

Figure 9.4: A buck/boost power factor correction circuit

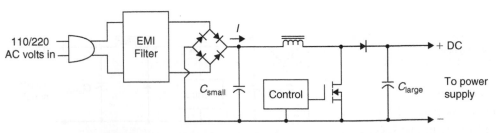

Figure 9.5: A boost power factor correction circuit

in output power, but suffer in severe rectifier reverse recovery losses, unless some zero-transition loss circuits are added, which can add cost.

Added cost is very important in this area since power factor correction is an unseen benefit to the customer who does not want to pay for anything he or she cannot directly see. An example is PFC corrected, electronic fluorescent lighting ballasts at two times the purchase price versus magnetic fluorescent ballasts. Only industrial customers are thoughtful enough to see the vast difference in their yearly electric bill 12 months later.

The basic PFC controller takes the form shown in Figure 9.6, which would include the critical and discontinuous-mode circuits and the continuous-mode circuits. There is a multiplier subcircuit inside the control IC which multiplies the instantaneous value of the input full-wave rectified voltage waveform with the output of the error amplifier. This produces a current limit signal which makes the input current follow the voltage sinusoidal waveshape. The AC input is filtered by the input EMI filter to produce a 50 to 60 Hz input current waveform that is free of switching artifacts.

The inductor operating mode is a major consideration in designing a PFC circuit. The discontinuous-mode of operation is typically used for power levels less than 300 to 600 W. It has high peak currents that limit its use at the higher input power levels. For powers greater than 300 W, the continuous-mode of operation is typically used. This lowers the peak currents seen by the power switch and output rectifier and is much easier to filter in the input EMI filter since there are no rapid transitions in the input switched current waveforms. The only disadvantage is that the diode-related switching losses rise significantly since the power switch must force the output rectifier to turn off at the beginning of each on-time period. The choice of output rectifier (low Trr) becomes critical to the operation of the PFC stage.

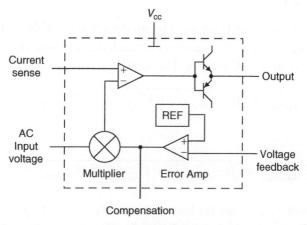

Figure 9.6: A generalized typical power factor control IC

9.1 How Power Factor and Harmonics Are Specified

I strongly recommend that your company engage a third-party EMI testing house to test your products. The minimum level of test equipment required to test for the discussed factors is very expensive and there is a long learning period involved.

The following discussion is primarily based upon EN61000-3-2, which is more industrial-based products. EN60555 is the newer version of IEC-555 and covers the emitted harmonics of household products. Knowing the product class is very important for designing the power factor interface for the AC line. Products that draw less than 75 W today and 50 W in the near future, do not have to comply with the relative limits, only to absolute maximums laid out by the specifications. The limits presented in Table 9.1 may change, so please refer to the latest released version. This is a developing field, so be aware of the most recent specifications at the time of your product's release.

The real power delivered to a load is given by

$$P_{in} = V_{in} \cdot I_{in} \cdot (\text{Power Factor}) \tag{9-1}$$

where

$$\text{Power Factor} = \frac{\text{Real Power}}{\text{Real Power} + \text{Reactive Power}} \tag{9-2}$$

In terms of strictly passive reactive loads, the power factor is the resulting phase between the voltage and the current waveforms. In power supplies, though, it is the distortion to the voltage waveform resulting from the time which input rectifiers conduct. Power factor is measured from 0 to 1 where 1 is where all the power is used by the load (purely resistive). The typical capacitive input filter found in power supplies has an average power factor of 0.5 to 0.7.

In running the tests, a power analyzer must be used such as the Voltech PM1000, PM1200, or PM3000. Also an audio spectrum analyzer is needed to measure the amplitude of the harmonic components of the AC current. The total input voltage and currents are given by

$$V_{RMS(total)} = \sqrt{V^2_{fund(RMS)} + V^2_{1(RMS)} + V^2_{2(RMS)} \cdots} \tag{9-3}$$

and

$$I_{RMS(total)} = \sqrt{I^2_{fund(RMS)} + I^2_{1(RMS)} + I^2_{2(RMS)} \cdots} \tag{9-4}$$

where the subscripts of 1, 2, . . . are the harmonics of 50 or 60 Hz. In power supplies the third harmonic is by far the next largest amplitude and therefore the largest problem. Harmonics

Table 9.1: IEC555-2 harmonic current limits

Harmonic	Class A RMS-Amps	Class D RMS-Amps
2	1.08	2.30
3	2.30	–
4	0.43	–
5	1.44	1.14
6	0.30	–
7	0.77	0.77
9	0.40	0.40
11	0.33	0.20
13	0.21	0.33
8 < n < 40	0.23 × 8/n	
11 < n < 39	0.15 × 15/n	0.15 × 15/n

cause problems because, in a pure sense, only the fundamental current frequency produces real power. So the reduction of harmonics produces a better power factor.

A term used in PFC is *total harmonic distortion*. This is defined as

$$T.H.D. = \frac{I_{1(\text{RMS})} + I_{2(\text{RMS})} \cdots}{I_{\text{RMS(total)}}}$$

(9-5)

and it is an indication as to the performance of a PFC circuit.

From the power analyzer or the spectrum analyzer, one can measure the amplitude values needed to verify compliance to the PFC specifications. EN61000-3-2 has the limits shown in Table 9.1. Class A and D are shown because they are common product categories.

These limits must be measured with a LISN (line impedance stabilization network) as specified by the regulatory agencies. This makes the input power line a 50-ohm impedance and serves the basis of all of these tests. The test results are highly dependent upon the AC line impedance.

9.1.1 Some Comments on the Design of PFC Circuits

First, the EMI filter is an integral part of any PFC circuit. It filters out the switching harmonics from the input current waveform. Without an EMI filter, your product will fail the EMI/RFI tests which are in addition to the power factor tests. Secondly, using a variac during the measurements will affect the input line impedance and thus affects the validity of the data you are trying to measure. Many units will pass testing without the use of a LISN, but fail when the LISN is used. The added impedance of a LISN distorts the waveforms more than the impedance of the typical raw AC line in use at that moment. Thirdly, all voltage measurements must be differential and use the specified current measuring apparatus.

9.2 A Universal Input, 180 W, Active Power Factor Correction Circuit

This design example demonstrates the design process of a 180 W discontinuous-mode boost PFC circuit. It can be scaled to provide output powers up to 200 W. The PFC stage is designed to work from every residential AC power system within the world; that is, from 85 to 270 VRMS at 50 and 60 Hz without the need for a jumper.

9.2.1 Design Specification

The design specification is as follows:

AC input voltage range: 85–270 VRMS
AC line frequencies   50–60 Hz
Output voltage: 400 VDC ± 10 V
Input power factor at rated load: >98%
Total harmonic distortion (THD)   under EN1000-3-2 limits

9.2.2 Predesign Considerations

Having a rating less than 200 W has many benefits for a power factor correction stage. The major benefit is that it can operate in the discontinuous-mode. Within higher power PFC designs the continuous-mode must be employed which presents a significant loss within the circuit due to the reverse recovery time of the output rectifier. In fixed frequency discontinuous-mode PFC controllers, there is still a period when the circuit operates in the continuous-mode ($V_{in} < 50$ V (approx.)). By employing a *critical conduction-mode* controller, the designer can guarantee that the continuous-mode is never entered.

The first consideration is to determine the peak AC input voltages.

110 V input:

$$V_{in(nom)} = 1.414(110\,V) = 155.5\,V$$
$$V_{in(hi)} = 1.414(130\,V) = 183.8\,V$$

240 V input (Britain—worst case):

$$V_{in(nom)} = 1.414(240 \text{ V}) = 339.4 \text{ V}$$
$$V_{in(hi)} = 1.414(270 \text{ V}) = 381.8 \text{ V}$$

The output voltage should be higher than the highest anticipated input peak crest voltage. The output voltage of the PFC stage is now chosen to be 400 VDC.

The maximum value for the peak inductor current will occur at the crest voltage of the minimum expected AC input voltage. This is

$$
\begin{aligned}
I_{pk(max)} &= 1.414(2)(P_{out(rated)}) / (eff_{est})(V_{in(min)RMS}) \\
&= 1.414(2)(180 \text{ W}) / (0.9)(85 \text{ V}_{RMS}) \\
&= 6.6 \text{ A}
\end{aligned}
$$

9.2.3 *Inductor Design*

In designing the boost inductor, one would designate the point of reference as the crest voltage of the minimum expected AC input voltage. For any set of operating conditions with this method of PFC control (i.e., fixed load and AC input voltage, the on time pulsewidth remains constant over the entire half-sinusoid waveform). To determine the on time at the minimum peak AC input voltage one would do the following operations:

$$
R = \frac{V_{out(DC)}}{\sqrt{2}V_{in\text{-}AC(min)}} = \frac{400 \text{ V}}{1.414(85 \text{ V}_{RMS})}
$$
$$
R = 3.3
$$

The maximum on time which occurs at this point is

$$
\begin{aligned}
T_{on(max)} &= \frac{R}{f(1+R)} = \frac{3.3}{(50 \text{ kHz})(1+3.3)} \\
&= 15.3 \text{ μs}
\end{aligned}
$$

The approximate maximum value of the boost inductor is

$$
\begin{aligned}
L &\approx \frac{T_{on(max)} \left(\sqrt{2}V_{in\text{-}AC(min)} \right)^2 (eff)}{2P_{out(max)}} \\
&\approx \frac{(15.3 \text{ μs})(1.414)(85 \text{ V}_{RMS})(0.9)}{2(180 \text{ W})} \approx 552 \text{ μH}
\end{aligned}
$$

The power winding of the inductor (transformer) not only must support the maximum average input current but the output current as well. So, the wire gauge of the winding should be

$$V_{w(max-av)} = \frac{P_{out}}{eff(V_{in(RMS)})} + \frac{P_{out}}{V_{out}}$$

$$= \frac{180\,W}{(0.9)(8.5\,V_{RMS})} + \frac{180\,W}{400\,V} = 2.8\,A$$

The wire gauge to accommodate this average current would then be #17 AWG. I will use three strands of #22 AWG (which adds up to the same wire cross-sectional area), which is more flexible during the winding process and will help reduce the AC resistance of the winding due to the skin effect. Also, due to the high voltages present within the same winding, I will be using quad-thickness insulation to reduce the threat of interturn arc-overs.

I am selecting a PQ core style. A major concern is the length of air-gap required for various core styles in unipolar applications. The larger air-gaps (>50 mils) cause excessive electromagnetic radiation into the immediate environment thus making it harder to RFI filter. To reduce the air-gap, one needs to find a ferrite core with the largest core cross-sectional area for a given core size. The PQ core has this characteristic. Referring to the WaAc *vs.* power charts provided by Magnetics, Inc., the resulting PQ core part number is P-43220-XX. (XX is the gap length in mils.)

The approximate air-gap needed in the core is

$$l_{gap} \approx \frac{0.4\pi L \cdot I_{pk}10^8}{A_c B_{max}^2}$$

$$\approx \frac{0.4\pi(552\,\mu H)(6.6\,A)10^8}{(1.70\,cm^2)(2000\,G)^2} \approx 66\,mils$$

Let us make the air-gap 50 mils, which is a custom air-gap. Magnetics has no problem with this practice and it usually adds only a couple of percent to the core cost. The inductance factor (AL) for this core with this gap is estimated at 160 mH/1000 T (using a linear extrapolation of AL reduction versus air-gap length).

The number of turns needed for this inductance is

$$N = 1000\sqrt{\frac{0.55\,mH}{160\,mH}} = 59\,turns$$

Checking to see if the core will support this many turns (neglecting the auxiliary winding area):

$$\frac{A_\mathrm{W}}{W_\mathrm{A}} = \frac{(59\,T)(.471\,\mathrm{mm}^2)}{47\,\mathrm{mm}^2} = 59\% \quad \mathrm{OK}$$

9.2.3.1 Designing the auxiliary winding

The auxiliary winding will have the low frequency (100 to 120 Hz) variation on its output peak rectified voltage, so the controller filter capacitor needs to be large to minimize the droop in the V_cc of the controller. The highest flyback-mode rectified voltage will occur at low input voltages and will be of the form

$$v_\mathrm{aux} \approx \frac{N_\mathrm{aux}(V_\mathrm{out} - V_\mathrm{in})}{N_\mathrm{pri}}$$

This AC waveform is seen in Figure 9.7. The PFC boost inductor construction is shown in Figure 9.8.

The MC34262 has a high-side driver clamp of 16 VDC, so in order to keep the high-side driver dissipation to a minimum, the peak voltage of the rectified auxiliary voltage should be around 16 V. Determine the turns ratio needed for this from

$$N_\mathrm{aux} = \frac{(59\,\mathrm{T})(16\,\mathrm{V})}{(400\,\mathrm{V} - 30\,\mathrm{V})} = 2.5 \text{ turns}$$

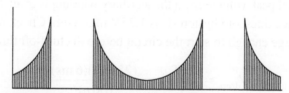

Figure 9.7: The rectified AC waveform present on the auxiliary winding

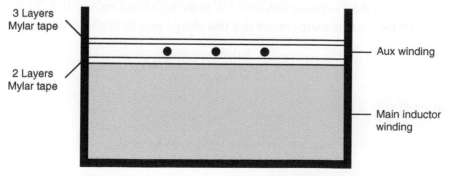

3 Layers Mylar tape

2 Layers Mylar tape

Aux winding

Main inductor winding

Figure 9.8: Construction of the PFC boost inductor

I will make this winding three turns because of concern about low AC line operation. I will use one strand of #28 AWG heavy insulated magnet wire.

The capacitor needed to filter this voltage with approximately 2 V of voltage ripple is

$$C_{aux} \approx \frac{I_{dd} T_{off}}{V_{ripple}} = \frac{(25 \, mA)(6 \, ms)}{2.0 \, V}$$
$$= 75 \, \mu F \quad make \, 100 \, \mu F \, @ \, 20 \, VDC$$

9.2.4 Transformer Construction

The two-winding transformer will be constructed by first winding the 59 turns of the three strands of #22 AWG quad-thickness magnet wire onto the bobbin. Then place two layers of Mylar tape. Then the three turns for the auxiliary winding, and lastly three layers of Mylar tape. The internal layers of tape are to discourage any arcing that may occur due to the high voltages between the primary winding and the auxiliary winding.

9.2.5 Designing the Start-up Circuit

I will use a passive resistor for starting up the control IC and to provide current to the gate drive of the MOSFET. For the resistor I need to use two resistors placed in series, since the 370 V peak on the rectified input is comparable to the breakdown voltages of the resistors themselves. The start-up resistors will charge the 100 μF bypass capacitor and the subsequent energy stored in the capacitor must be sufficient to operate the control IC for the 6 ms before the worst-case rectified peak voltage from the auxiliary winding is available to operate the IC. The start-up voltage threshold hysteresis is 1.75 V minimum. Checking whether the bypass capacitor is large enough to start the circuit before the turn-off threshold is reached:

$$V_{drop} = \frac{I_{dd} T_{off}}{C} = \frac{(25 \, mA)(6 \, ms)}{100 \, \mu F}$$
$$= 1.5 \, V \quad OK$$

I would like to keep the dissipation less than 1 W at the high input voltage line. To do this one needs to determine the maximum current that that should pass through the start-up resistors.

$$I_{start} < \frac{1.0 \, W}{270 \, V_{RMS}} = 3.7 \, mA$$

The total resistance is then:

$$R_{start} = \frac{270 \, V - 16 \, V}{3.7 \, mA} = 68 \, K(min)$$

Make the total resistance about 100 K or two 47 Kohm, 1/2 W resistors.

9.2.6 Designing the Voltage Multiplier Input Circuit

The minimum specified maximum linear limit of the input to the multiplier (pin3) is 2.5 V. This level should be the peak value of the divided rectified input waveform at the highest expected AC input voltage at the crest of the sinusoid (370 V). If a sense current of 200 μA is selected at this point the resistor divider becomes

$$R_{\text{bottom}} = \frac{2.5\,\text{V}}{200\,\mu\text{A}} = 12.5\,\text{K} \quad \text{make it 12 K}$$

The true sense current is 2.5 V/12 K = 208 μA.

The top resistor becomes

$$R_{\text{top}} = \frac{370\,\text{V} - 2.5\,\text{V}}{208\,\mu\text{A}} = 1.76\,\text{Mohms}$$

Make this two resistors in series each with a value of 910 Kohms.

The power rating of these resistors are $P = (370\,\text{V})2/1.76\,\text{M}\Omega$ or 0.8 W. Each resistor should have 1/2 W power rating.

9.2.7 Design of the Current Sensing Circuit

The current sense resistor should be sized in order to reach the 1.1 V current sense threshold voltage at the low AC input voltage. The value then becomes

$$R_{\text{CS}} = \frac{1.1\,\text{V}}{6.6\,\text{A}} = 0.3\,\text{ohms}$$

A leading edge spike filter of 1 K and 470 pF will also be added before inputting the current signal to pin 4.

9.2.8 Designing the Voltage Feedback Circuit

For the output voltage sense resistor divider, selecting the sense current as 200 μA, the lower resistor becomes

$$R_{\text{bottom}} = \frac{V_{\text{ref}}}{I_{\text{sense}}} = \frac{2.5\,\text{V}}{200\,\mu\text{A}} = 12.5\,\text{K} \quad \text{Make 12.0 K}$$

This makes the true sense current 2.5V/12 K = 208 μA. The upper resistor is

$$R_{\text{upper}} = \frac{(400\,\text{V} - 2.5\,\text{V})}{208\,\mu\text{A}} = 1.91\,\text{Mohms}$$

Make this resistor a 1-Mohm and a 910-Kohm resistor in series, each with a 1/2 W rating.

The compensation of the voltage error amplifier should be a single-pole rolloff with a unity gain frequency of 38 Hz. This is required to reject the fundamental line frequencies of 50 and 60 Hz. The feedback capacitor around the voltage error amplifier becomes

$$C_{fb} = \frac{1}{2\pi f R_{upper}} = \frac{1}{2\pi (38\,Hz)(1.82\,M)}$$
$$= 0.043\,\mu F \text{ or } .05\,\mu F$$

9.2.9 Designing the Input EMI Filter Section

I will be using a second order, common-mode filter. The difficulty in considering an input conducted EMI for this power factor correction circuit is its variable frequency of operation. The lowest instantaneous frequency of operation occurs at the crests of the sinusoid voltage waveform. This is where the core requires the longest time to completely discharge the core. The estimated frequency of operation has been 50 kHz, so I will use this as an assumed minimum frequency.

A good starting point is to assume that I will need 24 dB of attenuation at 50 kHz. This makes the corner frequency of the common-mode filter

$$f_C = f_{SW} \cdot 10^{\left(\frac{A_{tt}}{40}\right)}$$

where A_{tt} is the attenuation needed at the switching frequency in negative dB.

$$f_C = (50\,kHz)10^{\left(\frac{-24}{40}\right)} = 12.5\,kHz$$

Assuming that a damping factor of 0.707 or greater is good and provides a −3 dB attenuation at the corner frequency and does not produce noise due to ringing. Also assume that the input line impedance is 50 ohms since the regulatory agencies use an LISN test which makes the line impedance equal this value. Calculating the values needed in the common-mode inductor and "Y" capacitors:

$$L = \frac{R_L \cdot \zeta}{\pi \cdot f_C} = \frac{(50)(0.707)}{\pi (12.5\,kHz)} = 900\,\mu H$$

$$C = \frac{1}{(2\pi f_C)^2 L} = \frac{1}{[2\pi (12.5\,kHz)]^2 (900\,\mu H)}$$
$$= 0.18\,\mu F$$

Real-world values do not allow a capacitor of this large a value. The largest value capacitor that will pass the AC leakage current test is 0.05 μF. This is 27 percent of the

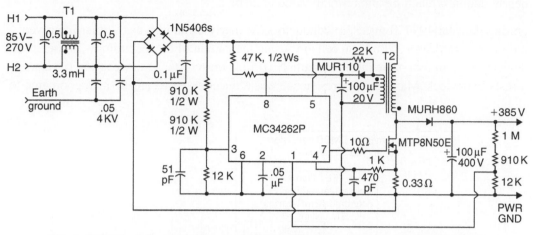

Figure 9.9: The schematic for the 180 W power factor circuit (with EMI filter)

calculated capacitor value, so the inductor must be increased 360 percent in order to maintain the same corner frequency. The inductance then becomes 3.24 mH and the resultant damping factor is 2.5 which is acceptable.

Coilcraft offers off-the-shelf common-mode filter chokes (transformers) and the part number closest to this value is E3493. With this filter design I can expect a minimum of −40 dB between the frequencies of 500 kHz and 10 MHz. If later during the EMI testing stage, I find I need additional filtering, I will add a third order to the filter design by using a differential-mode filter.

The resulting schematic of the power factor correction circuit is given in Figure 9.9.

9.2.10 Printed Circuit Board Considerations

The unit in which this power factor correction circuit resides is going to be marketed everywhere in the world. The toughest safety requirements are issued by VDE in Germany. Here the creepage distance, or the distance that an arc must travel over a surface, is 3.2 mm for those signals that are opposite phases of an AC power line up to 300 VRMS. This means that there must be 3.2-mm spacing between traces of H1 and H2 (Hot and Neutral), and their rectified DC signals. Also there must be a 3.2-mm (minimum) surface distance between the windings on the input common-mode filter transformer and between high and low pins of the flyback inductor. The spacing of the 400 V output must be more than 4.0 mm from all other traces carrying less voltage. The creepage between any earth ground trace and the other traces must be more than 8.0 mm.

All current-carrying traces should be as wide and as short as possible. One-point grounding practices between the input, output, and low-level grounds should be done at the ground side of the current sense resistor.

CHAPTER 10

Off-line Converter Design and Magnetics

Sanjaya Maniktala

The design of the switching power supply section of an AC/DC converter is more of an art than an electrical design. The schematic is only a portion of the whole design, since the physical design of all of the portions of the product also enter into consideration. In addition to the high voltage switching power supply design, switching noise and dielectric isolation are the primary concerns.

The AC mains is capable of delivering many kilowatts of power to any load, so many transformer-isolated topologies are available, depending on how much output power is required. The determining factor is the peak current that the power switches must sustain. A good maximum level is less than 30A per semiconductor switch. This makes the discontinuous flyback topology the appropriate choice for loads less than 100–150W. The one-transistor forward is a good choice for loads less than 300W, the half-bridge topology until about 800W and the full-bridge to beyond 1kW.

The primary concern is the dielectric isolation between the primary circuits and the output circuits. This requires very special practices in the construction of the transformer, gaps in the printed circuit board layout, and isolated feedback circuits. There are many specifications as to how this should be done issued by the safety regulatory agencies around the world.

The transformer is larger in off-line switching power supplies due to the added requirements of spacing between the windings and the added insulating tape needed between winding layers within the transformer. Also, with so much additional empty space—that is, nonmetal volume—factors such as leakage inductance increase drastically. Noise control circuits, such as snubbers and clamps, are now greatly needed. Always rely upon a knowledgeable magnetics designer and transformer supplier for the design of the transformer. His knowledge can make a large difference in the efficiency of the supply and the noise emitted from the supply, not to mention passing the safety and noise testing.

Sanjaya Maniktala does a very thorough job of illustrating the electrical and physical considerations in the design of AC/DC transformers. This chapter will also give you a good understanding of the needs of the design.

—**Marty Brown**

Off-line converters are derivatives of standard DC-DC converter topologies. For example, the flyback topology, popular for low-power applications (typically <100W), is really a buck-boost, with its usual single-winding inductor replaced by an inductor with multiple windings. Similarly, the forward converter, popular for medium to high powers, is a buck-derived topology, with the usual inductor ("choke") supplemented by a transformer. The flyback inductor actually behaves both as an inductor and a transformer. It stores magnetic energy as any inductor would, but it also provides "mains isolation" (mandated for safety reasons), just like any transformer would. In the forward converter, the energy storage function is fulfilled by the choke, whereas its transformer provides the necessary mains isolation.

Because of the similarities between DC-DC converters and off-line converters, most of the spadework for this chapter is in fact contained in Chapter 3. The basic magnetic definitions have also been presented therein. Therefore, the reader should read that chapter before attempting this one.

Note that in both the flyback and the forward converters, the transformer, besides providing the necessary mains isolation, also provides another very important function—that of a *fixed-ratio down-conversion step*, determined by the "turns ratio" of the transformer. The turns ratio is the number of turns of the input ("primary") winding, divided by the number of turns of the *output* ("secondary") winding. The question arises—why do we even feel the need for a transformer-based step-down conversion stage, when in principle, a switching converter should by itself have been able to up-convert or down-convert at will? The reason will become obvious if we carry out a sample calculation; we will then find that without any additional "help," the converter would require impractically low values of duty cycle to down-convert from such a high input voltage to such a low output. Note that the worst-case AC mains input (somewhere in the world) can be as high as 270V. So when this AC voltage is rectified by a conventional bridge-rectifier stage, it becomes a DC rail of almost $\sqrt{2} \times 270 = 382$V, which is fed to the input of the switching converter stage that follows. But the corresponding output voltage can be very low (5V, 3.3V, or 1.8V, and so on), so the required DC transfer ratio (*conversion ratio*) is extremely hard to meet, given the minimum on-time limitations of any typical converter, especially when switching at high frequencies. Therefore, in both the flyback and forward converters, we can intuitively think of the transformer as performing a rather coarse fixed-ratio step-down of the input to a more amenable (lower) value, from which point onwards the converter does the rest (including the regulation function).

10.1 Flyback Converter Magnetics

10.1.1 Polarity of Windings in a Transformer

In Figure 10.1, the *turns ratio* is $n = n_P/n_S$, where n_P is the number of turns of the primary winding, and n_S is the number of turns of the secondary winding.

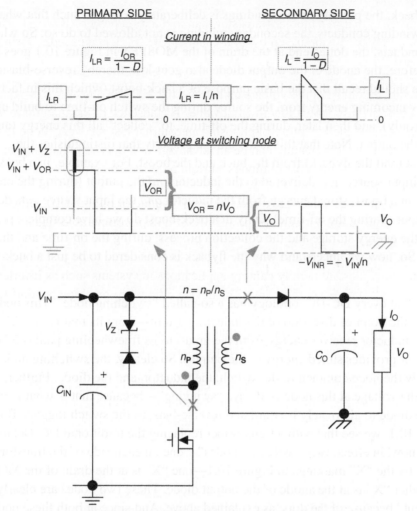

Figure 10.1: Voltage and currents in a flyback

We have also placed a *dot* on one end of each of the windings. All dotted ends of a transformer are considered to be mutually "equivalent." All non-dotted ends are also obviously mutually equivalent. That means that when the voltage on a given dotted end goes "high" (to whatever value), so does the voltage on the dotted ends of all other windings. That happens because all windings share the same magnetic core, despite the fact that they are not physically (galvanically) connected to each other. Similarly, all the dotted ends also go "low" at the same time. Clearly, the dots are only an indication of relative polarity. Therefore, in any given schematic, we can always swap the dotted and non-dotted ends of the transformer, without changing the schematic in the slightest way.

In the flyback, the polarity of the windings is deliberately arranged such that when the primary winding conducts, the secondary winding is not allowed to do so. So when the switch conducts, the dotted end at the drain of the MOSFET in Figure 10.1 goes low. And therefore, the anode of the output diode also goes low, thereby reverse-biasing the diode. We should recall that the basic purpose of a buck-boost (which this in fact also is) is to allow incoming energy from the source during the switch on-time to build up in the inductor (only), and then later, during the off-time, to "collect" all this energy (and no more) at the output. Note that this is the unique property that distinguishes the buck-boost (and the flyback) from the buck and the boost. For example, in a buck, energy from the input source gets delivered to the inductor *and* the output (during the on-time). Whereas, in a boost, stored energy from the inductor *and* the input source gets delivered to the output (during the off-time). Only in a buck-boost do we have complete separation between the energy storage and the collection process, during the on-time and the off-time. So, now we understand why the flyback is considered to be just a buck-boost derivative.

We know that every DC-DC topology has a so-called "switching node." This node represents the point of diversion of the inductor current—from its main path (i.e., in which the inductor receives energy from the input) to its freewheeling path (i.e., in which the inductor provides stored energy to the output). So clearly, the switching node is necessarily the node common to the switch, the inductor, and the diode. Further, we will find that the voltage at this node is always "swinging"—because that is what is required to get the diode to alternately forward and reverse-bias, as the switch toggles. But looking at Figure 10.1, we see that with a transformer replacing the traditional DC-DC inductor, there are now, in effect, two "switching nodes"—one on each side of the transformer, as indicated by the "X" markings in Figure 10.1—one "X" is at the drain of the MOSFET, and the other "X" is at the anode of the output diode. These two nodes are clearly "equivalent" because of the dots, as explained above. And since at both these nodes, the voltage is swinging, therefore both are considered to be "switching nodes" (of the transformer-based topology). Note that if we had, say, three windings (e.g., an additional output winding), we would have had three (equivalent) switching nodes.

10.1.2 Transformer Action in a Flyback, and Its Duty Cycle

Classic "transformer action" implies that the voltages across the windings of the transformer, and the currents *through* each of them, *scale* according to the turns ratio, as described in Figure 10.1. But it is perhaps not immediately apparent why the flyback inductor exhibits transformer action.

When the switch turns ON, a voltage V_{IN} (the rectified AC input) gets impressed across the primary winding of the transformer. And at the same time, a voltage equal to $V_{INR} = V_{IN}/n$

("R" for reflected) gets impressed across the secondary winding (in a direction that causes the output diode to get reverse-biased). Therefore, there is no current in the secondary winding when the primary winding is conducting.

Let us calculate what V_{INR} is. This voltage translation across the isolation boundary follows from the induced voltage equation applied to each winding

$$V_P = -n_P \frac{d\phi}{dt} \quad \text{and} \quad V_s = -n_s \frac{d\phi}{dt} \tag{10-1}$$

Note that both windings enclose the same magnetic core, so ϕ the flux is the same for both, and so is the rate of change of flux $d\phi/dt$ for each winding. Therefore

$$V_S = -n_S \times \left(\frac{V_P}{-n_P} \right) \tag{10-2}$$

or

$$V_S = n_S \times \left(\frac{V_{IN}}{n_P} \right) = \frac{V_{IN}}{n} \equiv V_{INR} \tag{10-3}$$

Also,

$$\frac{V_P}{n_P} = \frac{V_S}{n_S}$$

$$\frac{V_P}{V_S} = n \tag{10-4}$$

This above equation represents classic "transformer action" with respect to the voltages involved. But we also learn from the preceding equation that the volts/turn for any winding (at any given instant) is the same for all the windings present on a given magnetic core—and this is what eventually leads to the observed voltage scaling.

Note also that voltage scaling in any transformer occurs irrespective of whether a given winding is passing current or not. That is because, whether a given winding is contributing to the net flux ϕ present in the core or not, each winding encloses this entire flux, and so the basic equation $V = -N \times d\phi/dt$ applies to all windings, and so does voltage scaling.

We know that energy is built up in the transformer during the on-time. When the switch turns OFF, this stored energy (and its associated current) needs to flyback/freewheel. We also know that the voltages will automatically try to adjust themselves in any possible way, so as to make that happen. So we can safely assume the diode will somehow conduct during the switch off-time. Now, assuming we have reached a "steady state," the voltage on the output capacitor has stabilized at some fixed value V_O. Therefore, the voltage at the secondary-side switching node gets clamped at V_O (ignoring the diode drop). Further, since one end of the secondary winding is tied to ground, the voltage *across* this winding is now equal to V_O. By transformer action, this reflects a voltage *across* the primary winding, equal to $V_{OR} = V_O \times n$. But the switch is OFF during this time. Therefore, under normal circumstances, the voltage at the primary-side switching node would have settled at V_{IN}. However, now this reflected output voltage V_{OR}, coming through the transformer, adds to that. Therefore, the voltage at the primary-side switching node eventually goes up to $V_{IN} + V_{OR}$ (for now, we are ignoring the turn-off spike encircled in Figure 10.1).

> **Note:** During the on-time, the primary side is the one determining the voltages across all the windings. And during the off-time, it is the secondary winding that gets to "call the shots"!

We can calculate duty cycle from the most basic equation (from voltseconds law)

$$D = \frac{V_{OFF}}{V_{OFF} + V_{IN}} \tag{10-5}$$

We have the option of performing this calculation, either on the primary winding, or the secondary winding. Either way, we get the same result, as shown in Table 10.1.

We should be always very clear that transformer action applies only to the voltages *across* windings. And "voltage *across*" is not necessarily "voltage *at*"! To measure the voltage *at* a given point, we have to consider what the reference level (i.e., "ground" by definition) is, with respect to which its voltage needs to be measured, or stated. In fact, the reference level (i.e., by definition, "ground") is called the "primary ground" on the primary side and the "secondary ground" on the secondary side. Note that these are indicated by different ground symbols in Figure 10.1.

To find out the (absolute) voltage at the swinging end of any winding, we can use the following level-shifting rule:

To get the absolute value of the voltage at the swinging end of any winding, we must add to the voltage across the winding, the DC voltage present at its "non-swinging" end.

Table 10.1: Derivation of DC transfer function of flyback

	Primary Winding	Secondary Winding
V_{ON}	V_{IN}	$V_{INR} \equiv V_{IN}/n$
V_{OFF}	$V_{OR} \equiv V_O \times n$	V_O
DC Transfer Function	$D = \dfrac{V_{OFF}}{V_{ON} + V_{OFF}}$	
	$D = \dfrac{V_{OR}}{V_{IN} + V_{OR}}$	$D = \dfrac{V_O}{V_{INR} + V_O}$
	$D = \dfrac{nV_O}{V_{IN} + nV_O}$	

So, for example, to get the voltage at the drain of the MOSFET (swinging end of primary winding), we need to add V_{IN} (voltage at other end of winding), to the voltage waveform that represents the voltage across the primary winding. That is how we got the voltage waveforms shown in Figure 10.1.

Coming to the question of how currents actually reflect from one side of the transformer to the other, it must be pointed out that even though the final current scaling equations of a flyback transformer are exactly the same as in the case of an actual transformer, this is not strictly "classic transformer action." The difference from a conventional transformer is, that in the flyback, the primary and secondary windings do not conduct at the same time. So in fact, it is a mystery why their currents are related to each other at all!

The current scaling that occurs in a flyback actually follows from *energy considerations*. The energy in a core is in general written as

$$E = \frac{1}{2}LI^2 \tag{10-6}$$

We know the windings of our flyback conduct at different times, but the energy associated with each of the current flows must be equal to the energy in the core, and must therefore be equal to each other (we are ignoring the ramp portion of the current here for simplicity). Therefore,

$$E = \frac{1}{2}L_P I_P^2 = \frac{1}{2}L_S I_S^2 \tag{10-7}$$

where L_P is the inductance measured across the primary winding—with the secondary winding floating (no current), and L_S the inductance measured across the secondary winding—with the primary winding floating. But we also know that

$$L = N^2 \times A_L \times 10^{-9} \text{ henry} \qquad (10\text{-}8)$$

where A_L is the *inductance index*, defined previously. Therefore, in our case we get

$$L_P = n_P{}^2 \times A_L \times 10^{-9}$$

$$L_S = n_S{}^2 \times A_L \times 10^{-9}$$

Substituting in the energy equation, we get the well-known current scaling equations

$$n_P I_P = n_S I_S \qquad (10\text{-}9)$$

or

$$\frac{I_P}{I_S} = \frac{1}{n} \qquad (10\text{-}10)$$

We see that analogous to the volts/turns rule, the *ampere-turns* also need to be preserved at all times. In fact, the core itself doesn't really "care" which particular winding is passing current at any given moment, so long as there is no sudden change in the *net* ampere-turns of the transformer. This becomes the "transformer version" of the basic rule we learned in Chapter 1—that the current through an inductor cannot change discontinuously. Now we see that the ampere-turns of a transformer cannot change discontinuously.

Summarizing, transformer action works as follows—when reflecting a voltage from primary to secondary side, we need to divide by the turns ratio. When going from the secondary to the primary side, we need to multiply by the turns ratio. The rule reverses for currents—so we multiply by the turns ratio when going from primary to secondary, and divide in the opposite direction.

10.1.3 The Equivalent Buck-boost Models

Because of the many similarities, and also because of the way voltages scale in the transformer, it becomes very convenient (most of the time) to study the flyback as an equivalent DC-DC (inductor-based) buck-boost. In other words, we separate out the coarse fixed-ratio step-down ratio and incorporate it into equivalent (reflected) voltages and currents. We thereby manage to reduce the flyback transformer into a simple energy-storage medium, just like any conventional DC-DC buck-boost inductor. In other words, for most practical purposes, the transformer goes "out of the picture." The advantage is that almost all the equations and design procedure we can write for a conventional buck-boost now apply to this *equivalent* buck-boost model. One exception to this is the leakage inductance issue (and everything related to it—the clamp, the loss in efficiency due to it, the turn-off voltage spike on the switch, and so on). We will discuss this exception later. But other than that, all other parameters, like the capacitor, diode, and switch currents for example, can be more readily visualized and calculated if we use this DC-DC model approach.

The *equivalent DC-DC model* is created essentially by reflecting the voltages and currents across the isolation boundary of the transformer to one side. But again, as in the case of the duty cycle calculation (see Table 10.1), we have two options here—we can either reflect everything to the primary side, *or* everything to the secondary side. We thus get the two equivalent buck-boost models as shown in Figure 10.2. We can use the primary-side equivalent model to calculate all the voltages and currents on the primary side of the original flyback and the secondary-side equivalent model for calculating all the currents and voltages on the secondary side of the original flyback.

We know that voltages and currents reflect across the boundary by getting either multiplied or divided by the turns ratio. In fact, the reflected output voltage, V_{OR}, is one of the most important parameters of a flyback, as we will see. As the name indicates, V_{OR} *is effectively the output voltage as seen by the primary side*. In fact, if we compare the switch waveform of the flyback in Figure 10.1 with that of a buck-boost, we will realize that to the switch, it seems as if the output voltage is really V_{OR}.

As an example, suppose we have a 50W converter with an output of 5V at 10A, and a turns ratio of 20. The V_{OR} is therefore $5 \times 20 = 100$V. Now, if we change the set output to say 10V and reduce the turns ratio to 10, the V_{OR} is still 100V. We will find that none of the primary-side *voltage* waveforms change in the process (assuming efficiency doesn't change). Further, if we have also kept the output *power* constant in the process, that is, by changing the load to 5A for an output at 10V, all the *currents* on the primary side will also be unaffected. Therefore, the switch will never know the difference. In other words, the switch virtually "thinks" that it is a simple DC-DC buck-boost, delivering an output voltage of V_{OR} at a load current of I_{OR}.

As mentioned, the only difference between a transformer-based flyback that "thinks" it is providing an output of V_{OR} at the rate of I_{OR}, and an inductor-based version that *really* is providing an output of V_{OR} at the rate of I_{OR}, is the *leakage inductance* of the flyback transformer. This is that part of the primary side inductance that is not coupled to the secondary side and therefore cannot partake in the transfer of useful energy from the input to the output. We can confirm from Figure 10.1, that the only portion of the primary-side (switch) voltage waveform that "doesn't make it" to the secondary side is

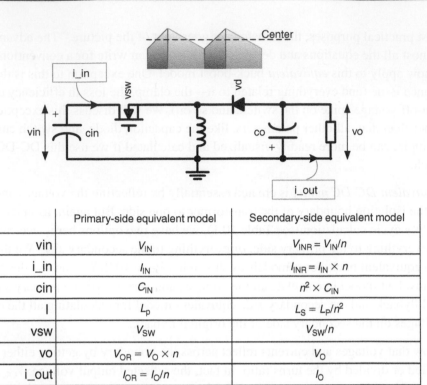

	Primary-side equivalent model	Secondary-side equivalent model
vin	V_{IN}	$V_{INR} = V_{IN}/n$
i_in	I_{IN}	$I_{INR} = I_{IN} \times n$
cin	C_{IN}	$n^2 \times C_{IN}$
l	L_p	$L_S = L_P/n^2$
vsw	V_{SW}	V_{SW}/n
vo	$V_{OR} = V_O \times n$	V_O
i_out	$I_{OR} = I_O/n$	I_O
center	$I_{OR}/(1 - D) = I_O/[n \times (1 - D)]$	$I_O/(1 - D)$
co	C_O/n^2	C_O
vd	$V_D \times n$	V_D
Duty Cycle	D	D
Current Ripple Ratio	r	r

Figure 10.2: The equivalent buck-boost models of the flyback

the *spike* occurring just after the turn-off transition. This spike comes from the uncoupled leakage inductance, as we will soon see.

Note that in the equivalent buck-boost models, the reactive component values also get reflected—though as the *square* of the turns ratio. We can understand this fact easily from energy considerations. For example, the output capacitor C_O in the original flyback was charged up to a value of V_O. So its stored energy was $1/2\ C_O V_O^2$. In the primary-side buck-boost model, the output of the converter is V_{OR}, that is, $V_O \times n$. Therefore, to keep the energy stored in this capacitor invariant (in the DC-DC model, as in the flyback), the output capacitance must get reflected to the primary side according to C_O/n^2. Note also from Figure 10.2 how the inductance reflects. This is consistent with the fact that $L \propto N^2$.

10.1.4 The Current Ripple Ratio for the Flyback

Looking at the equivalent buck-boost models in Figure 10.2, the center of the ramp on the secondary side (average inductor current, "I_L") must be equal to $I_O/(1 - D)$, as for a buck-boost (because the average diode current must equal the load current). This secondary-side "inductor" current gets reflected to the primary side, and so the center of the primary-side inductor current ramp is "I_{LR}," where $I_{LR} = I_L/n$. Equivalently, it is equal to $I_{OR}/(1 - D)$, where I_{OR} is the *reflected load current*, that is, $I_{OR} = I_O/n$. Similarly, the current swings on the primary and secondary sides are also related, via scaling (turns ratio n). Therefore, we see that the *ratio* of the swing to the center of the ramp is *identical* on both sides (primary-and secondary-side DC-DC models). We are thus in a position to define a *current ripple ratio r* for the flyback topology too—just as we did for a DC-DC converter. We just need to visualize r in a slightly different manner this time—*in terms of the center of the ramp (switch or diode), rather than the DC inductor level* (because there is no inductor present really). And as for DC-DC converters, we should normally try to set it to around 0.4.

The value of r *for a flyback is the same for either the primary and secondary DC-DC equivalent models.*

10.1.5 The Leakage Inductance

The leakage inductance can be thought of as a *parasitic inductance* in series with the primary-side inductance of the transformer. So just at the moment the switch turns OFF, the current flowing through both these inductances is "I_{PKP}," that is, the peak current on the primary side. However, when the switch turns OFF, the energy in the primary inductance has an available freewheeling path (through the output diode), but the leakage inductance energy has nowhere to go. So it expectedly "complains" in the form of a huge voltage spike (see Figure 10.1). This spike (or a scaled version of it) is *not* seen on the secondary side, simply because this is not a coupled inductance, like the primary inductance.

If we don't make any effort to collect this energy, the induced spike can be very large, causing switch destruction. Since we certainly can't get this energy to transfer to the secondary side, we have just two options—either we can try to *recover* it and cycle it back into the input capacitor, or we can simply burn it (dissipation). The latter approach is usually preferred for the sake of simplicity. It is commonly accomplished by means of a straightforward "zener diode clamp," as shown in Figure 10.1. Of course the zener voltage must be chosen according to the maximum voltage the switch can tolerate. Note that for several reasons, in particular that of efficiency, it is usually considered preferable to connect this zener *across* the primary winding (via a blocking diode in series with it). The alternative is to connect it from the switching node to primary ground.

We can ask—where does the leakage inductance really reside? Most of it is inside the primary winding of the transformer, though some of it lies in the PCB trace sections and transformer terminations, especially with those associated with the secondary winding, as we will see below.

10.1.6 Zener Clamp Dissipation

If we burn the energy in the leakage, it is important to know how this affects the efficiency. It is sometimes intuitively felt the energy dissipated every cycle is just $1/2 \times L_{LKP}I_{PK}^2$, where I_{PK} is the peak switch current, and L_{LKP} is the primary-side leakage. That certainly is the energy residing *in the leakage inductance* (at the moment the switch turns OFF), but it is *not* the *entire* energy that eventually gets dissipated in the zener clamp on account of the leakage.

The primary winding is in series with the leakage, so during the small interval that the leakage inductance is trying, in effect, to *reset*, by freewheeling into the zener, the primary winding is *forced to follow suit* and continue to provide this series current. Though the primary winding is certainly trying (and managing partly) to freewheel into the secondary side, a part of its energy also gets diverted into the zener clamp—until the leakage inductance achieves full reset (zero clamp current). In other words, some energy from the primary inductance gets literally "snatched away" by the series leakage inductance, and this also finds its way into the zener, along with the energy residing in the leakage itself. A detailed calculation reveals that the zener dissipation actually is

$$P_Z = \frac{1}{2} \times L_{LK} \times I_{PK}^2 \times \frac{V_Z}{V_Z - V_{OR}} \text{ W} \tag{10-11}$$

So the energy in the leakage $1/2 \times L_{LK} \times I_{PK}^2$ gets *multiplied* by the term $V_Z/(V_Z - V_{OR})$ (this additional term from the primary inductance).

Note that if the zener voltage is too close to the chosen V_{OR}, the dissipation in the clamp goes up steeply. V_{OR} *therefore always needs to be picked with great care*. That simply means that *the turns ratio has to be chosen carefully!*

10.1.7 Secondary-side Leakages Also Affect the Primary Side

Why did we use the symbol "L_{LK}" in the dissipation equation above? Why didn't we identify it as the *primary-side* leakage ("L_{LKP}")? The reason is that L_{LK} represents the *overall* leakage inductance as seen by the switch. So, it is partly L_{LKP}—but it also is influenced by the *secondary-side* leakage inductance. This is a little hard to visualize, since by definition, the secondary-side leakage inductance is not supposed to be coupled to the primary side (and vice versa). So how could it be affecting anything on the primary side?

The reason is that just as the primary-side leakage prevents the primary-side current from freewheeling into the output immediately (and thereby causes an increase in the zener dissipation), any secondary-side inductance also prevents the freewheeling path from becoming available immediately (following switch turn-off). Basically, the secondary-side inductance insists that we ("politely" and) slowly build up the current through it—respecting the fact that it is an inductance after all! However, until the current in the bona fide freewheeling path can build up to the required level, the primary-side current still needs to freewheel somewhere (because the switch is turning OFF)! The path the inductor current therefore seeks out is the one containing the zener clamp (being the only path available). The zener can therefore see significant dissipation, even assuming zero primary-side leakage.

In brief, the secondary-side leakage has created much the same effect as a primary-side leakage.

When both primary-and secondary-side leakages are present, we can find the effective primary-side leakage (as seen by the switch and zener clamp) as

$$L_{LK} = L_{LKP} + n^2 L_{LKS} \qquad (10\text{-}12)$$

So, like any other reactive element, the secondary-side leakage also reflects onto the primary side according to the square of the turns ratio, where it adds up in series with any primary-side leakage present.

For a given V_{OR}, if the output voltage is "low" (for example 5 V or 3.3 V), the turns ratio is much greater. Therefore, if the chosen V_{OR} is very high, the reflected secondary-side leakage can become even greater than any primary-side leakage. This can become quite devastating from the efficiency standpoint.

10.1.8 Measuring the Effective Primary-side Leakage Inductance

The best way to know what L_{LK} really is, is by measuring it! Commonly, a leakage inductance measurement is done by shorting the secondary winding pins and then measuring the inductance across the ends of the (open) primary winding. By shorting, we virtually cancel out all coupled inductance. And so what we measure is just the primary-side leakage inductance in this case.

However, the best method to measure leakage is actually an *in-circuit* measurement—so that we include the secondary-side PCB traces in the measurement. The recommended procedure is as follows.

On the given application board, a thick piece of copper foil (or a thick section of braided copper strands), with as short a length as possible, is placed directly across the diode solder pads on the PCB. A similar piece of conductor is placed across the output capacitor solder pads. Then, if we measure the inductance across the (open) primary winding pins, we will measure the effective leakage inductance L_{LK} *(not just* L_{LKP}*).*

We will find that the contribution from the secondary-sides traces can in fact make LLK several times larger than L_{LKP}. L_{LKP} can of course be measured, if desired, by placing a thick conductor across the secondary pins of the transformer.

The PCB used in the above procedure can be just a bare board with no components mounted on it, other than the transformer. Or it can even be a fully assembled board (though sometimes, we may need to cut the trace connecting the drain of the MOSFET to the transformer).

If we want to *mathematically* estimate the inductance of the secondary-side traces, the rule-of-thumb we can use is *20 nH per inch*. But here, we need to include the *full* electrical path of the high-frequency output current—starting from one end of the secondary winding, returning to its other end, through the diode and output capacitor(s). We will be surprised to calculate, or measure, that even an inch or two of trace length can dramatically decrease the efficiency by 5 to 10% in low output voltage applications.

10.1.9 Worked Example—Designing the Flyback Transformer

A 74 W universal input (90 VAC to 270 VAC) flyback is to be designed for an output of 5 V @ 10 A and 12 V @ 2 A. Design a suitable transformer for it, assuming a switching frequency of 150 kHz. Also, try to use a cost-effective 600 V-rated MOSFET.

10.1.9.1 Fixing the V_{OR} and V_Z

At maximum input voltage, the rectified DC to the converter is

$$V_{\text{INMAX}} = \sqrt{2} \times \text{VAC}_{\text{MAX}} = \sqrt{2} \times 270 = 382 \text{ V}$$

With a 600 V MOSFET, we must leave at least 30 V safety margin when at V_{INMAX}. So in our case, we do not want to exceed 570 V on the drain. But from Figure 10.1, the voltage on the drain is $V_{\text{IN}} + V_Z$. Therefore

$$V_{\text{IN}} + V_Z = 382 + V_Z \leq 570$$

$$V_Z \leq 570 - 382 = 188 \text{ V}$$

We pick a standard 180 V zener.

Note that if we plot the zener dissipation equation presented earlier, as a function of V_Z/V_{OR}, we will discover that in all cases, *we get a "knee" in the dissipation curve at around* $V_Z / V_{OR} = 1.4$. So here, too, we pick this value as an optimum ratio that we would like to target. Therefore

$$V_{OR} = \frac{V_Z}{1.4} = 0.7 \times V_Z = 0.71 \times 180 = 128 \text{ V}$$

10.1.9.2 Turns ratio
Assuming the 5 V output diode has a forward drop of 0.6V, the turns ratio is

$$n = \frac{V_{OR}}{V_O + V_D} = \frac{128}{5.6} = 22.86$$

Note that the 12V output may sometimes be regulated by a post-linear-regulator. In that case, we may have to make the transformer provide an output 3 to 5V higher (than the final expected 12V)—to provide the necessary "headroom" for the linear regulator to operate properly. This additional headroom not only caters to the dropout limits of the linear regulator, but in general also helps achieve a regulated 12V under all load conditions. However, there are also some clever *cross-regulation* techniques available that allow us to omit the 12V linear regulator, particularly if the regulation requirements of the 12V rail are not too "tight," and also if there is some minimum load assured on the outputs. In our example, we are assuming there is no 12V post regulator present. Therefore, the required turns ratio for the 12V output is 128/(12 + 1) = 9.85, where we have assumed the diode drop is 1V in this case.

10.1.9.3 Maximum duty cycle (Theoretical)
Having verified the selection of V_Z and V_{OR} at highest input, now we need to get back to the lowest input voltage, because we know from the previous discussions about the buck-boost (see the "general inductor design procedure" in the previous chapter) that V_{INMIN} is the worst-case point we need to consider for a buck-boost inductor/transformer design.

The minimum rectified DC voltage to the converter is

$$V_{INMIN} = \sqrt{2} \times VAC_{MIN} = \sqrt{2} \times 90 = 127 \text{ V}$$

We are ignoring the voltage ripple on the input terminals of the converter, and therefore we will take this as the DC input to the converter stage. So the duty cycle at minimum input voltage is

$$D = \frac{V_{OR}}{V_{OR} + V_{INMIN}} = \frac{128}{128 + 127} = 0.5 \text{ (flyback)}$$

This is clearly a "theoretical" estimate—implying 100% efficiency. We will in fact ignore this value ultimately, as we *will be estimating D more accurately by another trick.*

Note however, that this is the *operating* D_{MAX}. When we "power down" our converter for example, the duty cycle will actually increase further in an effort to maintain regulation (unless current limit and/or duty cycle limit is encountered along the way). Then depending upon the number of missing AC cycles for which we may need to ensure regulation (the *holdup time* specification), we will need to select a suitable input capacitance and also the maximum duty cycle limit, D_{LIM}, of our controller. Typically, D_{LIM} is set around 70%, and the capacitance is selected on the basis of the *3 µF/W rule-of-thumb*. For example, for our 74W supply with an estimated 70% efficiency at low line, we will draw an input power of 74/0.7 = 106W. Therefore we should use a $106 \times 3 = 318\,\mu F$ (standard value 330 µF) input capacitor. However, note that the ripple current rating of this capacitor (and its life expectancy) must be verified.

10.1.9.4 Effective load current on primary and secondary sides
Let us *lump* all the 74 W output power into an equivalent *single* output of 5 V. So the load current for a 5 V output is

$$I_O = \frac{74}{5} \approx 15\,A$$

On the primary side, the switch "thinks" its output is V_{OR} and the load current is I_{OR}, where

$$I_{OR} = \frac{I_O}{n} = \frac{15}{22.86} = 0.656\,A$$

10.1.9.5 Duty cycle
The actual duty cycle is important because a slight increase in it (from the theoretical 100% efficiency value) may lead to a significant increase in the operating peak current and the corresponding magnetic fields.

The input power is

$$P_{IN} = \frac{P_O}{\text{Efficiency}} = \frac{74}{0.7} = 105.7\,W$$

The average input current is therefore

$$I_{IN} = \frac{P_{IN}}{V_{IN}} = \frac{105.7}{127} = 0.832\,A$$

The average input current tells us what the actual duty cycle D is, because I_{IN}/D is also the center of the primary-side current ramp, and must equal I_{LR}; that is,

$$\frac{I_{IN}}{D} = \frac{I_{OR}}{1-D}$$

and solving,

$$D = \frac{I_{IN}}{I_{IN} + I_{OR}} = \frac{0.832}{0.832 + 0.656} = 0.559$$

We thus have a more accurate estimate of duty cycle.

10.1.9.6 Actual center of primary and secondary current ramps

The center of the secondary-side current ramp (lumped)

$$I_L = \frac{I_O}{1-D} = \frac{15}{1 - 0.559} = 34.01\,A$$

The center of the primary-side current ramp is

$$I_{LR} = \frac{I_L}{n} = \frac{34.01}{22.86} = 1.488\,A$$

10.1.9.7 Peak switch current

Knowing I_{LR}, we know the peak current for our selected current ripple ratio

$$I_{PK} = \left(1 + \frac{r}{2}\right) \times I_{LR} = 1.25 \times 1.488 = 1.86\,A$$

We may need to set the current limit of the controller, for example, based on this estimate.

10.1.9.8 Voltseconds

We have at V_{INMIN}

$$V_{ON} = V_{IN} = 127\,V$$

The on-time is

$$t_{ON} = \frac{D}{f} = \frac{0.559}{150 \times 10^3} \Rightarrow 3.727\,\mu s$$

So the voltsµseconds is

$$Et = V_{ON} \times t_{ON} = 127 \times 3.727 = 473\,V\mu s$$

10.1.9.9 Primary-side inductance

Note that when we come to designing off-line transformers, for various reasons like reducing high-frequency copper loss, reducing size of transformer, and so on, it is more common to set *r* at *around 0.5*. So the primary-side inductance must then be (from the "$L \times I$" rule)

$$L_\mathrm{P} = \frac{1}{I_\mathrm{LR}} \times \frac{\mathrm{Et}}{r} = \frac{473}{1.488 \times 0.5} = 636\,\mu\mathrm{H}$$

10.1.9.10 Selecting the core

Unlike made-to-order or off-the-shelf inductors, when designing our own magnetic components, we should not forget that adding an air gap dramatically improves the energy storage capability of a core. Without the air gap, the core could saturate even with very little stored energy.

Of course, we still need to maintain the desired *L*, corresponding to the desired *r*! So if we add too much of a gap, we will also need to add many more turns—thus increasing the copper loss in the windings. At one point, we will also run out of window space to accommodate these windings. So a practical compromise must be made here, one that the following equation actually takes into account (applicable to *ferrites* in general, for any topology):

$$V_\mathrm{e} = 0.7 \times \frac{(2+r)^2}{r} \times \frac{P_\mathrm{IN}}{f}\,\mathrm{cm}^3 \qquad (10\text{-}13)$$

where *f* is in kHz.

In our case we get

$$V_\mathrm{e} = 0.7 \times \frac{(2.5)^2}{0.5} \times \frac{105.7}{150} = 6.17\,\mathrm{cm}^3$$

We start looking for a core of this volume (or higher). We find a candidate in the EI-30. Its effective length and area are given in its datasheet as

$$A_\mathrm{e} = 1.11\,\mathrm{cm}^2$$
$$l_\mathrm{e} = 5.8\,\mathrm{cm}$$

So its volume is

$$V_\mathrm{e} = A_\mathrm{e} \times l_\mathrm{e} = 5.8 \times 1.11 = 6.438\,\mathrm{cm}^3$$

which is a little larger than we need, but close enough.

10.1.9.11 Number of turns

The voltage dependent equation

$$B = \frac{LI}{NA} \text{ tesla} \tag{10-14}$$

connects B to L. However we also know that a statement about r is equivalent to a statement about L—for a given frequency (the "$L \times I$ equation"). So combining these equations, and also connecting the swing in the B-field to its peak (through r), we get a very useful form of the voltage dependent equation, in terms of r (expressed in MKS units):

$$N = \left(1 + \frac{2}{r}\right) \times \frac{V_{ON} \times D}{2 \times B_{PK} \times A_e \times f} \text{ (voltage dependent equation, any topology)} \tag{10-15}$$

So even with *no* information about the permeability of the material, air gap, and so on, we already know the number of turns required on a core with area A_e that will produce a certain B-field. We also know that with or without an air gap, the B-field should not exceed 0.3 T for most ferrites. So solving the equation for N (N is n_P here, number of primary turns),

$$n_P = \left(1 + \frac{2}{0.5}\right) \times \frac{127 \times 0.559}{2 \times 0.3 \times 1.11 \times 10^{-4} \times 150 \times 10^3} = 35.5 \text{ turns}$$

We will have to verify that this can be accommodated in the window of the core—along with the bobbin, tape insulation, margin tape, secondary windings, sleeving, and so on. Usually, that is no problem for a flyback.

Note that if we want to reduce N, the only possible ways are to allow for a larger r, or decrease the duty cycle (i.e., pick a lower V_{OR}), or allow for a higher B (new material!?), or increase the area of the core—the latter, hopefully, without increasing the volume, because that would amount to overdesign. But certainly, just playing with the permeability and air gap is not going to help!

The number of secondary turns (5V output) is

$$n_S = \frac{n_P}{n} = \frac{35.5}{22.86} = 1.55 \text{ turns}$$

But we want an integral number of turns. Further, approximating this to one full turn is not a good idea since there will be more leakage. We therefore prefer to set

$$n_S = 2 \text{ turns}$$

So, with the same turns ratio (i.e., V_{OR} unchanged)

$$n_P = n_S \times n = 2 \times 22.86 \approx 46 \text{ turns}$$

The number of turns for the 12V output is obtained by the scaling rule

$$n_{S\_AUX} = \frac{12+1}{5+0.6} \times 2 = 4.64 \approx 5 \text{ turns}$$

where we have assumed the 5V diode has a drop of 0.6V and the 12V diode has a drop of 1V.

10.1.9.12 Actual B-field

So now we can use the voltage dependent equation again, to solve for B

$$B_{PK} = \left(1 + \frac{2}{r}\right) \times \frac{V_{ON} \times D}{2 \times n_P \times A_e \times f} \text{ teslas} \qquad (10\text{-}16)$$

But in fact we don't have to use this equation anymore! We realize that B_{PK} is inversely proportional to the number of turns. So if, with a calculated 35.5 turns, we had a peak field of 0.3T, then with 46 turns we will have (keeping L and r unchanged!)

$$B_{PK} = \left(\frac{35.5}{46}\right) \times 0.3 = 0.2315 \text{ teslas}$$

The swing is related to the peak by

$$\Delta B \equiv 2 \times B_{AC} = \frac{2r}{r+2} \times B_{PK} = \frac{1}{2.5} \times 0.2315 = 0.0926 \text{ T}$$

Note that in CGS units, the peak is now 2315 gauss, and the AC component is half the swing, that is, 463 gauss (since $r = 0.5$).

Note: If we start with a B-field target of 0.3T, we are likely to reach a lesser B-field after rounding up the secondary turns to the nearest higher integer, as we did above. That of course is not only expected, but acceptable. However note that on power-up or power-down, for example, the B-field will increase further, as the converter tries to continue regulating. That is why we need to set the maximum duty cycle limit and/or current limit accurately, or the switch can be destroyed due to inductor/transformer saturation. Cost-effective flyback designs with fast-acting current limit and fast switches (especially those with an integrated MOSFET) generally allow for a peak B-field of up to 0.42T, so long as the operating field is 0.3T or less.

10.1.9.13 Air gap

Finally, we need to consider the *permeability* of the material! L is related to permeability by the equation

$$L = \frac{1}{z} \times \left(\frac{\mu\mu_0 A_e}{l_e} \right) \times N^2 \text{ H} \qquad (10\text{-}17)$$

Here z is the *gap factor*

$$z = \frac{l_e + \mu l_g}{l_e} \qquad (10\text{-}18)$$

Note that z can range from 1 (no gap) to virtually any value. A z of 10, for example, increases the energy-handling capability of an ungapped core set by a factor of 10 (its A_L value falls by the same factor, and so does its *effective permeability*—$\mu e = \mu\mu_0/z$). So large gaps certainly help, but since we are still interested in maintaining L to a certain value based on our choice of r, we will have to increase the number of turns substantially. As mentioned, at some point, we just may not be able to accommodate these windings in the available window, and further, the copper loss will also increase greatly. So z in the range of 10 to 20 is a good compromise for gapped transformers made out of ferrite material. Let us see what it comes out to be, based on our requirements

$$z = \frac{1}{L} \times \left(\frac{\mu\mu_0 A_e}{l_e} \right) \times N^2 = \frac{1}{636 \times 10^{-6}} \times \left(\frac{2000 \times 4\pi \times 10^{-7} \times 1.11 \times 10^{-4}}{5.8 \times 10^{-2}} \right) \times 46^2$$

So,

$$z = 16$$

Finally, solving for the length of air gap,

$$z = 16 = \frac{5.8 + (2000)l_g}{5.8} \Rightarrow l_g = 0.0435 \, \text{cm (or 0.435 mm)}$$

> **Note:** In general, if we use a center-gapped transformer, the total gap in the center must be equal to the above calculated value, whether each center limb has been ground or not. But if spacers are being inserted on both side limbs (say on an EE or EI type of core), the thickness of the spacer on each outer limb must be half of the above-calculated value, because the total air gap is then as desired.

10.1.10 Selecting the Wire Gauge and Foil Thickness

In an inductor, the current undulates relatively smoothly. However, in a transformer, the current in one winding stops completely, to let the other winding take over. Yes, the core doesn't care (and doesn't even know) which of its windings is passing current at a given moment, as long as the ampere-turns is maintained—because only the net ampere-turns determine the field (and energy) inside the core. But as far as the windings themselves are concerned, the current is now pulsed—with sharp edges, and therefore with significant high-frequency content. Because of this, skin depth considerations are necessary for choosing the appropriate wire thickness of the windings of a flyback transformer.

> **Note:** We had ignored this for DC-DC inductors, but in high-frequency DC-DC designs too (or with high r) we may need to apply these concepts there too.

At high frequencies, the electric fields between the electrons become strong enough to cause them to repel each other rather decisively, and thereby cause the current to crowd on the *exterior* (surface) of the conductor (see exponential curve in Figure 10.3). This crowding worsens with frequency as per $\sqrt{f}$. There is thus the possibility that though we may be using thick wire in an effort to reduce the copper loss, a good part of the cross-section of the wire (its "innards") just may not be *available* to the current. The resistance presented to the current flow is inversely proportional to the area through which the current is flowing, or is able to flow. So this current crowding causes an increase in the effective resistance of the copper (as compared to its DC value). The resistance now presented to the current is called the *AC resistance* (see lower half of Figure 10.3). This is a function of frequency, because so is the skin depth. Instead of thus wasting precious space inside the transformer and losing efficiency too, we must try to use more optimum diameters of wire, in which the cross-sectional area is better-utilized. Thereafter, if we need to pass more current than the chosen cross-sectional area can handle, we need to *parallel* several such strands.

So how much current can a given wire strand handle? That depends purely on the heat buildup and the need to keep the overall transformer with an acceptable temperature rise. For this a good guideline/rule-of-thumb for the current density of flyback transformers is 400 *circular mils (cmils) per ampere*, and that is our goal too in the analysis that follows.

> Note: Expressing "current density" in the North American way of cmils/A needs a little getting used to. It is actually area per unit ampere, not ampere per unit area (as we would normally expect a "current density" to be)! So a higher cmils/A value actually is a lower current density (and vice versa)—and will produce a lower temperature rise.

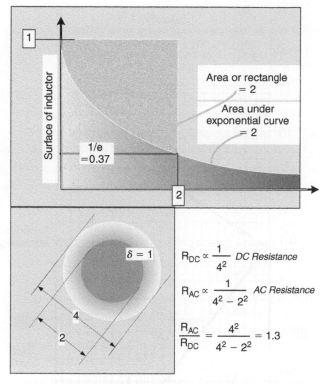

$R_{DC} \propto \dfrac{1}{4^2}$ *DC Resistance*

$R_{AC} \propto \dfrac{1}{4^2 - 2^2}$ *AC Resistance*

$\dfrac{R_{AC}}{R_{DC}} = \dfrac{4^2}{4^2 - 2^2} = 1.3$

Figure 10.3: Skin depth and AC resistance explained

We define the *skin depth* δ as the distance from the surface of a conductor at which the current density falls to $1/e$ times the value at the surface. Note that the current density at the surface is the same as the value it would have had all through the copper, were there no high-frequency effects. As a good approximation to the exponential curve, we can also imagine the current density remaining unchanged from the value at the surface, until the skin depth is reached, falling abruptly to zero thereafter. This follows from an interesting property of the exponential curve that the area under it from 0 to ∞ is equal to the area of a rectangle passing through its $1/e$ point (see Figure 10.3).

Therefore, when using round wires, if we choose the diameter as *twice* the skin depth, no point inside the conductor will be more than one skin depth away from the surface. So no part of the conductor is unutilized. In that case, we can consider this wire as having an AC resistance equal to its DC resistance—there is no need to continue to account for high-frequency effects so long as the wire thickness is chosen in this manner.

If we use copper foil, its thickness too needs to be about twice the skin depth.

In Figure 10.4 we have a simple nomogram for selecting the wire gauge and thickness. The *upper* half of this is based on the current-carrying capability as per the usual

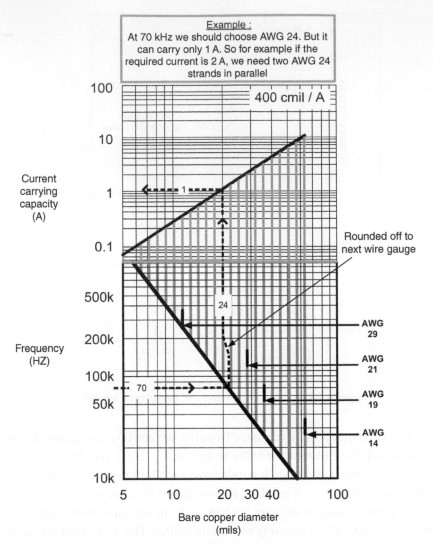

Figure 10.4: Nomogram for selecting wires and foils thicknesses, based on skin depth considerations

requirement of 400 cmil/A. But the readings can obviously be linearly scaled for any other desired current density. The vertical grid on the nomogram represents wire gauges. An example based on a switching frequency of 70 kHz is presented within the figure. In a similar manner, for our previous worked example, we see that for 150-kHz operation, we should use AWG 27. But its current-carrying capacity is only 0.5 A at 400 cmils/A (and only 0.25A at a *lower* current density of 800 cmils/A!). Therefore, since the center of the primary current ramp was iterated and estimated to be 1.488 A, we need *three* strands of AWG 27 (twisted together) to give a combined current capability of 1.5 A (which is slightly better than what we need).

Coming to the secondary side of the worked example, we remember we had lumped all the current as a 5 V equivalent load of 15 A. But in reality it is only 10 A, two-thirds of that. So the center of its current ramp, which we had calculated was about 34 A, is actually $(2/3) \times 34 = 22.7$ A. The balance of this, that is, $34 - 22.7 = 11.3$ A, reflects as $(5.6/13) \times 11.3 = 4.87$ A into the 12 V winding. So the center of the 12 V output's current ramp is 4.87 A. We can choose the 12 V winding arrangement using the same arguments we present below for the 5 V winding.

For the 5 V winding, we can consider using copper foil, since we have only two turns, and we need a high current capability. The center of the 5 V secondary-side current ramp is about 23 A. The appropriate thickness (2δ) at this frequency is found by projecting downward along the AWG 27 vertical line. We get about 14 mils thickness. But we still don't know if the current through it will follow our guideline of 400 cmil/A, since it is a foil. We need to check this out further.

One circular mil (cmil) is equal to 0.7854 square mils. Therefore 400 cmils is $400 \times 0.7854 = 314$ sq. mils (note $\pi/4 = 0.7854$). So for 23 A we need $23 \times 314 = 7222$ sq. mils. But the thickness of the foil is 14 mils. Therefore we need the copper foil to be $7222/14 = 515$ mils wide—that is, about half an inch. Looking at a bobbin for the EI-30 in Figure 10.5, we see it can accommodate a foil 530 mils wide. So this is just about acceptable. Note that if the available width is insufficient, we would need to look for another core altogether—one with a "longer" (stretched out) profile. Cores like that are available as American "EER" cores. Or we can again consider using several paralleled strands of round wire. The problem is that a bunch of 46 twisted strands (of AWG 27) is going to be bulky, difficult to wind, and will also increase the leakage inductance. So we may like to use, say, 11 or 12 strands of AWG 27 twisted together into one bunch, and then take four of these bunches (all electrically in parallel), laid out side by side to form one layer of the transformer. For a two-turns secondary, therefore, we would wind two layers of this.

10.2 Forward Converter Magnetics

The procedure presented in this section applies explicitly to the *single-switch* forward converter. However, the general procedure remains unchanged for the *two-switch* forward converter as well.

10.2.1 Duty Cycle

The duty cycle of a forward converter is

$$V_{\mathrm{O}} = V_{\mathrm{IN}} \times D \times \frac{n_{\mathrm{S}}}{n_{\mathrm{P}}}$$

(10-19)

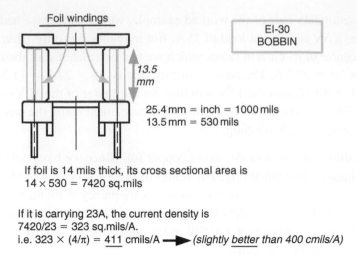

Foil windings

EI-30
BOBBIN

13.5 mm

25.4 mm = inch = 1000 mils
13.5 mm = 530 mils

If foil is 14 mils thick, its cross sectional area is
14 × 530 = 7420 sq.mils

If it is carrying 23A, the current density is
7420/23 = 323 sq.mils/A.
i.e. 323 × (4/π) = 411 cmils/A ⟶ (slightly *better* than 400 cmils/A)

Figure 10.5: Checking to see if a 23A foil can be accommodated on an EI-30 bobbin

Comparing this with the duty cycle of a buck, we see that the only difference is the term nS/nP. As mentioned, this is the coarse fixed-ratio step-down function available due to transformer action. We can therefore visualize that the input voltage V_{IN} gets reflected to the secondary side. This reflected voltage $V_{INR} = V_{IN}/n$ (where $n = n_P/n_S$) gets impressed at the secondary-side switching node. *From there on, we have in effect a simple DC-DC buck stage, with an input voltage of V_{INR} and an output of V_O* (see Figure 10.6). Therefore, the design of the forward converter's choke is *not* going to be covered here, as it is designed using the same procedure as that of any buck inductor. However, the forward converter's transformer is another story altogether!

> **Note:** Regarding choke design, we should keep in mind that for high-current inductors, as would be found in a typical forward converter, the calculated wire gauge may be too thick (and stiff) for winding easily over the core/bobbin. In that case, several thinner wire gauges may be twisted together to make the winding more flexible and easier to handle in production. Further, since choke and inductor design has usually little to do with high-frequency skin depth considerations, we can choose strands of almost any practical diameter, so long as we have enough net copper cross-sectional area to keep the temperature rise to within about 40 to 50°C.

Unlike a flyback transformer, the forward converter's secondary winding conducts at the same time as the primary winding. This leads to an *almost* complete flux cancellation inside the core. But there is one component of the primary current waveform which remains the

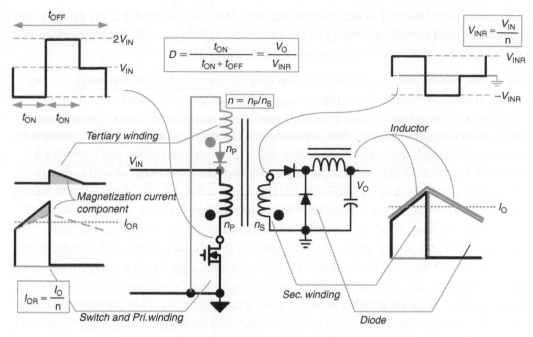

Figure 10.6: The single-ended forward converter

same, irrespective of the load. This is the *magnetization current* component—shown in gray on the left side of Figure 10.6. At zero load, this is the entire current through the primary winding and switch (assuming duty cycle remains fixed). As soon as we try to draw some load current, the secondary winding current increases, and so does the primary winding current. Each current increases proportionally to the load current, and so their increments too are mutually proportional—the proportionality constant being the turns ratio. But more significantly, they are of *opposite sign*—that is, looking at Figure 10.6, we see that the current enters the dotted end of the transformer on the primary side, and on the secondary side, it leaves by the dotted end at the same time. Therefore, the net flux in the core of the transformer remains unchanged from the zero load condition (assuming *D* is fixed)— because the core just never "sees" any change in the net ampere-turns flowing through its windings. All conditions inside the core, i.e. the flux, the magnetic fields, the energy stored, and even the core loss, are only dependent on the magnetization current. Of course the windings themselves have a different story to tell—they bear the entire brunt, not only of the actual load current, but the sharp edges and consequent high-frequency content of the pulsed current waveforms.

The magnetization current component is not coupled by transformer action to the secondary. In that sense, it is like a "parallel leakage inductance." We need to subtract this component from the total switch current, and only then will we find that the primary and secondary currents scale according to the turns ratio. In other words, the magnetization current does not scale—it stays confined to the primary side.

But in fact, the magnetization current is the only current component that is storing any energy in the transformer. So in that sense, it is like the flyback transformer! But, if we are to achieve a steady state, even the transformer needs to be "reset" every cycle (along with the output choke). But unfortunately, the magnetization energy is effectively "uncoupled," simply because of the output diode direction, and so we can't transfer it over to the secondary side. If we don't do anything about this energy, it will certainly destroy the switch by a spike similar to the leakage in a flyback. We don't want to burn it either, for efficiency reasons. Therefore, the usual solution is to use a "tertiary winding" (or "energy recovery winding"), connected as shown in Figure 10.6. Note that this winding is in flyback configuration with respect to the primary winding. It conducts only when the switch turns OFF, and thereby freewheels the magnetization energy back into the input capacitor. There is some loss associated with this "circulating" energy term, because of the diode drop and resistance of the tertiary winding. Note however, that any bona fide leakage inductance energy also gets recycled back into the input by the tertiary winding. So we don't need an additional clamp for it.

For various subtle reasons, like being able to ensure the transformer resets predictably under all conditions, and also for various production-related reasons, the number of turns of the tertiary winding is usually kept exactly the same as the primary winding. Therefore by transformer action, the voltage at the primary-side switching node (drain of the MOSFET) must rise to $2 \times V_{IN}$ when the switch turns OFF. Therefore, in a universal-input off-line *single-ended* (i.e., single-switch) forward converter, we need a switch rated for at least 800 V.

As soon as the transformer is reset (i.e. the current in the tertiary winding returns to zero), the drain voltage suddenly drops to V_{IN}—that is, *no voltage* is then present across the primary winding—and therefore there is no voltage across the secondary winding either. The catch diode of the output stage (i.e., the diode connected to the secondary ground in Figure 10.6) then freewheels the energy contained in the choke. Note that there is actually some ringing on the drain of the MOSFET for a while, around an average level of V_{IN}, just after transformer reset occurs. This is attributable to various undocumented parasitics (not displayed in the figure). The ringing, however, does contribute significantly to the radiated EMI.

Note that even prior to transformer reset, the secondary winding has not been conducting for a while—simply because the output diode (i.e., the one connected to swinging end of the secondary winding) has been reverse-biased during the time the tertiary winding was conducting.

Note also that the duty cycle of such a forward converter can under no circumstances ever be allowed to exceed 50%. The reason for that is we have to unconditionally ensure that transformer reset will always occur, every cycle. Since we have no direct control on the transformer current waveforms, we have to just leave enough time for the current in the tertiary winding to ramp down to zero on its own. In other words, we have to allow voltseconds balance to occur naturally in the transformer. However, because the number of turns in the tertiary winding is equal to the primary turns, the voltage across the tertiary winding is equal to V_{IN} when the switch is ON, and is also equal to V_{IN} (opposite direction) when the switch is OFF. Reset will therefore occur when t_{OFF} becomes equal to t_{ON}. So, if the duty cycle exceeds 50%, t_{ON} would certainly always exceed t_{OFF}, and therefore transformer reset would never be able to occur. That would eventually destroy the switch. Therefore, just to allow t_{OFF} to be large enough, the duty cycle must always be kept to less than 50%.

We realize that the forward converter transformer is always in DCM (its choke is usually in CCM, with an r of 0.4). Further, since the flux in the transformer remains unchanged for all loads, we can logically deduce that no part of the energy flowing through it into the output must be being stored in the transformer. So the question really is—what does the power-handling capability of a forward converter transformer depend on? We intuitively realize that we can't use any size transformer for any output wattage! So what governs the size? We will soon see that it is determined simply by how much copper we can squeeze into the available window area of the core (and more importantly, how well we can utilize this available area), without getting the transformer *too hot*.

10.2.2 Worst-case Input Voltage End

The most basic question in design invariably is—what input voltage represents the worst-case point at which we need to start the design of the magnetics (from the viewpoint of core saturation)? For the forward converter choke, this should be obvious—as for any buck converter, we need to set its current ripple ratio at around 0.4 at V_{INMAX}. But coming to the transformer, we need some analysis before we can make a proper conclusion.

Note that the transformer of a forward converter is in discontinuous mode (DCM), but the duty cycle is determined by the choke, which is in CCM. Therefore, the duty cycle of the transformer also gets "slaved" at the CCM duty cycle of $D = V_O/V_{INR}$, despite the fact that it is in DCM. This rather coincidental CCM + DCM interplay leads to an interesting observation—the voltseconds across the forward converter transformer is a constant, irrespective of the input voltage. The following calculation makes that clear, by the fact that V_{IN} cancels out completely:

$$\text{Et} = V_{IN} \times \frac{D}{f} = V_{IN} \times \frac{V_O}{V_{INR} \times f} = V_{IN} \times \frac{V_O \times n}{V_{IN} \times f} = \frac{V_O \times n}{f} \qquad (10\text{-}20)$$

So in fact, the swing Δ of the current or the field is the same at high input or at low input, or in fact at any input (as long as the choke is in CCM). Since the transformer is in DCM, its peak is equal to its swing, and so the peak too does not depend on V_{IN}. Of course, the peak *switch* current I_{SW_PK} is the sum of the peak of the magnetization current I_{M_PK}, and the peak of the secondary-side current waveform reflected onto the primary side, that is,

$$I_{SW\_PK} = I_{M\_PK} + \frac{1}{n}\left[I_O\left(1 + \frac{r}{2}\right)\right] \qquad (10\text{-}21)$$

So although the *current limit of the switch must be set high enough to accommodate* I_{SW_PK} *at* V_{INMAX} (since that is where the maximum peak of the reflected output current component occurs), *as far as the transformer core is concerned, the peak current (and corresponding field) is just* I_{M_PK}, *which does not depend on* V_{IN}! This is indeed an interesting situation. Note, also, that as far as the choke is concerned, the peak inductor current is no longer equal to the (reflected) peak switch current (as in a DC-DC buck topology), though the peak diode current still is. Yes, if we subtract the magnetization current from the switch current, and then scale (reflect) it to the secondary side according to the turns ratio, then the peak of that waveform will be equal to the peak inductor current.

So effectively I_M has the property of *input voltage rejection*. We can intuitively understand this in the following way—as the input increases, it tries to increase the slope of the transformer current ramp and thereby get ΔI to increase. However, the output choke, sensing a higher V_{INR}, decreases its duty cycle, and therefore also that of the transformer, and in effect tries to thereby reduce the current swing in the transformer. Coincidentally, these two opposing forces virtually counterbalance each other perfectly, and so there is no net change in the resultant current swing in the transformer.

As a corollary, the *core loss* in the transformer is independent of the input voltage. *The copper loss, on the other hand, is always worse at low inputs* (except for the DC-DC buck)—simply because the average input current has to increase so as to continue to satisfy the basic power requirement $P_{IN} = V_{IN} \times I_{IN} = P_O$.

Though we can pick any specific input voltage point for assuring ourselves that the core does not saturate *anywhere within its input range*, since the copper loss is at its worst at V_{INMIN}, we conclude that *the worst-case for a forward converter transformer is at* V_{INMIN}. For the choke, it is still V_{INMAX}.

10.2.3 Window Utilization

Looking at a typical winding arrangement on an ETD-34 core and bobbin in Figure 10.7, we see that the plastic bobbin occupies a certain part of the space provided by the core—thus

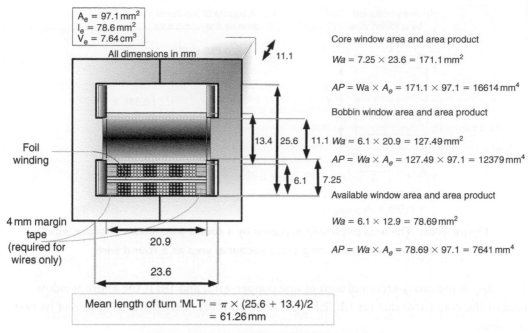

$A_e = 97.1\,mm^2$
$l_e = 78.6\,mm^2$
$V_e = 7.64\,cm^3$

All dimensions in mm

11.1

Foil winding

13.4 25.6 11.1

6.1 7.25

4 mm margin tape (required for wires only)

20.9

23.6

Core window area and area product

$Wa = 7.25 \times 23.6 = 171.1\,mm^2$

$AP = Wa \times A_e = 171.1 \times 97.1 = 16614\,mm^4$

Bobbin window area and area product

$Wa = 6.1 \times 20.9 = 127.49\,mm^2$

$AP = Wa \times A_e = 127.49 \times 97.1 = 12379\,mm^4$

Available window area and area product

$Wa = 6.1 \times 12.9 = 78.69\,mm^2$

$AP = Wa \times A_e = 78.69 \times 97.1 = 7641\,mm^4$

Mean length of turn 'MLT' = $\pi \times (25.6 + 13.4)/2$
= 61.26 mm

Figure 10.7: An ETD-34 bobbin analyzed

reducing the available window Wa from 171 mm² to 127.5 mm²—that is, by 74.5%. Further, if we include the 4-mm *margin tape* that needs to be typically provided on either side (to satisfy international safety norms regarding clearance and creepage requirements between primary and secondary sides), we are left with an available window of only 78.7 mm²— that's a total reduction of 78.7/171 = 46%. In addition to this, looking at the left side of Figure 10.8, we see that for any given wire, only 78.5% of the square area it "physically occupies" (or will occupy in the transformer) is actually conducting (copper). So in all, this leads to a total reduction of the available window space by 0.46 × 0.785 = 36%.

We realize some more space will also be lost to interlayer insulation (and any EMI screens if present), and so on. Therefore, finally, we estimate that perhaps only 30 to 35% of the available *core window area* will actually be occupied by copper. That is the reason why we need to introduce a *window utilization factor K* (later we will set it to an estimated value of 0.3). So

$$K = \frac{N \times A_{\mathrm{CU}}}{Wa} \tag{10-22}$$

and

$$N = \frac{K \times Wa}{A_{\mathrm{CU}}} \tag{10-23}$$

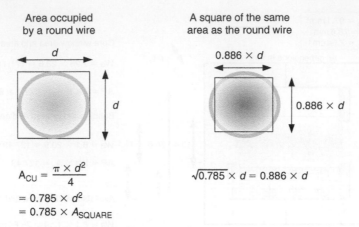

Figure 10.8: The area physically occupied by a round wire, and a "square wire" of the same conducting cross-sectional area as a round wire

Here A_{CU} is the cross-sectional area of *one* copper wire, and Wa is the entire window area of the core (note that for EE, EI types of cores this is only the area of *one* of its two windows!).

10.2.4 *Relating Core Size to Its Power Throughput*

We remember that the original form of the voltage dependent equation is

$$\Delta B = \frac{V_{IN} \times t_{ON}}{N \times A} \text{ teslas} \tag{10-24}$$

Substituting for N, the number of primary turns, we get

$$\Delta B = \frac{V_{IN} \times t_{ON} \times A_{CU}}{K \times Wa \times A} \text{ teslas} \tag{10-25}$$

Performing some manipulations,

$$\begin{aligned} \Delta B &= \frac{V_{IN} \times I_{IN} \times t_{ON} \times A_{CU}}{I_{IN} \times K \times Wa \times A} = \frac{P_{IN} \times (D/f) \times A_{CU}}{I_{IN} \times K \times Wa \times A} \\ &= \frac{P_{IN} \times (D/f) \times A_{CU}}{I_{SW} \times D \times K \times Wa \times A} \end{aligned} \tag{10-26}$$

$$\Delta B = \frac{P_{IN}}{(I_{SW}/A_{CU}) \times K \times f \times Wa \times A} = \frac{P_{IN}}{(J_{A/m^2}) \times K \times f \times AP} \tag{10-27}$$

where J_{A/m^2} is the current density in A/m^2, and AP is called the *area product* ($AP = Ae \times Wa$). Let us now convert into CGS units for greater convenience. We get

$$\Delta B = \frac{P_{IN}}{(J_{A/cm^2}) \times K \times f \times AP} \times 10^8 \text{ gauss} \qquad (10\text{-}28)$$

where AP is also in cm^2 now. Finally, converting the current density into cmils/A by using

$$J_{cmils/A} = \frac{197,353}{J_{A/cm^2}} \qquad (10\text{-}29)$$

we get

$$\Delta B = \frac{P_{IN} \times J_{cmils/A}}{197,353 \times K \times f \times AP} \times 10^8 \text{ gauss} \qquad (10\text{-}30)$$

Solving for the area product

$$AP = \frac{506.7 \times P_{IN} \times J_{cmils/A}}{K \times f \times \Delta B} \text{ cm}^4 \qquad (10\text{-}31)$$

Let us do some substitutions here. Assuming a typical current density of 600 cmil/A, utilization factor K of 0.3, ΔB equal to 1500 gauss, we get the following fundamental core-selection criterion:

$$AP = 675.6 \times \frac{P_{IN}}{f} \text{ cm}^4 \qquad (10\text{-}32)$$

> **Note:** In a typical forward converter, it is customary to set the swing in the B-field of the transformer at $\Delta B \approx 0.15$ teslas. This helps reduce core loss, and usually also leaves enough safety margin for avoiding hitting BSAT under say power-up condition at high line. Note that in a flyback, the core loss tends to be much less, because ΔI is a fraction of the total current (40% typically). But since the transformer of a forward converter is always in DCM, therefore the swing in B is now more significant—equal to its peak value, i.e., $B_{PK} = \Delta B$. So, if we set the peak field at 3000 gauss, ΔB would be 3000 gauss too, roughly twice that of a flyback set to the same peak. That is why we must reduce the peak field in a forward converter to about 1500 gauss.

10.2.5 Worked Example—Designing the Forward Transformer

We are building a 200-kHz forward converter for an AC input range of 90 to 270 volts. The output is 5 V at 50 A, and the estimated efficiency is 83%. Design its transformer.

10.2.5.1 Input power
We have

$$P_{IN} = \frac{P_O}{\text{Efficiency}} = \frac{5 \times 50}{0.83} \approx 300 \text{ W}$$

10.2.5.2 Selection of core
We use the criterion calculated previously:

$$AP = 675.6 \times \frac{P_{IN}}{f} = 675.6 \times \frac{300}{2 \times 10^5} = 1.0134 \text{ cm}^4$$

The area product of the ETD–34 shown in Figure 10.7 is

$$AP = W \frac{\left[\dfrac{25.6 - 11.1}{2}\right] \times 23.6 \times 97.1}{10^4} = 1.66 \text{ cm}^4$$

This is, in theory, probably a little larger than required. But it is the closest standard size in this range. Later we will see it is in fact just about adequate.

10.2.5.3 Skin depth
The skin depth is

$$\delta = \frac{66.1 \times [1 + 0.0042(T - 20)]}{\sqrt{f}} \text{ mm}$$

where f is in hertz and T is the temperature of the windings in °C. Therefore assuming a final temperature of $T = 80°C$ (40°C rise over a maximum ambient of 40°C), we get at 200 kHz

$$\delta = \frac{66.1 \times [1 + 0.0042 \times (60)]}{\sqrt{2 \times 10^5}} = 0.185 \text{ mm}$$

10.2.5.4 Thermal resistance
An empirical formula for EE-EI-ETD-EC types of cores is

$$\text{Rth} = 53 \times V_e^{-0.54} \text{ °C/W}$$

where V_e is in cm³. Therefore since $V_e = 7.64 \text{ cm}^3$, for the ETD-34

$$\text{Rth} = 53 \times 7.64^{-0.54} = 17.67°\text{C/W}$$

10.2.5.5 Maximum B-field

For a 40°C estimated rise in temperature, the maximum allowed dissipation is

$$P \equiv P_{CU} + P_{CORE} = \frac{\deg C}{Rth} = \frac{40}{17.67} = 2.26 \text{ watts}$$

Let's *divide this loss equally into copper and core losses* (typical first-cut assumption). So

$$P_{CU} = 1.13 \text{ watts}$$
$$P_{CORE} = 1.13 \text{ watts}$$

Therefore, the allowed *core loss per unit volume* is

$$\frac{\text{core loss}}{\text{volume}} = \frac{1.13}{7.64} \Rightarrow 148 \text{ mW/cm}^3$$

Using "System B" of Table 3.5 we get

$$\frac{\text{core loss}}{\text{volume}} = C \times B^p \times f^d$$

where B is in gauss and f in hertz. Therefore, solving for B

$$B = \left[\frac{\text{core loss}}{\text{volume}} \times \frac{1}{C \times f^d} \right]^{1/P}$$

If we are using the ferrite grade "3C85" (from Ferroxcube), we see from Table 3.6 that $p = 2.2$ and $d = 1.8$ and $C = 2.2 \times 10^{-14}$. Therefore

$$B = \left[148 \times \frac{1}{2.2 \times 10^{-14} \times 2^{1.8} \times 10^{5 \times 1.8}} \right]^{1/2.2} = 720 \text{ gauss}$$

We note that the "B" referred to here is actually, by convention, B_{AC}. So, we get the total allowed swing as

$$\Delta B = 2 \times B = 2 \times 720 = 1440 \text{ gauss}$$

10.2.5.6 Voltμseconds

Earlier, we had presented the following form of the voltage dependent equation

$$\Delta B = \frac{100 \times Et}{Z \times A} \text{ gauss}$$

where A is the effective area in cm^3. The *duty cycle of a typical forward converter is set to about 0.35 at low line* so as to meet the typical 20-ms holdup time requirement, without requiring an inordinately-sized input capacitor. The rectified input at low line is $90 \times 2 = 127$ V. The applied voltseconds is therefore (at any line voltage)

$$\text{Et} = V_{IN} \times \frac{D}{f} = 127 \times \frac{0.35}{2 \times 10^5} = 222.25 \text{ V}\mu\text{s}$$

10.2.5.7 Number of turns

Since $\Delta B = 1440$ gauss, we solve the following equation for N

$$\Delta B = \frac{100 \times \text{Et}}{Z \times A} \text{ gauss}$$

$$n_P = \frac{100 \times \text{Et}}{\Delta B \times A} = \frac{100 \times 222.25}{1440 \times 0.97} = 15.9 \text{ turns}$$

Note that this says nothing about the required *inductance*. We need these many number of turns, irrespective of the (primary) inductance. Yes, changing the inductance will affect the peak magnetization and the switch current, because it changes the proportionality constant connecting B and I. However, B still remains fixed, independent of the inductance!

Assuming a 0.6 V forward drop across the diode, the required turns ratio is

$$n = \frac{n_P}{n_S} = \frac{V_{IN}}{V_{INR}} = \frac{V_{IN}}{\left(\dfrac{V_O + V_D}{D} \right)} = \frac{127 \times 0.35}{5 + 0.6} = 7.935$$

Therefore, the number of secondary turns is

$$n_S = \frac{15.9}{7.935} = 2.003 \text{ turns}$$

Note that this could have turned out to be significantly different from an integer. In that case, we would round it off to the nearest (higher) integer, and then recalculate the primary turns, the new flux density swing, and the core loss—similar to what we did for the flyback. But at the moment, we can simply use

$$n = 8 \text{ (turns ratio)}$$

$$n_P = 16 \text{ turns}$$

$$n_S = 2 \text{ turns}$$

10.2.5.8 Secondary foil thickness and losses

The concept of skin depth presented earlier actually represents a single wire standing freely in space. For simplicity, we just ignored the fact that the field from the *nearby windings* may be affecting the current distribution significantly. In reality *even the annular area we were hoping was fully available for the high frequency current, is not.* Every winding has an associated field, and when this impinges on nearby windings, the charge distribution changes, and eddy currents are created (with their own fields). This is called the *proximity effect*. It can greatly increase the *AC resistance* and thus the copper losses in the transformer.

The first thing we need to do to improve the situation is have *opposing* flux lines cancel each other. In a forward converter, that is in fact something that tends to happen automatically, because the secondary windings pass current at the same time as the primary, and in the opposite direction. However, even that can prove totally inadequate, especially at the higher power levels that a forward converter is more commonly associated with. So a further reduction in these proximity losses is achieved by *interleaving* as shown in Figure 10.9.

Basically, by splitting the sections, and trying to get primary and secondary layers adjacent to each other as much as possible, we can increase cancellation of local adjoining fields. In effect, we are trying to prevent the ampere-turns from cumulating as we go from one layer to the next. Note that the ampere-turns are proportional to the local fields that are causing the proximity losses. However, it is impractical to interleave too much—because we will need several more layers of primary-to-secondary insulation, more terminations, and also more EMI screens at every interface (if required)—all of which will add up to higher cost and eventually lead to possibly higher, rather than lower, leakage. Therefore, *most medium-power off-line supplies just split the primary into two sections, one on either side of a single-section secondary.*

The other way to reduce losses is to decrease the thickness of the conductor. But there are several ways we can do this. If, for example, we take a winding made up of single-strand wire, and split the wire into several paralleled finer strands in such a way that *the* overall DC resistance does not change in the process, we will find that the AC resistance goes up first before it reduces. On the other hand, if we take a foil winding, and decrease its thickness, the AC resistance falls before it rises again.

In Figure 10.9, we have also defined p, the *layers per portion*. Note how p gets reassigned when we interleave.

But how do we go about actually estimating the losses? Dowell reduced a very complex multidimensional problem into a simpler, one-dimensional one. Based on his analysis, we can show that there is an optimum thickness for each layer. Expectedly, this turns out to be much less than $2 \times \delta$, where δ is the skin depth defined earlier.

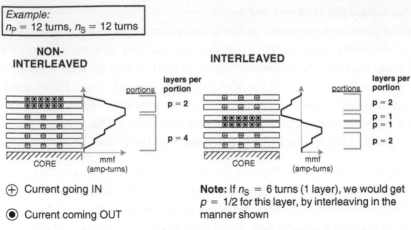

Figure 10.9: How proximity losses are reduced by interleaving

> **Note:** In the flyback, we had ignored the proximity effect for the sake of simplicity. But in any case, since the primary and secondary windings do not conduct at the same time, interleaving won't help. But interleaving is still carried out in the flyback, in a manner similar to the forward converter. However, the purpose then is to increase coupling between primary and secondary, and thereby reduce the leakage inductance. However, this also increases the capacitive coupling—unless grounded screens are placed at the primary-secondary interface. Screens are in general helpful in reducing high-frequency noise from coupling into the output, and suppressing common-mode conducted EMI. But they also increase the leakage inductance, which is of great concern particularly in the flyback. Note also that screens must be very thin, or they will develop very high eddy current losses of their own. Further, the ends of a primary-secondary screen should not be connected together, or they will constitute a shorted turn for the transformer.

In Figure 10.10, we have plotted out Dowell's equations in a form applicable to a square current waveform (unidirectional) in a transformer with foil windings. Note that the original Dowell curves actually plot F_R versus X. But we have plotted F_R/X versus X, where

$$F_R = \frac{R_{AC}}{R_{DC}} \tag{10-33}$$

and

$$X = \frac{h}{\delta} \tag{10-34}$$

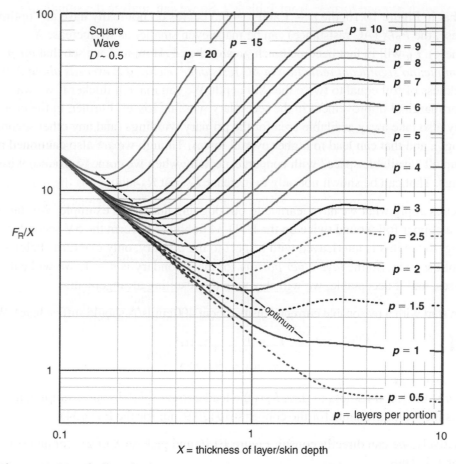

Figure 10.10: Finding the lowest AC resistance, as the thickness of a foil is varied

h being the thickness of the foil. The reason why we have not plotted F_R versus X is that F_R is only the ratio of the AC to DC resistance. It is not F_R, but R_{AC} that we are really interested in minimizing. So the "optimum R_{AC}" point need not necessarily be the point of the lowest F_R.

Let us try to understand this for a stand-alone foil (similar to what we did in Figure 10.3). If we slowly increase the thickness of the foil, once the foil thickness exceeds 2δ, the AC resistance won't change any further, since the cross-sectional area available for the high-frequency current remains confined to δ on each side of the foil. But the DC resistance continues to decrease as per $1/h$—and as a result F_R will increase. So the relationship between R_{AC} and F_R is not necessarily obvious. Therefore, since $F_R = R_{AC}/R_{DC}$, with $R_{DC} \propto 1/h$, we get $R_{AC} \propto F_R/h$. And this is what we really need to minimize (for a foil). Further, since we always like to write any frequency-dependent dimension in reference to the skin depth, we have plotted not F_R/h, but F_R/X, versus X, in Figure 10.10.

Note that in Figure 10.10, the $p = 1$ and $p = 0.5$ curves do not really have an "optimum." For these, the F_R/X (AC resistance) can be made even smaller as we increase X (thickness). F_R will in fact become much greater than 1. However, we see that *for p = 1 for example, no significant reduction in AC resistance occurs if X exceeds about 2*, that is, thickness of foil equal to twice the skin depth. We can make it thicker if we want, but only for marginal improvement in the secondary winding losses. Further, in the process, we may also take away available area for the primary windings (and any other secondary windings), and that can lead to higher overall losses. Though we are also cautioned not to fill up all "available space" with copper, especially when we come to (round) wire windings. That can be shown not only to increase F_R, but R_{AC} too.

Now let us apply what we have learned to our ongoing numerical example. We start by taking a copper foil wound twice on the ETD-34 bobbin—to form the 5V secondary winding. Since this is interleaved with respect to the primary, only one turn "belongs" to each split section. So the *layers per portion* for the secondary is $p = 1$. We will calculate the losses, and if acceptable, we will stay with the resulting arrangement.

We can start with a reasonable current density (about 400 cmils/A should suffice here). We use

$$h = \frac{I_O \times J_{\text{cmils/A}} \times 10^2}{\text{width} \times 197,353} \text{ mm}$$

where h is the foil thickness in mm, I_O is the load current (50A in our example), and *width* is the width available for the copper strip (20.9 mm for the ETD-34).

Alternatively, we can directly consult Figure 10.10 and pick an X of 2.5 for an estimated F_R/X of 1.4. Thus

$$h = X \times \delta = 2.5 \times 0.185 = 0.4625 \text{ mm}$$

The *mean length per turn* (MLT) of the ETD-34 is 61.26 mm (see Figure 10.7), the ("hot") resistivity of copper (ρ) is 2.3×10^{-5} ohms-mm, so we get the resistance of the secondary winding in ohms as

$$R_{\text{AC\_S}} = \left(\frac{F_R}{X}\right) \times \frac{\rho \times \text{MLT} \times n_S}{\text{width} \times \delta} = (1.4) \times \frac{2.3 \times 10^{-5} \times 61.26 \times 2}{20.9 \times 0.185} = 1.02 \times 10^{-3}$$

Note that since F_R/X is set to 1.4, the corresponding F_R is

$$F_R = 1.4 \times \frac{h}{\delta} = 1.4 \times \frac{0.4625}{0.185} = 3.5$$

This is fairly high, but as explained, it is actually helpful here, because R_{AC} goes down. Now, the current in the secondary looks like a typical switch waveform, with its center equal to the load current (50A), and a certain current ripple ratio set by the output choke. Its RMS value is

$$I_{RMS\_S} = I_O \times \sqrt{D \times \left(1 + \frac{r^2}{12}\right)} \, A \qquad (10\text{-}35)$$

However, we do not yet know what the current ripple ratio of the choke, r, is at 90 VAC. The r has probably been set to 0.4 *at* V_{INMAX}, *not at* V_{INMIN}. Nevertheless, it is easy to work out the new r as follows. The duty cycle is inversely proportional to input voltage. Therefore, if D is 0.35 at 270 VAC, then at 90 VAC it is 0.35/3 = 0.117. *Further, r varies as per (1 – D) for a buck stage.* Therefore the value of r at 90 VAC is

$$r = \frac{1 - 0.35}{1 - 0.117} \times 0.4 = 0.294$$

So the RMS current in the secondary winding is

$$I_{RMS\_S} = I_O \times \sqrt{D \times \left(1 + \frac{r^2}{12}\right)} = 50 \times \sqrt{0.35 \times \left(1 + \frac{0.294^2}{12}\right)} = 29.69 \, A$$

The heat dissipated in the secondary windings is finally

$$P_S = I_{RMS\_S}^2 \times R_{AC\_S} = 29.69^2 \times 1.02 \times 10^{-3} = 0.899 \, W$$

If the losses are not acceptable, we may need to look for a bobbin that will allow a *wider width* of foil. Or we can consider paralleling several thinner foils to increase p. For example, if we take four paralleled (thinner) foils in parallel (each insulated from the others), we will get four effective layers for the secondary, and the layers per portion will then become 2.

10.2.5.9 Primary winding and losses

For the secondary, we have finally chosen copper foil of thickness 0.4625 mm (i.e., $0.4625 \times 39.37 = 18$ mil). Let us assume each foil is covered on both sides by a 2-mil thick mylar tape. Since 1 mil is 0.0254 mm, we have effectively added 4×0.0254 mm to the foil thickness. In addition there will be three layers of tape between each of the two primary-secondary boundaries. So in all, the thickness occupied by the secondary and the insulation, h_S, is

$$h_S = (n_S \times h) + (n_S \times 4 \times 0.0254) + (12 \times 0.0254) \, mm$$

or,

$$h_S = n_S \times (h + 0.102) + 0.305 \, mm$$

So in our case

$$h_S = 2 \times (0.4625 + 0.102) + 0.305 = 1.434 \, mm$$

The ETD-34 has an available height inside the bobbin of 6.1 mm. That now leaves 6.1 − 1.434 = 4.67 mm. Therefore each section of the split primary has an available winding height of 2.3 mm only. We should ultimately check that we can accommodate the primary winding we decide on, within this space.

Note that for the primary, the available width is only 12.9 mm (since there is 4-mm margin tape on each side—for the secondary, since we have a foil with tape wrapped over it, we do not need the margin tape). We need to find how best to accommodate eight turns into this available area, with minimum losses.

> **Note:** It is not mandatory to use a particular thickness of insulating tape, provided it is safety-approved to withstand a specified voltage. We can, for example, use 1 mil approved tape or even 1/2 mil, if it suits our production, helps lower the cost, and/or improves performance in some way.

Let us first understand the basic concept for winding wires here. For a *stand-alone* wire, as in Figure 10.3, as we increase the diameter of the wire, the cross-sectional area available for the high-frequency current is $(\pi \times d) \times \delta$. And since resistance is inversely proportional to cross-sectional area, we get $R_{AC} \propto 1/d$. Similarly, $R_{DC} \propto 1/d^2$. So $F_R \propto d$. Therefore, $R_{AC} \propto 1/F_R$. This actually means that a higher F_R (bigger diameter) will decrease the AC resistance! That is not surprising, because the annulus available for the high-frequency current does increase if the diameter increases. However, this is not the way to go when dealing with "non-standalone" wire. Because, by increasing the diameter, we will inevitably move on to higher number of layers, and Dowell's equations then tell us that the losses will increase, not decrease.

On the top left side of Figure 10.11, we have Dowell's *original* curves, which show how F_R varies with respect to X (i.e., h/δ). The parameter for each curve is layers per portion (i.e., p). Note that Dowell's curves talk in terms of foils only. They don't care about the actual number of turns in the primary or secondary (i.e. from the electrical point of view), but only the effective layers per portion (from the field point of view). So, when we consider a layer of round wires of diameter d, we need to convert this into an equivalent foil. Looking back at the right side of Figure 10-8, we see that that this amounts to replacing a wire of diameter d with a foil slightly thinner (i.e., with the same amount of copper, but in a square shape). Alternatively, if we want to get a foil of $X = 4$, for example, we need to start with a wire of diameter 1/0.886 = 1.13 times X. Finally, as indicated, all these copper squares then merge (from the field point of view), to give an equivalent layer of foil.

In Figure 10-11, we are also conducting a certain "experiment"—as an alternative way of laying out wires optimally. Suppose we have several round wires laid out side by side

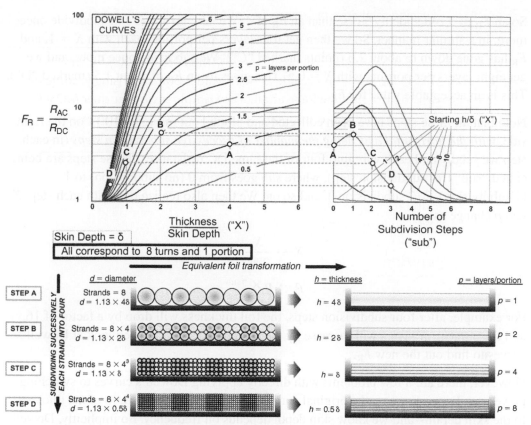

$$F_R = \frac{R_{AC}}{R_{DC}}$$

Figure 10.11: Understanding the process of "subdivision," keeping the DC resistance unchanged, and how the equivalent foil transformation process takes place

with a diameter $1.13 \times 4\delta$. Suppose also, that this constitutes one layer per portion in a given winding arrangement. This is therefore equivalent to a single-layer foil of thickness 4δ, that is, $X = 4$. Now using Dowell's curves, the FR is about 4 (points marked "A" in Figure 10-11). Suppose we then divide each strand into four strands, where each strand has a diameter half the original. Therefore, the cross-sectional area occupied by copper remains the same because

$$A = 4 \times \frac{\pi \times \left(\dfrac{d}{2}\right)^2}{4} = \frac{\pi \times d^2}{4} \tag{10.36}$$

However, the equivalent foil thickness is now half of what it was —2δ (i.e., $X = 2$). And we also now have *two layers per portion* from Dowell's standpoint. Consulting Dowell's curves, we get an F_R of about 5 now (marked "B"). Since we are keeping R_{DC} fixed in the process, $R_{AC} \propto F_R$. Therefore now, decreasing F_R is a sure way to go, to decrease R_{AC}.

So an F_R of 5 is decidedly worse than an F_R of 4. We now go ahead and subdivide once more, in a similar manner. So we then get four layers per portion, each with $X = 1$, and F_R has gone down to about 2.6 (points marked "*C*"). We subdivide once more, and we get eight layers per portion, with $X = 0.5$. This gives us an F_R of about 1.5 (marked "*D*"). This is an acceptable value for F_R.

Note that all these steps have been collected and plotted out in Figure 10.11 on the right side, *with the horizontal axis being the number of successive subdivision steps* (in each step we subdivided each wire into four of the same DC resistance). These steps are being called "sub" (for subdivision step), where *sub* goes from 0 (no subdivision) to 1 (1 subdivision), 2 (2 subdivisions), and so on. We then also realize that with each step, X and p change as per

$$X \rightarrow \frac{X}{2^{\text{sub}}}$$

$$p \rightarrow p \times 2^{\text{sub}}$$

For example, after four subdivision steps, the foil thickness will drop by a factor of 16, and the number of layers will increase by the same factor. We can then look at Dowell's curves to find out the new F_R.

However, there are a few problems with directly applying Dowell's curves to switching power regulators. For one, the original curves only talked about the ratio of the thickness to the skin depth—and we know skin depth depends on frequency. So implicitly, Dowell's curves provide the F_R for a *sine wave*. Further, Dowell's curves do not assume the current has any DC value. So engineers, who adapted Dowell's curves to power conversion, would usually first break up the current waveform into its *AC and DC components*, apply the F_R obtained from the curves to the AC component only, compute the DC loss separately (with $F_R = 1$), and then sum as follows:

$$P = I_{\text{DC}}{}^2 \times R_{\text{DC}} + I_{\text{AC}}{}^2 \times R_{\text{DC}} \times F_R \qquad (10\text{-}37)$$

However, in our case, we have preferred to follow the more recent approach of using the actual (unidirectional) current waveform, splitting it into Fourier components, and summing to get the *effective* F_R. The losses are expressed in terms of the thickness of the foil *as compared to the δw at the fundamental frequency* (first harmonic). We also include the DC component in computing this *effective* F_R. That is the reason, when calculating the secondary winding losses, we were able to use the simple equation

$$P = I_{\text{RMS}}{}^2 \times R_{\text{AC}} \equiv I_{\text{RMS}}{}^2 \times (F_R \times R_{\text{DC}}) \qquad (10\text{-}38)$$

In that case, the F_R was actually the effective F_R (computed for a square wave with DC level included), though not explicitly stated. However, note that the graphs in

Figure 10.11 are still based on the original sine-wave approach, and the purpose here was only to demonstrate the *subdivision technique* through the original curves.

But in Figure 10.12, we have finally modified Dowell's original sine-wave curves. Fourier analysis has been carried out while constructing these curves, and so the designer can apply them *directly* to the typical (unidirectional) current waveforms of power conversion. We will now use these curves to do the calculations for the primary winding of our ongoing numerical example.

But one question may still be puzzling the reader—why are we not using the previous F_R/X curves (see Figure 10.10) that we used for the secondary? The reason is the situation is different now. The curves in Figure 10.10 are Dowell's curves for a square wave, except that on the vertical axis we have used F_R/X, not F_R. That is useful only when we are varying h and seeing when we got the lowest R_{AC}. But for the primary windings, we are going to fix the height of the windings in each step of the iterations that follow. We will be using the subdivision technique in each iteration, and therefore *keep the DC resistance constant*. So now the minimum R_{AC} (for a given iteration step) *will be achieved at the minimum F_R, not at the minimum F_R/X.*

The subdivision method was originally presented in Figure 10.11, except that now we will use the modified curves in Figure 10.12.

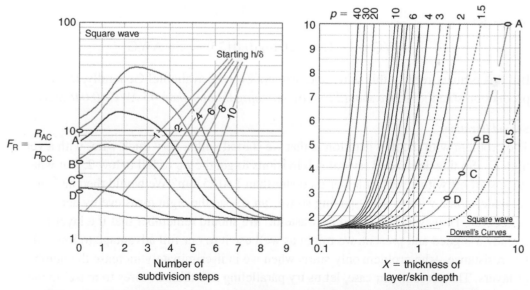

Figure 10.12: Dowell's curves modified for square current waveforms, and the corresponding FR curves for the subdivision method

First Iteration:

Let us plan to try and fit eight turns on one layer. *Lesser* number of layers will usually be better. We remember that we have 12.9 mm available width on the bobbin. So if we stack eight turns side by side (no gap between them) we will require each of these eight round wires to have a diameter of

$$d = \frac{\text{width}}{\text{turns per layer}} = \frac{12.9}{8} = 1.6125 \, \text{mm}$$

We can check that the available height of 2.3 mm is big enough to accommodate this diameter of wire. The *penetration ratio X* is (using the equivalent foil transformation)

$$X = \frac{0.886 \times d}{\delta} = \frac{0.886 \times 1.6125}{0.185} = 7.723$$

The p is equal to 1. From either of the graphs in Figure 10.12, we can see that the F_R will be about 10 in this case (marked "A"). Further, from the graph on the left side, we can see that we need to subdivide the "$X = 7.7$" curve (imagine it close to the $X = 8$ curve) seven times to get the F_R below 2. That would give strands of diameter

$$d \rightarrow \frac{d}{2^{\text{sub}}} = \frac{1.6125}{2^7} = 0.0125 \, \text{mm}$$

The corresponding AWG can be calculated by rounding off

$$\text{AWG} = 18.154 - 20 \log(d)$$

So we get

$$\text{AWG} = 18.154 - 20 \log(0.0125) \Rightarrow 56 \, \text{AWG}$$

But this is an extremely thin wire, and may not even be available! Generally, from a production standpoint, *we should not use anything thinner than 45 AWG (0.046 mm).*

Second Iteration:

The problem with the first iteration is that we started with a very thick wire, with a very high F_R. So this demanded several subdivisions to get the F_R to fall below 2. But what if we start off with a wire of lesser diameter than 1.6125 mm? We would then need to introduce some wire-to-wire spacing so we can spread the eight turns evenly across the bobbin. However, that would be wasteful! We should remember that if a layer is already assigned and present, we might as well use it to our full advantage to lower the DC resistance—the problem only starts when we indiscriminately increase the number of layers. Therefore in our case, let us try paralleling *two* thinner wires to make up the primary. We still want to keep to one layer (without spacing). That means we will now have 16 wires placed side by side in one layer. We then define a *bundle* as the number of wires paralleled to make the primary winding (we will be subdividing each of these further). So in our case

$$\text{bundle} = 2$$

The diameter we are starting off with is

$$d = \frac{\text{width}}{\text{turns per layer}} = \frac{12.9}{16} = 0.806 \, \text{mm}$$

The penetration ratio X is

$$X = \frac{0.886 \times d}{\delta} = \frac{0.886 \times 0.806}{0.185} = 3.86$$

The p is still equal to 1. From both the graphs in Figure 10.12, we can see that the F_R will be about 5.3 in this case (marked "B"). Further, from the graph on the left side, we can see that we need to subdivide five times to get the F_R below 2. That would give strands of diameter

$$d \rightarrow \frac{d}{2^{\text{sub}}} = \frac{0.806}{2^5} = 0.025 \, \text{mm}$$

This is still thinner than the practical AWG limit of 0.046 mm.

Third Iteration:

So we now parallel *three* wires to make up the primary. That means we will have 24 wires side by side in one layer.

$$\text{bundle} = 3$$

The diameter we are starting off with is

$$d = \frac{\text{width}}{\text{turns per layer}} = \frac{12.9}{24} = 0.538 \, \text{mm}$$

The penetration ratio X is

$$X = \frac{0.886 \times d}{\delta} = \frac{0.886 \times 0.538}{0.185} = 2.58$$

The p is still equal to 1. From both the graphs in Figure 10.12, we can see that the F_R will be about 3.7 in this case (marked "C"). Further, from the graph on the left side, we can see that we need to subdivide four times to get the F_R below 2. That would give strands of diameter

$$d \rightarrow \frac{d}{2^{\text{sub}}} = \frac{0.538}{2^4} = 0.034 \, \text{mm}$$

But this is still too thin!

Fourth Iteration:

Let us now parallel four wires to start with. We will have 32 wires in one layer.

$$bundle = 4$$

The diameter we are starting off with is

$$d = \frac{width}{turns\ per\ layer} = \frac{12.9}{32} = 0.403\ mm$$

The penetration ratio X is

$$X = \frac{0.886 \times d}{\delta} = \frac{0.886 \times 0.403}{0.185} = 1.93$$

The p is still equal to 1. From both the graphs in Figure 10.12, we can see that the F_R will be about 2.8 in this case (marked "D"). Further, from the graph on the left side, we can see that we need to subdivide three times to get the F_R below 2. That would give strands of diameter

$$d \rightarrow \frac{d}{2^{sub}} = \frac{0.403}{2^3} = 0.05\ mm$$

This corresponds to AWG 44, and would be of acceptable thickness.

Note that by the process of subdivision, the number of layers per portion has gone up as

$$p \rightarrow p \times 2^{sub}$$

So with three subdivisions we have

$$p \rightarrow p \times 2^{sub} = 1 \times 2^3 = 8\ (layers\ per\ portion)$$

that is, eight layers. The penetration ratio has similarly now become

$$X \rightarrow \frac{X}{2^{sub}} = \frac{1.93}{2^3} = 0.241$$

The F_R is now about 1.8 as can also be confirmed from the graph on the right side of Figure 10.12 (for $X = 0.241$, $p = 8$).

The number of strands each original "bundle" has been divided into is

strands $= 4^{sub} = 4^3 = 64$

So finally, the primary winding consists of four bundles in parallel, each bundle consisting of 64 strands, side by side in one layer, with an F_R of about 1.8.

We can continue the process if we want to get a slightly lower F_R. But at some point we will find the F_R *will start to go up again.* For our purpose, we will take an F_R of less than 2 as acceptable to proceed with the loss estimates.

Note that further tweaking will always be required since when we bunch wires together to form a bundle, they will "stack" in a certain manner that will affect the dimensions from what we have assumed. Further, the diameter of the wire we used was for bare wire, and is slightly less than the coated diameter. Note that in general, *if after winding several layers evenly, we are left with a few turns that seem to need another layer to complete, we are better off reducing the primary number of turns and sticking to the existing completed layers, because even a few turns extra will count as a new layer from the field point of view, and increase proximity losses.*

We can now calculate the losses for the two primary sections combined, since they can be considered to be identical and with the same F_R. The AC resistance in ohms of the entire primary winding is

$$R_{AC\_P} = (F_R) \times \frac{\rho \times \text{MLT} \times n_P}{\pi \times \dfrac{d^2}{4} \times \text{bundles} \times \text{strands}} = (1.8) \times \frac{2.3 \times 10^{-5} \times 61.26 \times 16}{\pi \times \dfrac{(0.05)^2}{4} \times 4 \times 64}$$

So the loss is

$$P_P = I_{RMS\_P}{}^2 \times R_{AC\_P} = \left(\frac{I_{RMS\_S}}{n}\right) \times R_{AC\_P} = \left(\frac{29.69}{8}\right)^2 \times 0.08 = 1.102 \, \text{W}$$

Had we gone further and divided the primary into five bundles and then subdivided three times, we would get eight layers with 64 strands of 0.04 mm diameter wire per bundle, and an F_R of 1.65—which seems better than the 1.8 we got in the last step. But since the wires are so thin to start with, the DC resistance now goes up, and the dissipation will rise to 1.26 W.

10.2.5.10 Total transformer losses

The total dissipation in the transformer is therefore

$$P = P_{CORE} + P_{CU} = P_{CORE} + P_P + P_S = 1.13 + 1.102 + 0.899 = 3.131 \, \text{W}$$

The estimated temperature rise

$$\text{degC} = R\text{th} \times P = 17.67 \times 3.145 = 55.3°\text{C}$$

What we are seeing is a typical practical situation! The temperature rise is 15°C higher than we were expecting! However, 55°C is perhaps still acceptable (even from the standpoint of getting safety approvals without special transformer materials). Admittedly, there is room for more optimization. However, the next time we do the process, we must note that the core loss is only a third the total loss, not half, as we had initially assumed.

Note also that methods in related literature may predict a smaller temperature rise. But the fact is that these are usually based on the sine-wave versions of Dowell's equations, and we know that will typically underestimate the losses significantly.

CHAPTER 11

A "True Sine Wave" Inverter Design Example

Raymond Mack

Working on this section and the primary side of the switchmode power supply are no less danger-ous than poking a screwdriver into an AC socket. If you do not understand what you are doing, you could end up on your back, or at least with some fireworks occurring in your hands. Graphic analogies aside, you must take the right precautions when probing the switching power supply.

One time after setting up the instrumentation on a bench and making sure I had all of the equip-ment I needed, I started my AC/DC converter. It was operating fine, until I went to view the drain voltage of the main power switch. Within a millisecond I had vaporized the ground clip of the oscilloscope voltage probe with a large flash and "pop." After a few expletives, I realized I had neglected to insert an earth ground isolator onto the AC power plug of the oscilloscope. Happily, the oscilloscope survived, having been designed for accidents or idiots such as myself. I had learned two lifelong lessons: make sure your life insurance is up to date and always keep several ground isolator plugs in your tool box. Today, my oscilloscope is always isolated from ground, no matter what type of circuit I am working on.

Several other habits I have developed over the years: don't work barefooted, know where the earth ground points are on the rest of your workbench, and know where the workbench's circuit breaker is located. It is not so much a matter of being afraid, but more a matter of having a healthy respect for the power you are working with. It is no different than operating your back-yard BBQ grill with its hot surfaces and its propane bomb (sorry…tank) just underneath it.

This chapter is an adjunct to Chapters 5 and 6. This topic is how to interface a switching power supply to the various AC power grids in the world. The AC/DC rectification stage has three or four major functions: Creating a high voltage DC voltage for the input to the switching power supply, filtering the noise from the input to the power supply from the AC mains, filtering the noise created by the switching power supply to the AC mains, protecting the power supply from adverse transients from the AC mains, and perhaps performing power factor correction of the AC input current. This is a lot to do for a small, often minimized, section of an AC/DC power supply.

The design of this section is important to the reliable operation of your final product and whether the safety and emissions regulating agencies will let you sell your product around the world. It can be the most frustrating portion of the design cycle. You will set your product on *a table in a test lab in front of an engineer who, frankly, neither understands your product or cares. He has his standards to test. These are GO/NO-GO tests and you may have to return to your workbench and add noise control components to your design or slightly modify the physical design of the product.*

California has recently passed a law that limits the amount of current that is drawn from the AC line during a product's OFF or sleep mode. This makes the use of inexpensive wall transformers (wall warts) forbidden. Only an off-line switching power supply can meet these standards. So paying attention to this chapter is important!

Ray Mack overviews the function and design of this circuit section. There are many resources that will help you design the input rectification stage on the web.

—**Marty Brown**

We will design a "true sine wave" uninterruptible power supply in this chapter. "True sine wave" for our design means 20% or less total harmonic distortion. This product is intended to give instant crossover from line power to battery power and provide power for devices that require sine wave operation. The description in this chapter is intended to show the iterative nature of designing a complicated switching system, so you will see several missteps at each major decision point.

It is imperative that you remember that all of the circuitry in this design is connected directly to the AC power mains. This presents a potentially life-threatening situation. Always use a suitable isolation transformer to isolate the circuit from the AC power mains while testing and analyzing this design.

11.1 Design Requirements

The following are the requirements for our design:

1. 115 VAC, 60 Hz, 650 VA maximum input power

2. Class B FCC EMI certification

3. 115 VAC, 60 Hz, 300 VA pseudo-sine wave output, less than 20% total harmonic distortion (THD). Operation over 0.5–1.0 power factor load

4. 300 W-h power capacity

5. instant switchover—zero dropped cycles

11.2 Design Description

The requirement for low harmonic distortion means that we will need to approximate a sine with numerous steps. Figure 11.1 shows the spectrum for two waveforms used by square-wave inverters. The first waveform is that of a square wave. The amplitude is equal to the RMS value of an equivalent sine wave. The second waveform has the value

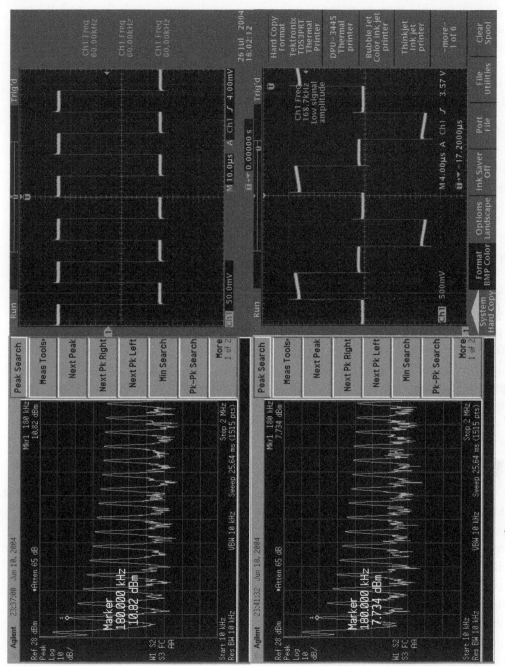

Figure 11.1: Spectra for two waveforms used by square wave converters

the null bus while the input supply is very low or capacitor is being depleted, as even input and output ripple less expressive square wave converters. The first spectrum is that of a square wave. The amplitude is equal to the RMS value of an equivalent sine wave. The second waveform has four steps that do a reasonable approximation of a sine wave for electronic loads that use full wave rectification. The RMS value is equal to the RMS of an equivalent sine wave and the peak is the same as an equivalent sine wave. Inductive loads, such as motors, need a cleaner sine wave to minimize losses. There are very few capacitive loads in the real world, but an electronic load like a computer will come close. A design that will supply an inductive load will most likely also supply a capacitive load with the same efficiency.

The first design step is to use an arbitrary waveform generator to simulate a stepped sine wave and measure THD. Figure 11.2 shows the test waveforms and the associated spectra for two test waveforms. The spectrum analyzer has a lower limit of 9 kHz, so the test waveforms are 60 kHz. The real design will scale the circuit values by a factor of 1000.

The spectra of the three sampled sine waves shows energy at the fundamental and an alias on each side of the sample frequency. The four-step waveform in Figure 11.1 is sampled at 240 kHz (four samples per cycle), so the first alias frequencies are at 180 kHz and 300 kHz. These are the same frequencies that occur for a simple square wave. The first waveform in Figure 11.2 is sampled at 480 kHz (eight samples per cycle), so the alias frequencies occur at 420 kHz and 540 kHz. The second signal in Figure 11.2 is sampled at 960 kHz (16 samples per cycle), so the alias frequencies are at 900 kHz and 1020 kHz. There are no harmonics of the 60 kHz fundamental anywhere in the spectrum for the signals of Figure 11.2. All the energy at higher frequencies is related to the sample frequency. Energy decreases for each alias as frequency increases. Table 11.1 lists the energy relative to the fundamental for each alias up to 3 MHz for the eight-step signal. The total harmonic distortion (THD) is 4.6%, which is well within the 20% THD goal. This means we can drive the load directly without filtering. The 16-step waveform has about 1% THD.

The next decision we must make is how to switch from mains power to battery power. We could use a relay to provide switching from mains power to battery power. Relays are not especially fast devices, so we would need to provide a reservoir capacitor to hold enough energy for the period that the relay is switching. It is not long, but tens of milliseconds are normal. We also require electronics to control the relay. If we use electronics to provide the switch, we will likely have approximately the same number and cost of components as the relay solution. Another reason to look at an electronic solution is that the relay has reliability issues. Current flow causes contact wear on the normally closed contacts and the normally open contacts oxidize over time. An advantage of the electronic solution is that we can likely produce a circuit that will power the output from both the input supply and the batteries while the input supply reservoir capacitor is being depleted. An even simpler and significantly less expensive transfer circuit uses diodes to isolate the battery

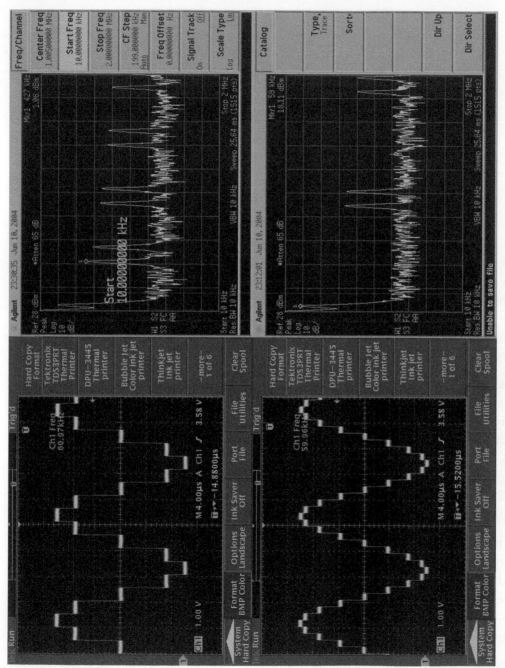

Figure 11.2: Test waveforms and their associated spectra

Table 11.1: Relative distortion power versus
alias frequency for eight-step waveform

Frequency (kHz)	Relative power
420	0.020
540	0.013
900	0.004
1020	0.003
1380	0.002
1500	0.002
1860	0.001
1980	0.001
2340	0.000
2460	0.000

pack from the main power supply. The output power conversion circuit will pull up to 15 A from the battery pack. A Schottky diode will have about 0.5 V forward voltage, so the power dissipation will be approximately 8 W. An electronic solution will likely have power dissipation near 2 W. The analysis of the battery pack reveals that there is no advantage to the extra complexity of the electronic transfer because there is ample reserve energy in the battery pack.

Sealed lead acid batteries are the only cost effective technology in this application. This market is very competitive, so prices are almost identical among manufacturers for equivalent batteries. We will almost certainly want at least 24 V, and perhaps as much as 48 V, to minimize the current draw. We must derate the capacity of a battery based on peak current draw. A 20-Ah battery is capable of supplying 1 A for 20 hours. The same battery will supply 20 A for only about 36 minutes. Our 300 W-h requirement means that the 20-hour rating must be 500 W-h. Table 11.2 lists several configurations and total cost (Digi-Key prices in case lots).

We have four candidate battery packs that will supply the required energy with roughly equivalent cost. With a difference of only $3.30 between the highest and lowest, we need to look at complexity and reliability to choose an appropriate system. The 36 V system appears to be the best choice because it gives a significant margin in energy. It also will cost less to manufacture because it uses three identical batteries, where the least expensive option mixes 12 V and 6 V batteries.

A lead acid battery has a float voltage of 2.40–2.42 V per cell when fully charged. Our system will require 44 V across the battery pack to maintain charge. Allowing a 6 V drop across the charging circuit means we need 50 V for the charging circuit and the power

Table 11.2: Battery composition versus total cost (2004 dollars)

Battery voltage	Total voltage	Capacity (Ah)	W-h	Total cost
6	120	4.2	504	141.75
6	72	7.2	518	135.64
6	42	12	504	115.45
12	72	7.2	518	97.34
12	48	12	576	106.82
12	36	17	612	98.64
12	24	28	672	99.90
12/6	42	12	504	96.60

conversion circuit. Likewise, a lead acid battery is significantly discharged when the voltage drops below 1.95 V per cell. This sets our lowest power conversion voltage at 35.1 V.

My original design goal was to use as many off-the-shelf components as possible. The current levels in the preregulator preclude using a single off-the-shelf inductor in the output filter. I ran a second design using the 72 V battery. This quick analysis revealed that the duty cycle would range from 50% to almost 100%. The current level decreases, but the change in duty cycle requires 600 μH at 7 A rather than 470 μH at 12 A. The current level for both inductors will require a special-order part.

The nominal output voltage is 120 VAC. The highest supply voltage needed corresponds to the 170 V peak of the AC output voltage. The second supply voltage is 120 V. We can use a boost converter or a transformer converter to step up the voltage from the battery. As mentioned in Chapter 5, the boost converter cannot limit current under fault conditions because the switch is not in the current path between input and output. A forward or push-pull converter allows us to shut off the output by shutting off control of the switch.

There are two options for supplying the two voltages. The first is to use a single output and use the PWM circuit to change the voltage between 120 V and 170 V. The second method is to use the converter to generate both voltages simultaneously and electronically switch between the voltages. For the first method, the PWM control loop must be fast enough to track the change from 120 V to 170 V during the 2-ms time period that the voltage needs to have the peak value. It is possible to design a control loop that will respond so quickly, but it is likely to be quite complex.

The second method uses more components but the control loop is much simpler. Our design will generate both the 120 V and 170 V sources from separate windings on the power transformer. We can only control one of the voltages, so we must provide a way to ensure the two voltages always have close to a 50 V difference. One solution is to control the 170 V supply and clamp the 120 V to 50 V below the 170 V supply. This is probably a reasonable solution, since the 120 V supply will provide the bulk of energy in a true sine wave application and will tend to drop below 120 V. In an electronic application, the 170 V supply will provide the majority of the energy.

The final stage to consider is the output driver. The output section is a standard H bridge that generates the alternating AC signal. The biggest difference between the H bridge here and an H bridge in a switching power supply is that the voltage and current may very well not be in phase. Out-of-phase operation is certain for a load with less than unity power factor. This requirement for current out of phase with voltage requires that we use MOSFETs so that current can flow in either direction in the on switches.

11.3 Preregulator Detailed Design

Figure 11.3 shows the schematic of the input power conversion circuit. Only one-third of the mechanical power switch is shown. The first section controls power to the power line supply. The second section disconnects the battery from the output circuit and the third section disconnects the low voltage tap of the battery. The input inductors and the differential mode capacitors are designed to increase the power factor to the greatest extent possible within a reasonable cost. Economical inductors will not be effective at reducing low-order harmonics. There are no harmonic requirements at this time in the United States, so reducing harmonics is not a requirement. We can reduce the harmonic content by reducing the size of the input reservoir capacitor and increasing the ripple voltage.

Figure 11.4 shows the input regulator that reduces the input voltage to 50 V. The peak battery voltage during float is 43.6 V. 50 V gives enough head room for the battery charging circuit. Keeping the input voltage low reduces the range for the final power conversion circuit. The lower difference between the input voltage and the float voltage reduces losses while charging the battery pack. The battery charger is a dissipating regulator, so higher input voltage will waste more power. Setting the maximum duty cycle for the input regulator to 90% allows the lowest voltage to be 56 V, and the 187 V

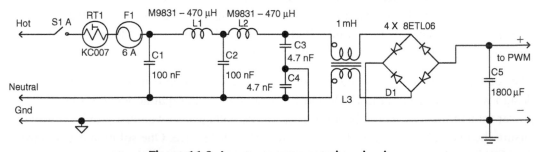

Figure 11.3: Input power conversion circuit

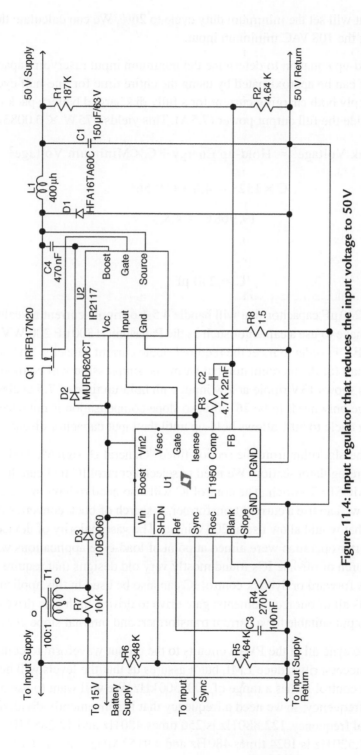

Figure 11.4: Input regulator that reduces the input voltage to 50 V

maximum input will set the minimum duty cycle to 26%. We can calculate the reservoir capacitor using the 108 VAC minimum input.

We use the hold-up equation to determine the minimum input reservoir capacitor. The hold-up energy can be approximated by using the entire time for one-half cycle (8.3 ms). We need to supply both charging current for a fully discharged battery pack (4 A) plus the current to provide the full output power (7.5 A). This yields $575 \, W \times 0.0083 \, s = 4.8 \, J$.

$$C \times \text{Peak Voltage}^2 = \text{Hold-up Energy} + C \times \text{Minimum Voltage}^2 \qquad (11\text{-}1)$$

$$C \times 152^2 = 4.8 + C \times 56^2 \qquad (11\text{-}2)$$

$$19,968 \, C = 4.8, \qquad (11\text{-}3)$$

so

$$C = 240 \, \mu F \qquad (11\text{-}4)$$

A search for a 240 μF capacitor that will handle 4.5 A of ripple current reveals no available parts. We will have to use a capacitor such as the Panasonic 1800 μF 200 WV model ECOS2DP182EX in order to meet the required ripple current capacity. The voltage ripple will be very small for the preregulator, but harmonic suppression will be more difficult. A new analysis shows 13 V ripple at low input with hold-up time of 7.5 cycles to 56 V. The duty cycle range only needs to be 26–50%, so slope compensation is not necessary. Setting maximum duty cycle to 50% allows hold-up until the input capacitor discharges to 100 V.

Next, we use the algorithm from the section titled "General Design Method" in Chapter 5 to design the preregulator section. We need to select a control IC that can drive a high side driver for the MOSFET switch. The control IC will also need to have an external current sense input so we can use a current transformer. A search of buck converter ICs resulted in no devices that would allow use at 200 V input. The vast majority of devices listed for buck converter operation were aimed at point of load-type applications with internal switches. A search of off-line ICs found mostly very old designs that require significant design work. A forward or flyback control IC can also be used in this application. The LTC1950 meets all of our requirements: gate drive to drive a high side drive circuit; a current sense input suitable for a current transformer; and internal slope compensation.

We will want to sync all of the PWM circuits to the output waveform to control EMI. This does not necessarily reduce EMI, but it assures us that the levels will be constant over time. The control IC has a range of 100–500 kHz. We will want to use a crystal to control the frequency, so we need a frequency that is conveniently divided from a standard crystal frequency. 122,880 Hz is 256 times 480 Hz and 12.288 MHz is a standard frequency. 491,520 Hz is 1024 times 480 Hz and 4.9152 MHz is a standard frequency.

Both of these frequencies will require a divide by 10 circuit and multiple binary divisions. 4.9152 MHz yields 153 kHz for switching frequency, so it is slightly better than the 122 kHz for the 12.288 MHz oscillator. 153 kHz is a 6.54 μs cycle time.

The duty cycle range is only 2:1, so transformer drive is feasible, but a high side driver IC is likely to be less expensive and more straightforward. The IR2117 is a good choice. The IR2117 will only drive 250 mA peak current, so we will have to verify that switching time is adequate. The required current capability of the high side drive depends on the gate charge of the MOSFET. The IRFB17N20 is reasonable because it has low on resistance and is the least expensive of the low r_{ds} 200 V MOSFETs. It has 30 nC of gate charge. The switching time will be 120 ns (30 nC/250 mA). A 470-nF capacitor will supply all of the current necessary for the boost circuit.

The current sense transformer can be a special design or we can use a standard toroid with a single turn primary. The secondary will not carry appreciable current, so a miniature toroid (around 1 mH) should be adequate. R7 needs to be large enough to generate sufficient volt-seconds to reset the core. R8 will be determined empirically in the lab to generate the required voltage at 7A peak inductor current. A reasonable start from the 100:1 guess is 1.5 Ω. The feedback resistors need to divide the 50.0 V output to 1.23 V. The actual values are arbitrary. We chose 4.64 K for R2, so R1 must be 187 K. The resistor divider on the shutdown pin sets the shutdown voltage to 100 V.

The next step is to pick the ripple current and design the inductor. This design is a cascaded system, so we need to use a longer transient response for one section than the other. The load on the input section will tend to vary less than the output load, so we can use a small amount of inductor ripple. The low ripple current will provide a long response time. Ripple current of 500 mA is a reasonable value. We use the inductor equation to calculate the inductor value:

$$L = V \times dt/di = (187\,V - 50\,V) \times (0.26 \times 6.54 \ \mu s) / 0.50\,A = 466\,\mu H \qquad (11\text{-}5)$$

Our prototype will use four 390 μH (Miller parts from Digi-Key) in series/parallel to result in 400 μH at 11 A and 780 μH at lower current. The ripple current at full output will be around 600 mA instead of the desired 500 mA, and the ripple current will be 300 mA at light loads. A 150 μF/63 V Panasonic EEUFC1J511 can handle 690 mA ripple current and has an impedance of 0.178 Ω. The ripple voltage will be on the order of 70 mV.

The HFA16TA60C is an adequate commutation diode. The average current at full load will be 8.5A and we need more than 200 V PRV to have enough margin. This diode blocks 600 V and has 16 A average current capability. D2 will have almost the full 187 V input when reverse biased, so an MURD620CT 200 V FRED will be appropriate. The second diode of the package is tied in parallel. D3 will have a short high voltage pulse during core reset, so a 10BQ060 60 V Schottky diode will be adequate.

We do not need an auxiliary power supply to run the control IC. The battery pack provides an "instant-on" power supply for the control IC. A single battery will not provide enough voltage to drive the switch when the battery is fully discharged, and two batteries supply too much voltage when the batteries are fully charged. We will use two batteries and reduce the voltage to provide the IC control voltage. We do not need soft start operation, since the load will be supplied by the battery until the output reaches the battery voltage.

11.4 Output Converter Detailed Design

Push-pull operation is a reasonable choice for the output converter, since it will derive its input from a relatively constant input voltage. The frequency doubling allows using much smaller inductors and filter capacitors. The National LM5030 is an excellent choice for this application. It meets all of our requirements: external synchronization, internal slope compensation, and large gate drive capability. We will set the maximum duty cycle (input side) to 40% to leave room for control and to stay away from the need to ensure that the devices do not conduct simultaneously. The lowest voltage is 35.1 V battery voltage minus the 0.7 V voltage drop of the switching diode, or 34.4 V. The maximum voltage is the 50 V preregulator voltage minus the 0.7 V drop of the switching diode, or 49.3 V. This sets the minimum duty cycle to 28%. The duty cycle on the output side will be double or a range of 56–80%.

The RMS AC output current is 2.5 A. However, the average current for the 170 V supply will be 900 mA and the 120 V supply will be 1.25 A. The peak currents will be 3.5 A and 2.5 A, respectively. The maximum rectifier diode voltage will occur at minimum duty cycle. The maximum input voltage for a 170 V supply will be 303 V plus the rectifier voltage drop. The maximum 120 V supply will be 214 V. PRV for both supplies will equal double the input voltage. One way to reduce the stress on the diodes for the 170 V supply is to put a 50 V supply in series with the 120 V supply instead of a single 170 V supply. This arrangement will force the two supplies to track more closely. Increasing the maximum duty cycle to 45% and putting two supplies in series gives 190 V and 79 V for the two PRV values. The minimum duty cycle changes to 63%. The HFA08TA60 is a reasonable diode for the 120 V supply, but the HFA16TA60 is actually less expensive and allows us to use the same part in two positions. The MURD620CT is a reasonable diode for the 50 V supply. The forward voltage drop at 1 A forward current is 1.2 V for both diode types, so the transformer voltages need to be 80 V and 191 V. Figure 11.5 shows the circuit for the power conversion circuit.

The output current is low even at full load, so we can start with 600 mA ripple current. The supply will transition to discontinuous mode at 300 mA output current, which corresponds to about 25 VA in the load. The 50 V inductor needs to be:

$$L = \text{V} \times dt/di = (79\text{V} - 50\text{ V}) \times (0.63 \times 3.27\,\mu s)/0.60\text{ A} = 100\,\mu\text{H} \qquad (11\text{-}6)$$

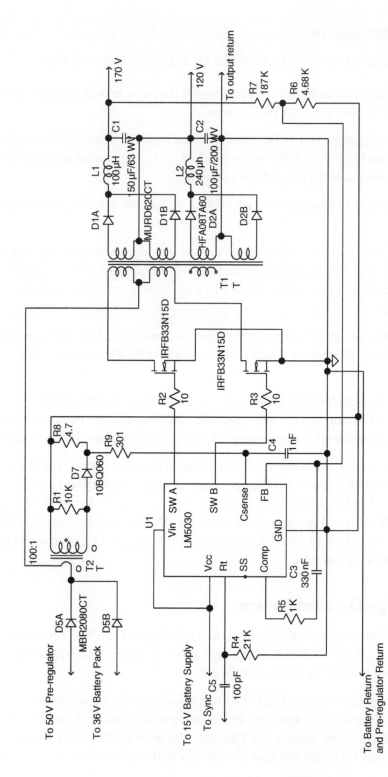

Figure 11.5: Circuit for the power conversion circuit

The 120 V inductor needs to be:

$$L = \text{V} \times dt/di = (190\,\text{V} - 120\,\text{V}) \times (0.63 \times 3.27\,\mu\text{s})/0.60\,\text{A} = 240\,\mu\text{H} \qquad (11\text{-}7)$$

Putting the supplies in series makes the peak current the same in both supplies, so both inductors need to be rated for 3.8 A peak current.

The lowest input voltage is 34.4 V. The switch current is on the order of 12 A, so the voltage drop for the IRFB33N15D is on the order of 0.7 V. The current sense resistor will drop 0.5 V. The lowest transformer voltage is then 33.2 V.

The transformer ratios for the windings are 80/33.2 = 2.41 for the 50 V supply and 191/33.2 = 5.75.

A reasonable set of windings will likely be four turns per side on the primary (eight turns total) of #10 wire. The 50 V secondary could be nine turns per side of #18 wire. The 120 V winding could be 23 turns per side of #18 wire. This is a reasonable place to start with our prototype.

Once again, the first design decision needs correction. Since the LM5030 has a 500 mV current sense voltage, the resistor would dissipate almost 6W. This indicates that a current sense transformer is a more reasonable design.

The IRFB33N15D is rated at 33 A drain current at 25°C. It can still handle the 12 A peak current at 150°C. The drain voltage rating must be 150 V because the maximum drain voltage will be double the 50 V input voltage. A 100 V device will not have any margin.

The output ripple voltage is not especially important in this application. The Panasonic 100 μF/200 WV EEU-EB2D101 has enough ripple current capability and a low dissipation factor. The 150 μF/63 WV EEU-FC1J151 is reasonable for the 50 V supply. The MBR2080CT is a good power transfer diode for D5. The average current at full load will be 11 A and we only need 60 V PRV to have enough margin. Only one diode in the package will conduct at any time.

11.5 H Bridge Detailed Design

The output H bridge is a standard design, as shown in Figure 11.6. We will have enough margin with 200 V MOSFET switches. The high side switches can be driven with an IR2117 such as we used in the input preregulator.

The 1-μF boost capacitors for the IR2117 drivers should be film capacitors so they have minimal leakage. Electrolytic capacitors are not acceptable because of the amount of time that the high side drive must be on. The high side drive for the 170 V supply can use a 60 V P-channel MOSFET because it must only withstand the voltage of the 50 V

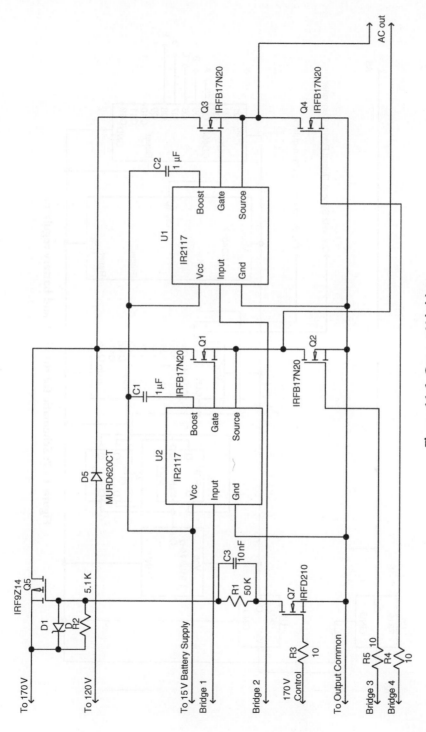

Figure 11.6: Output H bridge

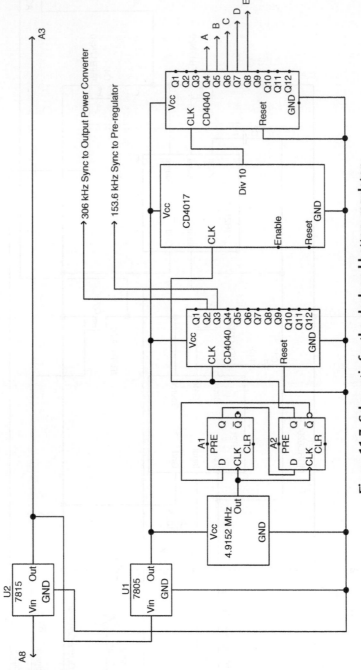

Figure 11.7: Schematic for the clock and battery regulator

supply. C3 supplies speed up current to turn on Q5 because the current supplied by R1 is insufficient by itself to turn on the switch. R1 must be a 500 mW resistor. The resistor will only dissipate about 250 mW because the duty cycle is 25% for the 170 V supply. R2 supplies the turn-off path for Q5.

11.6 Bridge Drive Detailed Design

Figure 11.7 shows the clock divider schematic, the logic power supply, the battery regulator supply, and the sync signal generation for the power conversion circuits. Figure 11.8 shows the logic that drives the H bridge.

The drive to the bottom switches (Q2 and Q4) must be on for 1.04 ms (one-eighth of each half-cycle) during the middle of the zero voltage period. This sequence allows the current to be steered from one bottom switch to the other during the zero time and guarantees no overlap of the bottom switches and the top switches. The top switches and the bottom switches must have nonoverlapping drive to ensure no current shoot-through.

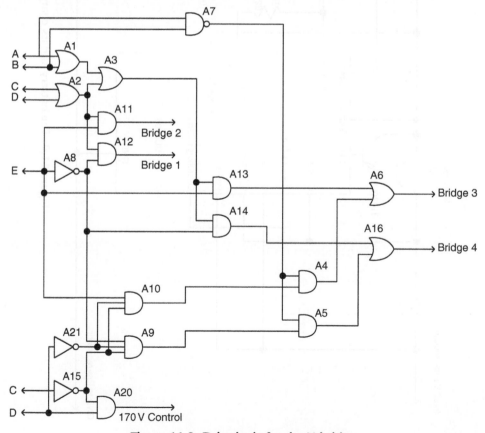

Figure 11.8: Drive logic for the H bridge

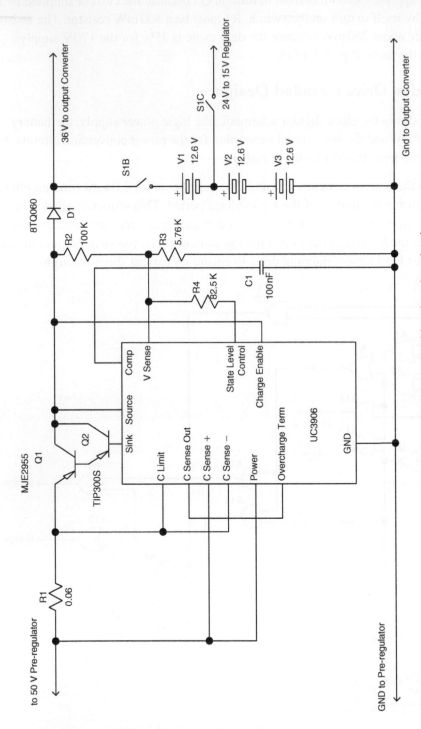

Figure 11.9: The battery and its charging circuit

Figure 11.9 shows the schematic of the battery and the charging circuit. The charging circuit follows the design method recommended in TI Application Note U-104. Q1 and Q2 implement a Darlington pass element because the IC can only supply 25 mA of drive. D1 is required to disconnect the IC from the battery when power fails. D1 guarantees that none of the IC circuit is reverse biased by the battery.

You should notice that the schematics in Figures 11-4 and 11-5 show a long connection from the GND pin of the control IC to the output filter capacitor. This is a reminder that the current sense resistor and components on the left side of each of the schematics should be connected to the GND pin of the IC and the GND pin should have a single small connection to the PGND connections of the power side of the circuit. The power line circuit and the 50 V pre-regulator should be built on one PC board with the output circuits built on a second PC board. The alternative is to build the input circuits on one side of the PC board and build the output circuits on the other side with very wide power traces in the middle for connection to the battery and charger.

CHAPTER 12

Thermal Analysis and Design

Marty Brown

It is no surprise that power supplies get hot. Sometimes they get very hot. You have done every-thing you can do to maximize the efficiency in your design. Now you are faced with getting rid of the resulting heat energy.

Marketing will always want a switching power supply that costs zero, measures zero and is 100 percent efficient. You, as the power supply engineer, will just have to get used to not living up to the expectations of marketing.

The drive to further miniaturize electronics and the use of more and more surface mount pack-aging present a real challenge. The use of hot components that cannot be bolted to a heatsink means that more nontraditional methods of transporting heat away from these parts must be employed.

The rule of thumb in the reliability field is that, for every increase in 10 degrees Celsius of a component's temperature, its life is cut in half. So one's goal as a designer is to keep the com-ponent's temperature as low as possible to maximize the product life.

This chapter presents the mathematical method of modeling a thermal system and illustrates some approaches for the design of mechanical heatsinking assemblies. The author (me) presents the material in an understandable manner to help you through your design.

—Marty Brown

Proper thermal design is essential to the overall design of a power supply. Overdissipation failures account for probably the largest portion of the failures. Therefore, it is essential that the designer understand its basic principles.

Thermal analysis is really no more difficult than Ohm's Law. There are similar parameters to voltage, resistance, nodes, and branches. For the majority of electronic applications, the thermal "circuit" models are quite elementary and, if enough is known of the thermal system, values can be calculated in a matter of minutes. If one has a temperature-measuring probe, the thermal components can also easily be measured and calculated.

There are two main goals in designing the thermal system: the first is to never allow any component to exceed its *maximum operating junction temperature* ($T_{J(max)}$); the second is to keep the components as cool as possible given the restrictions for space and weight. Failure to maintain the first condition will cause a component to fail within minutes. The second consideration affects the long-term life of the system. MIL-217, a reliability prediction tool for high-reliability applications, makes the following generalization: *"The life of a component will be halved every +10°C rise above room temperature."* In most applications, the designer should be concerned if any component's case temperature exceeds +60°C.

12.1 Developing the Thermal Model

Thermal system analysis is actually a variation of Ohm's Law. There are equivalent circuit elements which map directly to the elements within the electrical domain (refer to Table 12.1).

These elements always form a loop, with the power source providing the driving force for the entire model. Each circuit element and node corresponds to a physical structure or surface within the actual physical design. The power source corresponds to a heat-producing element within the circuit which creates a calculable or measurable power. The power semiconductors are typically the major heat-producing elements within a power supply. The power may be measured by graphically multiplying the terminal voltage and current waveforms from an oscilloscope and normalizing the energy product to one second (power = energy/second) or by measuring voltage and current directly using a digital voltmeter (DVM), if it is a DC application. The result is expressed in watts.

The thermal resistance can represent two physical situations. The first is the resistance to heat flow across a surface boundary, such as a power transistor bolted to the surface of a heatsink. The second situation is how well the heat spreads through a body from the heat-emitting surface to a radiating surface. Both of these physical situations are simply represented by a single thermal element—the thermal resistor, which is represented by the Greek symbol theta. Its units are measured in degrees Celsius per watt (°C/W), which

**Table 12.1: Analogous elements between the
thermal and electrical domains**

Electrical element	Thermal equivalent
Voltage source	Power (heat) source
Resistance	Thermal resistance
Node voltage	Temperature of element
Current loop	Thermal loop
Circuit ground	Ambient air temperature

represents the temperature difference across a boundary given a certain power dissipation. Some of the thermal resistances related to the semiconductor are as follows.

Power packages

$R_{\theta\text{JA}}$ Thermal resistance from the junction to the air
$R_{\theta\text{JC}}$ Thermal resistance from the junction to the case
$R_{\theta\text{CS}}$ Thermal resistance from the case to the heatsink
$R_{\theta\text{SA}}$ Thermal resistance from the heatsink to the air

Diodes

$R_{\theta\text{JL}}$ Thermal resistance from the junction to the lead
$R_{\theta\text{LA}}$ Thermal resistance from the lead to the air

All of the semiconductor case-related parameters are published by the semiconductor manufacturers. The sink-to-air parameter is published by the heatsink manufacturers, if one buys a heatsink. If one makes his or her own, it is easy to measure these resistances from any model.

Every thermal model has as its ground the ambient air temperature, unless the heat-removing medium is water or a refrigerant, in which case the ambient temperature of that medium is used. This must be the case, since the power producing device can be no cooler than the coolest media around it and since heat flows from the warmer to the cooler body.

The nodes in the model are the respective surfaces of bodies along the path of flow of the heat. These can be transistor cases, heatsink surfaces, the semiconductor die, etc. (Refer to Figure 12.1). The calculated temperatures of these surfaces can actually be measured using a temperature probe at their respective surfaces. If the power dissipation is not known but all the thermal resistances are known, one can extrapolate backwards within the model and determine the power being dissipated within the die by simply measuring the temperature difference across one of the thermal boundaries.

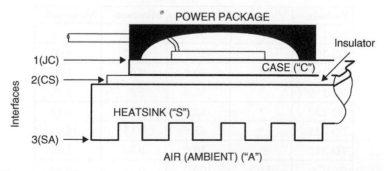

Figure 12.1: Development of the thermal model for power packages

12.2 Power Packages on a Heatsink (to-3, to-220, to-218, etc.)

This physical situation can be modeled as shown in Figure 12.2. The thermal equation would look like:

$$T_{j(\max)} = P_D(R_{\theta JC} + R_{\theta CS} + R_{\theta SA}) + T_A \qquad (12\text{-}1)$$

Since the heatsink performs the vast majority of the heat radiation, it is assumed that all the power flows through all the other thermal elements.

Temperature tests can be conducted at ambient room temperature, but the designer must remember that the typical product is enclosed in a case and its internal temperature rise must be added to the readings. Another consideration is the highest external ambient temperature the product may experience. In the desert, where this book was written, daytime temperatures may reach $+43°C$ in the shade and exceed $55°C$ inside an automobile.

Some typical thermal resistances associated with the different power packages are given in Table 12.2.

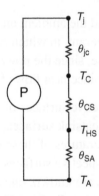

Figure 12.2: The thermal model for a transistor on a heatsink

Table 12.2: Thermal resistances of common thru-hole power packages

Package	Minimum	Maximum	Minimum	Maximum
TO-3	*	30.0	0.7	1.56
TO-3P	*	30.0	0.67	1.00
T0–218	*	30.0	0.7	1.00
TO-218FP	*	30.0	2.0	3.20
TO-220	*	62.5	1.25	4.10
TO-225	*	62.5	3.12	10.0
TO-247	*	30.0	0.67	1.00
DPACK	71.0	100.0	6.25	8.33

These thermal estimates are minimums and maximums for those types of packages. The thermal resistance values are highly dependent on the size of the die inside the package, so refer to the data sheet for the exact maximum value.

The insulating pad also adds to the thermal resistance of the case-to-heatsink. Choosing the proper insulating pad can minimize this thermal resistance. Two common technologies are mica and silicone. There are also some ceramic technologies but these are for highly specialized applications. In addition, some insulators require thermal grease to attain a good thermal contact, such as mica.

12.3 Power Packages Not on a Heatsink (Free Standing)

Power packages not mounted on a suitable heatsink can expect to dissipate less than five percent of the maximum specified power capability of the package. So 100 W devices will only dissipate 1 to 2 W when they are free standing. This also includes using the PC board copper plating as a heatsink. Thus, great discretion should be used when cost is the most important issue.

The thermal model for the case in Figure 12.3 is shown in Figure 12.4. The thermal equation becomes

$$T_{j(max)} = P_D \cdot R_{\theta JA} + T_A \qquad (12\text{-}2)$$

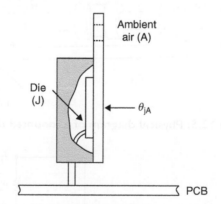

Figure 12.3: A free-standing power package

Figure 12.4: Thermal model of a free-standing power package

As one can see by the typical values of the junction-to-air thermal resistance, it doesn't require much power to result in very high junction temperatures. If the designer can possible mount the power package on any metal surface to increase the radiating surface area, it will only improve the junction temperature.

12.4 Radial-leaded Diodes

The diodes within the power supply typically dissipate a large amount of power. These are the input rectifiers and the output rectifiers. In a bipolar centered switching power supply, the output rectifiers dissipate as much power as the bipolar power switches, so their contribution to the heat within the system is significant. The physical situation is shown in Figure 12.5.

As one can see, the thermal parameters define a physically different situation. For a radial-leaded diode, the heat can only be conducted from the die, via the leads. The thermal resistance would then change as a function of lead-length and is published this way in data sheets. The thermal expression (see Figure 12.6a) is

$$T_{j(max)} = P_D \cdot R_{\theta LA} + T_A \tag{12-3}$$

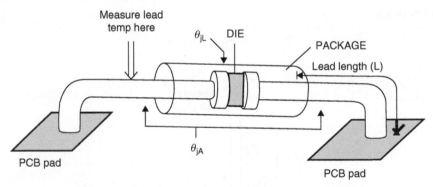

Figure 12.5: Physical diagram of a mounted diode

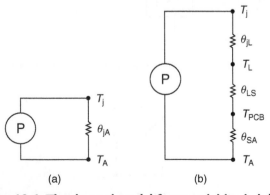

Figure 12.6: The thermal model for an axial-leaded diode

This is for the typical PC board-mounted application where only the PC board traces are used to conduct the heat away from the diode. The typical value range of the lead-to-air thermal resistance is between 30 to 40°C/W and is a variable which is dependent on the lead length.

There are some heatsinks for radially-leaded diodes that solder to one of the leads. These are also available from the transistor heatsink manufacturers. In this situation the thermal equation (see Figure 12.6b) becomes

$$T_{j(max)} = P_D(R_{\theta JL} + R_{\theta SA}) + T_A \tag{12-4}$$

These heatsinks will help a marginal heat situation. The alternative is to use a rectifier in a power transistor package such as a TO-220, TO-218, etc., and place it on a heatsink or to investigate a different technology of diode that exhibits a lower forward voltage drop such as a Schottky.

12.5 Surface Mount Parts

The use of surface mount parts is widespread. Surface mount parts can rid themselves of heat only through their leads which are soldered to a printed circuit board. The thickness and surface area of the copper island become the heatsink system. The thermal resistances in surface mount devices are much higher, therefore their designs have much less margin and room for error. Table 12.3 has nominal values for thermal resistances of common surface mount packages. Please refer to the individual part data sheet for the exact value.

It is very important to select the package that is appropriate for the function being performed. For switching signals, which are signal currents of less than 50 mA, the SOT23, SOD123, and other simple packages with gull-wing, J-leaded, and solder bump

Table 12.3: Typical surface mount package thermal resistances

Package	J-A[1]	J-C[2]
SOD123	340	150
SOT23	556	75
SOT223	159	7.5
SO-8	63	21
SMB		13
SMC		11
DPAK	80	6
D2PAK	50	2

[1] Thermal resistance for a reference pad size.
[2] Thermal resistance for a very large pad size.

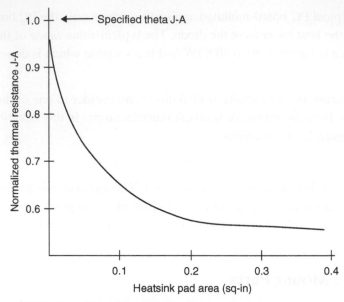

Figure 12.7: Example of the effect of increasing pad area versus θJA

leaded packages are very compact and economical. For currents of 100 mA through amperes, the package must have a tab or multiple leads connected directly to the die. This is typically the drain, collector, or cathode. The common packages are the SOT223, DPAK, SMB, and SMC. These tabs offer a very low resistance channel to remove the heat from the die and get it onto the PC board for dissipation.

In surface mount printed circuit board applications, more than one issue usually must be considered. Heatsinking must be considered, along with signal and EMI/RFI considerations. The trace that must dissipate the greatest heat within a switching power supply is also the node that has the largest *dv/dt*s which couple very easily to the surrounding traces.

Laying out heatsinking systems for surface mount packaging technology systems is still an uncertain process. Semiconductor manufacturers still do not offer adequate information for each power package to feel confident about the adequacy of the heatsinking design. The graph in Figure 12.7 is a normalized plot based upon a SOT223 package. The curve is 2-oz. copper on the top of the PCB only. Curves such as that shown in Figure 12.7 are needed to properly size the PC board heatsink island.

12.6 Examples of Some Thermal Applications

These examples will show the reader a typical application of thermal analysis but with common application variations. These variations are useful in defining thermal boundaries within a design.

12.6.1 Determine the Smallest Heatsink (or Maximum Allowed Thermal Resistance) for the Application

This approach is useful for determining the smallest possible heatsink that an application can use before the thermal limit of a power device is exceeded. This is an example of a consumer market approach to designing a heatsink system.

12.6.1.1 Specification

The device is an FDP6670 (Fairchild MOSFET) in a switching power supply. Convection cooling.

$$PD = 10 \text{ watts}$$
$$T_{A(max)} = +50°C$$
$$\theta_{JC} = 2.0°C/W$$
$$\theta_{SA} = 0.53°C/W \text{ (Thermalloy P/N 53-77-5)}$$
$$T_{J(max)} = 175°C$$

The thermal model is shown in Figure 12.8.

Rearranging Equation 12-1 and solving for the thermal resistance of the heatsink,

$$\theta_{SA(max)} < (T_J - T_A)/P_D - \theta_{JC} - \theta_{CS} \qquad (12-5)$$

θ_{CS} is assumed at 1.0°C/W. Being conservative by not requiring the junction at its maximum temperature makes the maximum allowable junction temperature 150°C. The result is

$$\theta_{SA(max)} = 7.0°C/W$$

The PC board-mounted heatsink choices are: Thermalloy part numbers 7021B through 7025B for low-cost sheet-metal type heatsinks.

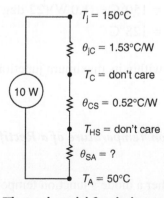

$T_j = 150°C$

$\theta_{jC} = 1.53°C/W$

$T_C = $ don't care

10 W

$\theta_{CS} = 0.52°C/W$

$T_{HS} = $ don't care

$\theta_{SA} = ?$

$T_A = 50°C$

Figure 12.8: Thermal model for design example 12.6.1

12.6.2 Determine the Maximum Power That Can Be Dissipated by a Three-Terminal Regulator at the Maximum Specified Ambient Temperature without a Heatsink

A three-terminal regulator's overcurrent protection is totally dependent upon the heatsinking system. When the die reaches approximately 165°C, the regulator shuts down. This example demonstrates the nonheatsink capabilities of a μA7805.

12.6.2.1 Specification

The desired three-terminal regulator is a mA7805KC (TO220) (Texas Instruments).

$T_{J(max)}$	150°C
$T_{A(max)}$	+50°C
$V_{in(max)}$	10.0 VDC
$I_{out(max)}$	200 mA

$$\theta_{JA} = 22°C/W$$

The power dissipated by the regulator is

$$P_D = (V_{in(max)} - V_{out}) \cdot I_{out(max)} \qquad (12\text{-}6)$$

or

$$P_D = 1.0W$$

The thermal model is that of Figure 12.4, and the thermal equation is Equation 12-2 rearranged to

$$
\begin{aligned}
T_{A(max)} &= T_{J(max)} - P_D \cdot \theta_{JA} \\
T_{A(max)} &= 150°C - (1.0\,W)(22\ \text{deg C/W}) \\
T_{A(max)} &= 128°C
\end{aligned}
\qquad (12\text{-}7)
$$

So the μA7805KC will operate within its maximum junction temperature ratings for this application.

12.6.3 Determine the Junction Temperature of a Rectifier with a Known Lead Temperature

This is useful in verifying whether a diode's junction temperature is within its safe operating temperature.

12.6.3.1 *Specification*

This is a zener diode, shunt regulator application. The diode is a 1N5240B ($10°V_{(nom)}$, $\pm5\%$).

$I_{Z(max)}$	50 mA
$T_{A(max)}$	$+50°C$
T_L	$+46°C$ (measured at TA $= +25°C$)
Lead length	3/8 inch (1.0 cm) each ($175°C/W$)

The worst-case power dissipation is

$$P_D = 1.05(10\,V)(50\,mA) = 525\,mW \text{ or } 0.525\,W$$

This situation would fit the thermal model as seen in Figure 12.7b. It does not matter in this case that not all the elements of the model are known, since all the elements above the lead temperature node are known for this first step. The thermal expression for the temperature rise above the measured lead temperature is

$$T_{J(rise)} = P_D \cdot \theta_{JL} \tag{12-8}$$

or

$$T_j = (0.525\,W)(175°C/W) = 92°C \text{ rise}$$

The junction temperature at the specified maximum local ambient temperature is

$$\begin{aligned} T_{J(max)} &= T_{J(rise)} + T_{A(max)} \\ T_{J(max)} &= 142°C \end{aligned} \tag{12-9}$$

The maximum junction temperature specified in the data sheet is $+200°C$, so the junction will be operating safely.

Index

Printed and bound by CPI Group (UK) Ltd, Croydon, CR0 4YY

03/10/2024

01040334-0004